AF361732

Physicochemical and Sensory Evaluation of Grain-Based Food

Physicochemical and Sensory Evaluation of Grain-Based Food

Editors

Luca Serventi
Charles Brennan
Rana Mustafa

MDPI • Basel • Beijing • Wuhan • Barcelona • Belgrade • Manchester • Tokyo • Cluj • Tianjin

Editors
Luca Serventi
Lincoln University
New Zealand

Charles Brennan
RMIT University
Australia

Rana Mustafa
University of Saskatchewan
Canada

Editorial Office
MDPI
St. Alban-Anlage 66
4052 Basel, Switzerland

This is a reprint of articles from the Special Issue published online in the open access journal *Foods* (ISSN 2304-8158) (available at: https://www.mdpi.com/journal/foods/special_issues/physicochemical_sensory_grain).

For citation purposes, cite each article independently as indicated on the article page online and as indicated below:

LastName, A.A.; LastName, B.B.; LastName, C.C. Article Title. *Journal Name* **Year**, *Volume Number*, Page Range.

ISBN 978-3-0365-4449-6 (Hbk)
ISBN 978-3-0365-4450-2 (PDF)

Contents

Editorial

Physicochemical and Sensory Evaluation of Grain-Based Food

Luca Serventi [1,*], Charles Brennan [2] and Rana Mustafa [3]

[1] Department of Wine, Food and Molecular Biosciences, Faculty of Agriculture and Life Sciences, Lincoln University, Lincoln 7647, New Zealand
[2] School of Science, RMIT University, P.O. Box 2474, Melbourne, VIC 3001, Australia; charles.brennan@rmit.edu.au
[3] Department of Plant Sciences, College of Agriculture and Bioresources, University of Saskatchewan, 51 Campus Drive, Saskatoon, SK S7N 5A8, Canada; rana.mustafa@gifs.ca
* Correspondence: luca.serventi@lincoln.ac.nz

Citation: Serventi, L.; Brennan, C.; Mustafa, R. Physicochemical and Sensory Evaluation of Grain-Based Food. *Foods* **2022**, *11*, 1237. https://doi.org/10.3390/foods11091237

Received: 14 April 2022
Accepted: 20 April 2022
Published: 26 April 2022

Publisher's Note: MDPI stays neutral with regard to jurisdictional claims in published maps and institutional affiliations.

Grain-based food is a staple of the human diet. Whether it is cereals, legumes or pseudocereals, grain-based diets provide nutritional benefits. This can be in the form of macronutrients (starch, fibre, protein, and lipids) and micronutrients (minerals and vitamins), as well as bioactive peptides and phytochemicals [1]. Grains are used to develop bakery products, such as savory (bread, gluten-free bread, crackers, and pasta) and sweet (cakes and muffins) [2] in addition to plant-based beverages (milk alternatives), fermented products (such as yoghurt and fermented paste), extrudates and other snacks [3]. Furthermore, grain-based ingredients offer emulsifying, foaming and thickening abilities [4]. Raw materials include cereals (barley, corn, millet, rice, rye, spelt, wheat), legumes (beans, chickpeas, lentils, peas, and soybeans) and pseudocereals (amaranth, buckwheat, quinoa, and sorghum). The functionalities are numerous, spanning from health to taste. In order to fully exploit the nutritional potential of grain-based foods, consumer acceptance must be achieved. This will guarantee compliance. The acceptability of food can be studied both instrumentally and via sensory science.

The physicochemical evaluation of food can be performed via numerous techniques, exploring a broad range of functionalities. Foaming, emulsifying and thickening abilities can be assessed with specific tests. These results offer valuable information on grain-based ingredients capability to incorporate air (thus increasing volume), stabilize emulsion systems (air in water, oil in water, and water in oil) and increase viscosity, offering mouthfeel, while preventing syneresis and phase separation upon storage. Food products can be assessed for texture, rheology (viscosity and pasting properties), thermal properties (through differential scanning calorimetry and thermogravimetric analysis) as well as water mobility (nuclear magnetic resonance) in addition to image analysis (microscopy and particle size) [5–7].

Sensory evaluation includes both consumer preference and trained panels. Consumer panels reveal human preferences for appearance, aroma, taste and texture. It can be performed traditionally in sensory booths, or with modern techniques such as immersive technologies and augmented reality. The goal is to predict consumers' acceptability of food products. Tests include hedonic scale, threshold, the triangle test and others [8]. Trained panels are used in focus groups, which allow us to study specific attributes with experts of each type of food. This technique is useful in describing new foods as well as in the investigation of their shelf-life stability [9].

Sustainable food supply is a contemporary issue of high relevance. Societies must be able to produce food sustainably, meaning with lower environmental impact (less carbon and water footprint, and minimized land use), high nutritional quality, safety and sensory quality. Local crops, plant-based foods and upcycling of processing side streams are three answers to this call. The application of grains to non-traditional foods (egg, dairy, meat alternatives) and traditional (bakery) offers new ways to deliver nutrition along with high taste. Examples of upcycling include *aquafaba* and *liluva* (the processing water of legumes) used in egg replacers or as alternatives to hydrocolloids [10–12].

In recent years, there has also been attention focused on the bioactive ingredients of cereal grains and their benefits in terms of nutritional well-being [13,14]. However, these bioactive ingredients (fibre and phenolic compounds for instance) can affect the physical nature of foods as well as their sensory quality [15]. This is particularly the case when considering the use of wholegrains in foods [16].

Therefore, the aim of this Special Issue is to illustrate the latest scientific advances in the field of grain-based foods, investigating their physicochemical properties and sensory qualities. The focus is on sustainable solutions such as local crops (amaranth, ancient grains, buckwheat, maize, quinoa, rice, and spelt), plant-based products (yoghurt and egg alternatives) and upcycled ingredients (*aquafaba*, *liluva*, and pomace).

Author Contributions: Conceptualization, L.S., C.B. and R.M.; methodology, L.S., C.B. and R.M., software, L.S., C.B. and R.M.; validation, L.S., C.B. and R.M.; formal analysis, L.S., C.B. and R.M.; investigation, L.S., C.B. and R.M.; resources, L.S., C.B. and R.M.; data curation, L.S., C.B. and R.M. writing—original draft preparation, L.S.; writing—review and editing, L.S. and C.B.; visualization, L.S., C.B. and R.M.; supervision; L.S., C.B. and R.M. Project administration, L.S., C.B. and R.M.; funding acquisition, L.S., C.B. and R.M. All authors have read and agreed to the published version of the manuscript.

Funding: This research received no external funding.

Conflicts of Interest: The authors declare no conflict of interest.

References

1. Carcea, M. Nutritional value of grain-based foods. *Foods* **2020**, *9*, 504. [CrossRef] [PubMed]
2. Hui, Y.H.; Corke, H.; De Leyn, I.; Nip, W.K.; Cross, N.A. (Eds.) *Bakery Products: Science and Technology*; John Wiley & Sons: New York, NY, USA, 2008.
3. Satija, A.; Hu, F.B. Plant-based diets and cardiovascular health. *Trends Cardiovasc. Med.* **2018**, *28*, 437–441. [CrossRef] [PubMed]
4. Woomer, J.S.; Adedeji, A.A. Current applications of gluten-free grains—A review. *Crit. Rev. Food Sci. Nutr.* **2021**, *61*, 14–24. [CrossRef] [PubMed]
5. Joyner, H.S. Explaining food texture through rheology. *Curr. Opin. Food Sci.* **2018**, *21*, 7–14. [CrossRef]
6. Schiraldi, A.; Dimitrios, F. Calorimetry and thermal analysis in food science. *J. Therm. Anal. Calorim.* **2019**, *138*, 2721–2732. [CrossRef]
7. Chen, Q.; Tai, X.; Li, J.; Li, C.; Guo, L. High Internal Phase Emulsions Synergistically Stabilized by Sodium Carboxymethyl Cellulose and Palm Kernel Oil Ethoxylates as an Essential Oil Delivery System. *J. Agric. Food Chem.* **2021**, *69*, 4191–4203. [CrossRef] [PubMed]
8. Aschemann-Witzel, J.; Ares, G.; Thøgersen, J.; Monteleone, E. A sense of sustainability?—How sensory consumer science can contribute to sustainable development of the food sector. *Trends Food Sci. Technol.* **2019**, *90*, 180–186. [CrossRef]
9. Djekic, I.; Lorenzo, J.M.; Munekata, P.E.; Gagaoua, M.; Tomasevic, I. Review on characteristics of trained sensory panels in food science. *J. Texture Stud.* **2021**, *52*, 501–509. [CrossRef]
10. Campos-Vega, R.; Oomah, B.D.; Vergara-Castaneda, H.A. (Eds.) *Food Wastes and By-Products: Nutraceutical and Health Potential*; John Wiley & Sons: New York, NY, USA, 2020.
11. Mustafa, R.; Reaney, M.J. Aquafaba, from food waste to a value-added product. In *Food Wastes By-Products: Nutraceutical Health Potential*; Wiley: Hoboken, NJ, USA, 2020; pp. 93–126. [CrossRef]
12. Serventi, L. (Ed.) *Upcycling Legume Water: From Wastewater to Food Ingredients*; Springer Nature: Berlin, Germany, 2020; p. 174.
13. Radoš, K.; Čukelj Mustač, N.; Drakula, S.; Novotni, D.; Benković, M.; Kraljić, K.; Štifter, S.; Voučko, B.; Ćurić, D. The effect of cryo-grinding and size separation on bioactive profile of buckwheat hulls. *Int. J. Food Sci. Technol.* **2022**, *57*, 1911–1919. [CrossRef]
14. Yin, R.; Fu, Y.; Yousaf, L.; Xue, Y.; Hu, J.; Hu, X.; Shen, Q. Crude and refined millet bran oil alleviate lipid metabolism disorders, oxidative stress and affect the gut microbiota composition in high-fat diet-induced mice. *Int. J. Food Sci. Technol.* **2022**, *57*, 2600–2610. [CrossRef]
15. Chumsri, P.; Chaijan, M.; Panpipat, W. A comparison of nutritional values, physicochemical features and in vitro bioactivities of Southern Thai short-grain brown rice with commercial long-grain varieties. *Int. J. Food Sci. Technol.* **2021**, *56*, 6515–6526. [CrossRef]
16. Allai, F.M.; Azad, Z.; Gul, K.; Dar, B.N. Wholegrains: A review on the amino acid profile, mineral content, physicochemical, bioactive composition and health benefits. *Int. J. Food Sci. Technol.* **2022**, *57*, 1849–1865. [CrossRef]

Article

Physicochemical Properties and Mouthfeel in Commercial Plant-Based Yogurts

Maija Greis [1,2,*], **Taru Sainio** [1], **Kati Katina** [1], **Alissa A. Nolden** [2], **Amanda J. Kinchla** [2], **Laila Seppä** [1] **and Riitta Partanen** [3]

[1] Department of Food and Nutrition, University of Helsinki, P.O. Box 66, 00014 Helsinki, Finland; taru.sainio@helsinki.fi (T.S.); kati.katina@helsinki.fi (K.K.); laila.seppa@helsinki.fi (L.S.)
[2] Department of Food Science, University of Massachusetts Amherst, Amherst, MA 01003, USA; anolden@umass.edu (A.A.N.); kinchla@umass.edu (A.J.K.)
[3] Valio Ltd., P.O. Box 10, FI-00039 Helsinki, Finland; riitta.partanen@valio.fi
* Correspondence: maija.greis@helsinki.fi

Abstract: There is a growing need for plant-based yogurts that meet consumer demands in terms of texture. However, more research is required to understand the relationship between physicochemical and mouthfeel properties in plant-based yogurts. The purpose of this study was to determine the physicochemical properties of five commercial plant-based yogurt alternatives with different chemical compositions, making comparisons to dairy yogurts and thick, creamy, thin, and watery mouthfeel sensations. The physicochemical parameters studied included large and small deformation rheology, particle size, soluble solids, acidity, and chemical composition. Significant differences in flow behavior and small deformation rheology were found between dairy- and plant-based yogurts. Among plant-based yogurts thick, creamy, thin, and watery mouthfeel sensations were strongly associated with steady shear rates and apparent viscosity. The results highlight the importance of large deformation rheology to advance the use of plant-based ingredients in the development of yogurt alternatives. Furthermore, this study demonstrates that dairy- and plant-based yogurts with a similar mouthfeel profiles may have different viscoelastic properties, which indicates that instrumental and sensory methods should not be considered substitutive but complementary methods when developing plant-based yogurts in a cost-effective and timely manner.

Keywords: physicochemical properties; rheology; sensory evaluation; dynamic mouthfeel perception; plant-based yogurt alternative; oat

Citation: Greis, M.; Sainio, T.; Katina, K.; Nolden, A.A.; Kinchla, A.J.; Seppä, L.; Partanen, R. Physicochemical Properties and Mouthfeel in Commercial Plant-Based Yogurts. *Foods* **2022**, *11*, 941. https://doi.org/10.3390/foods11070941

Academic Editors: Luca Serventi, Charles Brennan and Rana Mustafa

Received: 24 February 2022
Accepted: 21 March 2022
Published: 24 March 2022

Publisher's Note: MDPI stays neutral with regard to jurisdictional claims in published maps and institutional affiliations.

1. Introduction

In terms of dairy alternatives, oat-based products are a popular substitute due to their mild flavor properties and potential positive health benefits [1]. The functional properties inherent to plant-based ingredients often include a lower gelling strength compared to animal-based systems; therefore, the gelling structures are enhanced through the use of hydrocolloids [2–5]. In previous work, we reported that the sensory properties among oat-based yogurts differ, some of them resembling their dairy counterparts, both in mouthfeel and pleasantness [6]. Due to the complexity and variety in the composition of these products, it is difficult to explain their mouthfeel differences through compositional factors alone. Therefore, rheology, with the help of acidity, soluble solids, and particle size measurements, was applied to better understand the mouthfeel sensations and pleasantness of these plant-based yogurts.

There is extensive prior literature exploring the relationship between the rheological properties and sensory attributes of dairy yogurts [4,7–14]. Other physicochemical parameters have also been successfully linked to mouthfeel in dairy yogurts. Particle size-related parameters have been shown to influence the creamy mouthfeel [14–18]. In addition, the reduction in sugar in dairy yogurt has been linked to a decrease in viscosity, resulting in a thin and watery mouthfeel [19]. According to another study, a watery mouthfeel is the opposite to a creamy one and relates to low-fat content in emulsion-filled gels [20].

An increasing number of studies are exploring the consumer acceptance and physicochemical properties of different plant-based yogurts [2,5,21,22]. A noteworthy study reports the rheological properties, sensory perception, and consumer acceptability of lactic acid fermented, oat-based gels [2]. They demonstrated that a gel with a higher total solids content was perceived as creamier compared to a gel with a lower total solids content. Another study reports the compositional and physicochemical properties with liking of different commercial plant-based yogurts [20]. They concluded that soy, coconut, and cashew yogurts scored similarly in terms of texture liking as dairy yogurts. A more recent study aims to understand the sensory acceptability and textural properties in Australian commercial dairy and plant-based samples [20]. The selected soy, coconut, and dairy yogurts showed wide variations in their microstructure and rheology. The results highlight that the protein content, gel firmness, and consistency coefficient displayed a positive relationship with overall liking [21]. Notably, these previous studies did not include oat-based yogurts in their experiments [21,22].

Our study aimed to determine the physicochemical properties of plant-based yogurts. The results were compared to dairy counterparts and previously studied mouthfeel properties. Our hypothesis is that oat-based structures are predominantly carbohydrate gels, and thus provide a more fine-stranded network compared to dairy yogurts. Instead, dairy yogurts provide a distinguished particle gels system attributed to the network of protein particles and protein-covered fat droplets. Our previous findings suggest that the dominant mouthfeel attributes perceived during the early stages of mastication have a larger impact on mouthfeel pleasantness than the dominant attributes perceived later during mastication [6]. Therefore, conventional rheological methods are expected to be relevant in determining factors that contribute to mouthfeel liking and disliking.

We will examine these questions using a variety of commercial products. They represent a wide range of mouthfeel properties that would not be achievable if using a simple, controlled model product. By choosing a set of unflavored commercial products from the same plant source, we limit the differences in flavor and thus focus only on the mouthfeel. In this experiment, our focus is on four following specific positive and negative mouthfeel sensations contributing to the liking of the products: thickness and creaminess (positive) and wateriness and thinness (negative) based on the findings in our previous study [6].

2. Materials and Methods

2.1. Samples

Five unflavored plant-based yogurt alternatives (P1-P5) and two unflavored dairy yogurts (D1-D2) were purchased from a local supermarket (Table 1) in Finland. The plant-based products were spoonable yogurt-like semisolid snacks labeled as "oat-based yogurts". Dairy-based references included two spoonable dairy yogurts (fat contents of 2.5% and 4%). All samples were fermented with the help of an added starter. These yogurt alternatives were selected due to their different structures. In addition, they represent the variety of oat-based yogurt alternatives in the market. The reference samples resembled typical dairy yogurts in the market. The products were sourced in duplicate so that analysis could be split for sensory [6], and physicochemical analyses. All samples were stored at 5 °C prior to the sensory and physicochemical analyses. All samples were analyzed both in the sensory analysis and instrumental measurements at 10 °C within their declared shelf-life period. The studied yogurt alternatives are referred to as "plant-based" instead of "oat-based yogurts", as they contain pea and potato protein in addition to oat protein. All instrumental measurements were performed in triplicate, apart from particle size assessment, where three separate measurements were conducted for each sample. A summary of the analysis is presented in Table 2.

Table 1. The bases, thickeners, stabilizers, and oils as declared on the labels of all the samples.

	Base	Thickener	Stabilizer or Preservative	Oil (g/100 mL)
D1	Dairy	None	None	Milk fat (2.5)
D2	Dairy	None	None	Milk fat (4)
P1	Oat base (water, oat 12%), potato protein	Potato starch	Calcium carbonate (E170), Tricalcium phosphate (E341)	Rapeseed oil (2.2)
P2	Oat base (water, oat 8.5%)	Modified starch, pectin	Potassium sorbate (E202)	Canola oil (2.4)
P3	Oat base (water, oat flakes 8%)	Starch (corn, potato), pectin	Tricalcium phosphate (E341)	Canola oil (2.5)
P4	Water, oat 12%, and potato protein	Starch (tapioca, potato), xanthan, and locust bean gum	None	Canola oil (0.8)
P5	Oat base (water, oat 8.2%), pea protein	Modified potato starch	None	Canola oil (0.9)

Table 2. Overview of the physicochemical parameters extracted from instrumental measurements.

Type of Measurement	Explanation	Codes
Large deformation test: Steady shear rate (SS)	η at $10\,s^{-1}$ at $t = 10\,s$. η at $50\,s^{-1}$ at $10\,s$.	SS10 SS50
Large deformation test: Flow curves (FCs)	The area of the hysteresis loop between the upward and downward curves Shear thinning index, n, and consistency, K, were calculated from the power law ($\eta = K*\dot{\gamma}n^1$) from the upward flow curve Apparent viscosities (η_{app}) from upward flow curve (Pa·s) calculated from Ostwald-de Waele $= K\dot{\gamma}^{\wedge}(n^{-1})$ at shear rates 1.5, 5, 10, 25, and 50 (1/s)	HL n, K ηapp10
Small deformation test: Dynamic strain sweeps (DSSs)	Stress (G') at the end point of LVER Strain (γ) at the end point of LVER	G'LVE γLVE
Small deformation test: Dynamic frequency sweep (DFS)	G' at 1 Hz, Pa (DFS G'1 Hz) G'' at 1 Hz, Pa (DFS G''1 Hz)	G' G''
Particle size	Surface weighted particle size Volume weighted particle size 90th percentile of the particles less than d[0.9]	d[3.2] d[4.3] d[0.9]
Chemical composition	Fat content Carbohydrate content Sugar content Fiber content Protein content Oat content	Fat Carboh. Sugar Fiber Proteins Oat
Soluble solids Acidity	°Brix pH Total titratable acidity	°Brix pH TTA

2.2. pH and Titratable Acidity

The pH and titratable acidity (TTA) were analyzed from both sets of samples (sensory and instrumental) in order to confirm the statistical similarity between experiments. Total titratable acidity was analyzed using instrumental analysis: 10 g of each sample was homogenized (1 min) with 10 mL of acetone and 90 mL of Milli-Q water using a Bamix blender (Switzerland), as described in [23]. The TTA was determined as the amount of 0.1 M NaOH required to adjust the end pH of samples to 8.5. A pH meter (Model HI 99161,

Hanna Instruments, Woonsocket, RI, USA) and TTA titrator (EasyPlus Titration, Mettler Toledo, Columbus, OH, USA) were used for measurements.

2.3. Soluble Solids

For soluble solid analysis samples were centrifuged for 10 min at $7200 \times g$ (Galaxy MiniStar, VWR, Radnor, PA, USA). Soluble solids were determined with a digital refractometer (Pocket Refractometer PAL-1, Atago, Tokyo, Japan) from the resulting supernatant. The results are given as degrees °Brix at 10 ± 0.2 °C.

2.4. Particle Size Measurement

The particle size distribution of the samples was determined by static light scattering using a Malvern Mastersizer 3000 (Malvern Instruments, Worcestershire, UK) with an absorption parameter value of 1.5 and refractive index ratio of 1.33. Each sample was diluted with Milli-Q water at 1:50 and mixed for 30–45 min with a magnet mixer. The average d[4.3] and Sauter mean (d[3.2]) corresponding to fine microgel particles are both reported to compare differences in the average volume-weighted and surface weighted particle sizes, respectively. The 90th percentile d[0.9] is also reported to represent the distribution of coarser particles and is used to interpret the sensory perception data as shown in [16,17,24].

2.5. Rheological Measurements

The rheological behavior of plant-based and dairy yogurts was characterized by using flow curve, steady shear, and dynamic shear measurements adopted from previous literature [11,16,17,25,26]. All measurements were conducted with a HAAKE MARS 40 Rheometer and monitored by a RheoWin software package, version 2.93 (Thermo Fisher Scientific, Waltham, MA, USA). Samples were analyzed at 10 °C. A cone-plate configuration (cone diameter 35 mm, angle 2°, and gap 0.100 mm) was used in steady shear measurements and flow curves. A plate-plate configuration (diameter 35 mm, gap 1.500 mm) was used in dynamic shear measurements.

2.5.1. Steady Shear Data

The sample (0.4 mL) was placed between cone and plate and then covered with a solvent trap to avoid water evaporation during the resting and measurement. Samples were allowed to rest for 5 min before measurement and a fresh sample was loaded for each measurement. The steady flow properties of each sample were measured at two steady shear rates 10 s^{-1} and 50 s^{-1} [11,16,17,25–27]. Viscosity was measured for 120 s while one data point per one second was collected (120 points). In order to understand the thixotropic behavior of the samples, viscosity was plotted against time (s) at constant share rates (10 s^{-1} and 50 s^{-1}).

2.5.2. Flow Curves

Flow curves (FCs) were obtained from stepped shear stress ramp between 0.01 s^{-1} and 1000 s^{-1} [25]. The shear rate increased logarithmically for 200 s and then decreased logarithmically for 200 s from 1000 s^{-1} to 0.01 s^{-1}. The apparent viscosity was plotted against shear rate to examine the shear thinning behavior. To analyze the recovery of the structure, the area of the hysteresis loop (HL) was determined. Based on the flow curves (between 0.01 and 1000 s^{-1}), the consistency index, K, and shear thinning index, n, were calculated using the power law equation (Table 2). Apparent viscosities (η_{app}) at shear rates of 10 (s^{-1}) from the upward flow curve (Pa·s) were calculated from the Ostwald-de Waele equation.

2.5.3. Dynamic Shear Data

The viscoelastic properties of the samples were studied by strain sweeps and frequency sweeps [11,16,17,25,26]. A plate-plate configuration (diameter 35 mm, gap 1.500 mm) was used in the measurements. The sample (1.5 mL) was placed between the plates and covered with a solvent trap to avoid water evaporation during the resting and measurement. To determine the linear viscoelastic region (LVER), strain sweeps were run at 1 Hz. For the strain sweeps, the step-wise γ increased logarithmically from 0.0001 to 1. The end point of the linear viscoelastic region, thus the point where G$'$ was 10% lower than the plateau phase of linear viscoelastic region, was measured as stress (G$'$) and strain (γ). All the frequency sweeps were then performed within the linear viscoelastic region at the following a constant deformation: γ = 0.001 and over the range of f = 0.01–10 Hz. The values of the storage modulus (G$'$) and the loss modulus (G$''$) were plotted.

2.6. Sensory Analysis

The dynamic mouthfeel perception of the samples was collected by temporal dominance of sensation (TDS) with a consumer test. The participants (n = 87) in the study reported consuming either yogurt or yogurt alternatives. A full description of the applied sensory methods, the statistical analysis, and the results can be found in detail in our previous study [6]. According to our previous results, the drivers of mouthfeel liking in plant-based yogurts are thickness and creaminess and the drivers of disliking are wateriness and thinness. These four characteristics were chosen for the present analysis to investigate the physicochemical-mouthfeel relationship. A product average of the dominance durations for each attribute was calculated from the temporal data. The dominance durations are not an approximate visual summary of the panel but represent the average durations of dominant attributes, i.e., for how long each attribute was selected during the mastication. The dominance duration is a recommended parameter to be used when testing product differences in multivariate analysis [28]. It represents the magnitude of the selected attribute among the consumers. Dominance durations have been extracted using left-right standardized individual TDS sequences. This was performed so that panelists with longer perception times would not have more weight in the product means.

2.7. Data Analysis

To compare the physicochemical properties between plant-based and dairy yogurts, different parameters were calculated. Steady shear data, flow curves, dynamic strain sweeps, and dynamic frequency sweeps were extracted from the rheological data. In addition, particle size diameters d[3.2], d[4.3], and d[0.9], soluble solids, acidity, and compositional parameters were taken into further analysis. The instrumental data for all parameters measured were examined and determined normally distributed using the Shapiro-Wilk test. One-way analysis of variance was performed on all the instrumental measurements. When the effect was significant, Tukey's test was applied to determine differences between samples. All analyses were performed in triplicate using SPSS version 25 (SPSS Inc., Chicago, IL, USA).

In order to visualize which of the physicochemical and previously studied mouthfeel sensations contribute most to the differences between plant-based and dairy yogurts, principal component analysis (PCA) was conducted. PCA is a procedure that examines the relationships among a set of correlated variables. The obtained results were visualized graphically by projecting the samples (scores) and physicochemical as well as mouthfeel variables (loadings) onto the space defined by the two first PCs.

To determine if the previously studied mouthfeel sensations (thick, thin, creamy, and watery) could be explained by physicochemical properties in plant-based yogurts, a relationship between two datasets among plant-based yogurts was summarized and visualized by partial least squares regression (PLS-R). In addition, Pearson correlations were analyzed to support the results of the PLS-R. All extracted physicochemical parameters and mouthfeel sensations (thick, creamy, thin, and watery) were included for the analysis. PLS regression is designed to determine relationships existing between dependent (Y, mouthfeel sensations) and explanatory (X, physicochemical properties) variables by seeking underlying factors common to both sets of variables [29]. The model was developed using internal cross-validation based on y, mouthfeel sensations, and X, physicochemical properties. PLS-R is a suitable model because it allows for small to medium sample sizes, a large number of independent variables, and is robust to multicollinearity. Both PCA and PLS were analyzed using Unscrambler (Unscrambler 7.6 SR-1, Camo Asa, Oslo, Norway).

3. Results

3.1. Acidity

The acidity differences between the instrumental and sensory batches were small, indicating similarities between the batches (Table 3) and thus validating their comparison. The pH in both dairy and nondairy samples ranged from 3.4 and 4.4, with one sample (P4) having a significantly lower pH (3.4) compared to the other samples. The total titratable acidity showed clear differences between dairy and plant-based samples. Dairy yogurts had significantly higher TTA compared to plant-based yogurts, and P2 and P3 had the lowest TTA, 2.20 and 2.18, respectively.

Table 3. pH and TTA of all the samples in the instrumental analysis (±standard deviation) and difference to the samples in the sensory analysis. Superscript letters indicate statistical difference between the samples, in the same column ($p < 0.05$).

	pH		TTA	
	Instrumental Analysis ± STD	**±Sensory Analysis**	**Instrumental Analysis ± STD**	**±Sensory Analysis**
D1	4.27 ± 0.12 [bc]	-0.06	10.83 ± 0.09 [a]	-0.18
D2	4.18 ± 0.12 [bc]	-0.12	10.86 ± 0.10 [a]	0.22
P1	4.16 ± 0.08 [c]	0.01	4.43 ± 0.14 [c]	-0.09
P2	4.26 ± 0.10 [b]	-0.07	2.00 ± 0.08 [d]	-0.05
P3	4.43 ± 0.11 [a]	-0.17	2.18 ± 0.16 [d]	0.14
P4	3.47 ± 0.12 [d]	0.06	5.36 ± 0.45 [b]	0.23
P5	4.26 ± 0.11 [bc]	-0.08	5.54 ± 0.33 [b]	0.21

3.2. Soluble Solids

Figure 1 shows the calculated °Brix with the carbohydrates, sugars, and proteins that are obtained from the label information. The soluble solids (°Brix, %) ranged from 7.0 to 10.3 between all the samples, with P1 and P4 having the highest while P3 and P5 having the lowest °Brix among the plant-based samples. The figure demonstrates that samples with higher total carbohydrate content (P1, P2, and P4) also have the highest °Brix values.

3.3. Particle Size Measurements

The smallest particles by diameter were discovered in sample P4 (d[3.2] = 14 µm) (Table 4). The d[3.2] values ranged from 14 to 36 µm and 20 to 21 µm in plant-based and dairy yogurts, respectively. The d[4.3] values ranged from 22 to 68 µm and 27 to 28 µm in plant-based and dairy yogurts, respectively. The d[0.90] values ranged from 42 to 151 µm and 47 to 52 µm in plant-based and dairy yogurts, respectively. Sample P5 had the greatest particle size (d[0.90] = 151 µm) among all the samples. Compared to other plant-based samples, P3 had the most similarities with dairy yogurts in particle size and diameters.

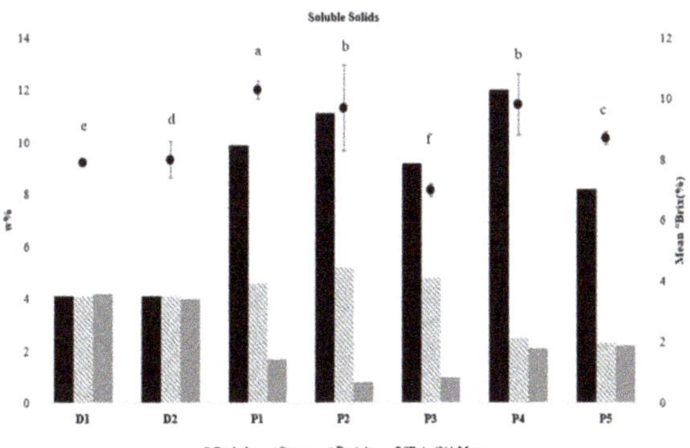

Figure 1. The final carbohydrate, sugar, and protein content as labelled in the products (w-%) and °Brix (%) with standard deviation. Superscript letters indicate statistical difference in °Brix (%) between the samples ($p < 0.05$).

Table 4. Particle size diameters (±standard deviation) of all the samples. Superscript letters indicate statistical difference in the same row ($p < 0.05$).

	D1	D2	P1	P2	P3	P4	P5
d[3.2] + s.d. (μm)	20 ± 0.2 [bc]	21 ± 0.3 [b]	15 ± 0.1 [e]	36 ± 0.4 [a]	20 ± 0.1 [c]	14 ± 0.1 [f]	19 ± 0.1 [d]
d[4.3] + s.d. (μm)	27 ± 0.2 [d]	28 ± 0.9 [c]	24 ± 0.5 [e]	48 ± 0.1 [b]	30 ± 0.4 [c]	22 ± 0.3 [f]	68 ± 1.2 [a]
d[0.9] + s.d. (μm)	47 ± 0.6 [e]	52 ± 2.2 [cd]	48 ± 1.4 [de]	76 ± 0.7 [b]	56 ± 0.7 [c]	42 ± 0.4 [f]	151 ± 3.5 [a]

3.4. Rheological Measurements

3.4.1. Steady Shear Data

Different parameters help to articulate discernable rheological differences among samples (Table 5). All the samples showed thixotropic behavior at steady shear rates ($10\ s^{-1}$ and $50\ s^{-1}$), thus demonstrating structural breakdown under flow. For most of the samples, the viscosity decreased rapidly at the beginning of the measurement and then decreased slowly, staying nearly constant (Table 5). The dairy yogurts had a stronger decline in their viscosity than in the plant-based samples. Particularly at shear rates of 5 and $10\ s^{-1}$, samples P2 and P3 showed similar behavior to dairy yogurts compared to other plant-based samples (Figure 2B). The viscosity of sample P2 remained nearly constant after the first drop at the beginning of the measurement (Figure 2A). Yet, a higher shear rate was associated with a lower viscosity also for P2.

Table 5. The mean value of the rheological parameters of all the samples. Superscript letters indicate statistical differences in the same row ($p < 0.05$).

	D1		D2		P1		P2		P3		P4		P5	
SS10 (Pa s)	3.92	±0.20 [b]	4.82	±0.17 [a]	1.99	±0.05 [d]	4.20	±0.06 [b]	4.20	±0.03 [b]	2.76	±0.06 [c]	2.50	±0.06 [c]
SS50 (Pa s)	1.53	±0.05 [b]	1.99	±0.11 [a]	0.52	±0.01 [e]	1.55	±0.00 [b]	1.01	±0.01 [c]	0.90	±0.01 [cd]	0.85	±0.01 [d]
HL (-)	57,416.48	±1479.05 [b]	59,720.44	±1242.04 [a]	10,937.42	±148.72 [d]	−4647.42	±152.23 [e]	17,678.42	±291.84 [c]	11,022.59	±177.27 [d]	16,278.60	±22.35 [c]
n (-)	0.31	±0.01 [b]	0.28	±0.01 [c]	0.15	±0.01 [d]	0.35	±0.00 [a]	0.15	±0.02 [d]	0.31	±0.00 [bc]	0.36	±0.01 [a]
K (Pa s^n)	21.15	±1.65 [b]	26.00	±2.06 [a]	14.02	±0.28 [c]	18.91	±0.15 [b]	27.94	±1.18 [a]	13.52	±0.08 [cd]	10.73	±0.09 [d]
η app10 (1/s)	4.35	±0.27 [b]	4.94	±0.29 [a]	1.99	±0.02 [d]	4.27	±0.03 [b]	3.97	±0.02 [b]	2.75	±0.01 [c]	2.45	±0.03 [c]
G′LVE (Pa)	302.00	±15.46 [b]	380.71	±37.11 [a]	59.08	±3.41 [de]	77.73	±1.39 [d]	195.05	±12.77 [c]	16.48	±0.92 [e]	24.90	±2.35 [e]
γ LVE (-)	0.01	±0.00 [c]	0.01	±0.00 [c]	0.02	±0.00 [c]	0.03	±0.00 [bc]	0.07	±0.02 [a]	0.06	±0.00 [ab]	0.06	±0.01 [a]
G′ (Pa)	303.30	±14.57 [a]	431.60	±13.47 [b]	61.22	±5.45 [e]	89.30	±1.01 [d]	226.15	±1.75 [c]	17.69	±1.58 [f]	25.67	±1.10 [f]
G″ (Pa)	74.60	±2.52 [a]	102.88	±0.81 [b]	7.21	±0.27 [de]	23.13	±0.12 [d]	14.81	±0.06 [c]	8.53	±0.67 [e]	10.51	±0.23 [e]

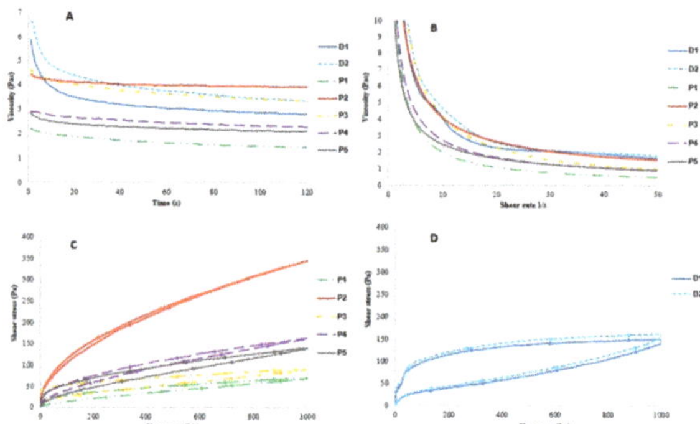

Figure 2. (**A**) Viscosity (Pa·s) during 120 s at a steady shear rate of $10\ \mathrm{s}^{-1}$. (**B**) Flow curve: viscosity (Pa·s) by shear rate (s^{-1}). (**C**) Hysteresis loops in plant-based yogurts. (**D**) Hysteresis loops in dairy yogurts.

3.4.2. Flow Curves

All the samples showed shear thinning behavior ($n < 1$) as the apparent viscosity decreased by increasing the shear rate (Figure 2B) in all the samples (Table 5). The thixotropic properties were measured by calculating the hysteresis loops, i.e., the area between the forward and backward curves (Figure 2C,D). A greater area within the hysteresis loops was reported with dairy yogurts (D1-D2) compared to other yogurts. Furthermore, for sample P2, the hysteresis loop showed the following different behavior compared to other samples: the forward and backward curves were partly overlapping (within 500–$1000\ \mathrm{s}^{-1}$), the backward curve being also partly higher than the forward curve, indicating reversible shear-thinning behavior (Figure 2C).

3.4.3. Dynamic Shear Data

Frequency sweeps showed that elastic properties dominated in the linear viscoelastic area. Examples of the viscoelastic properties of samples P2 and P3 as well as the dairy samples are shown in Figure 3. There are significant disparities among the samples in the storage modulus, indicating that the samples represent a wide range of texture properties, particularly in rigidity. All samples had $G' > G''$ and thus can be described as soft fluid gels (Table 5). The storage modulus of the dairy samples as well as samples P2 and P3 was significantly higher than the storage modulus for other samples, indicating a more rigid structure compared to other products. This could be due to a high fat content in samples P2 and P3, 2.4 and 2.4 g/100 mL, respectively. Sample P3, however, had the lowest storage modulus, while also the lowest fat content, at 0.8 g/100 mL.

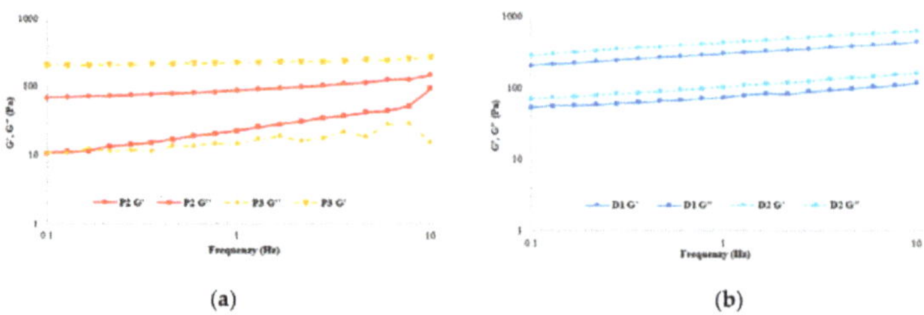

(a)   (b)

Figure 3. An example of the viscoelastic properties of both types of the following samples: plant-based samples P2 and P3 in Figure (**a**) and dairy samples in Figure (**b**).

3.5. Physicochemical Differences in Dairy- and Plant-Based Yogurts

A PCA analysis was applied to demonstrate the positioning of the plant-based and dairy yogurts when the average values of the instrumental and mouthfeel properties were applied. The first two components accounted for 70% of the total variability. The biplot graph (Figure 4) visualizes the similarities and differences between the products in physicochemical and mouthfeel properties. The first component, which explained the higher percentage of variability (53%), separated the samples clearly according to their viscosity, including both large and small deformation tests. Products P1, P4, and P5 were in the negative part of the first component, dairy products were in the positive part of the component, and products P2 and P3 were in the middle. The second component, which accounted for 17% of the variability, separated P2 and P3 from the other samples, at least according to the differences in steady shear viscosity compared to the other samples. In addition, samples were separated by negative and positive mouthfeel sensations, which were placed on opposite sides of the scale in both of the PCs. Furthermore, the rheological parameters and the PCA graph indicate a pattern between the following large and small deformation tests: Large deformation tests correlate positively with plant-based yogurts P2 and P3, whereas small deformation tests represent dairy yogurts.

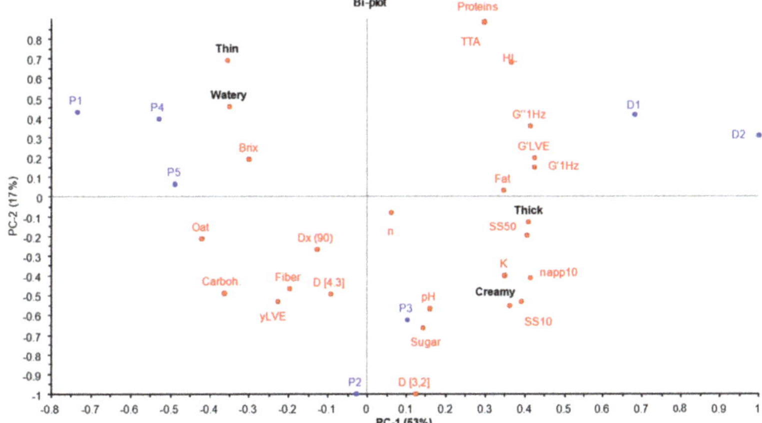

Figure 4. PCA biplot (scores and loadings) of the physicochemical properties for plant-based and dairy yogurts plus the following mouthfeel sensations: thick, creamy, thin, and watery. The abbreviations of the physicochemical parameters are in accordance with Table 2.

3.6. Physicochemical and Mouthfeel Properties among Plant-Based Yogurts

The relationship between the mouthfeel attributes and the physicochemical properties among plant-based yogurts was studied and visualized by PLS regression (Figure 5). Mouthfeel sensations used in the PLS regression figure represent the dominance durations for each attribute and thus describe the magnitude of each attribute. The first factor of the PLS regression model explained 43% of the variation in the physicochemical results and 78% of the variation in the sensory data within the five samples analyzed. The second factor explained 23% and 16%, respectively. Altogether, the PLS regression model explained 66% of the variation in the instrumental and 94% of the variation in the mouthfeel using the two first principal components, i.e., the variation in the samples is explained more specifically by sensory results than by instrumental results, according to the two components. However, the relatively large explanation rate in both results suggests that instrumental and sensory analyses are complementary methods. To determine which of the parameters had a significant correlation with the mouthfeel sensations, Pearson correlations were studied among the plant-based samples (Table A1).

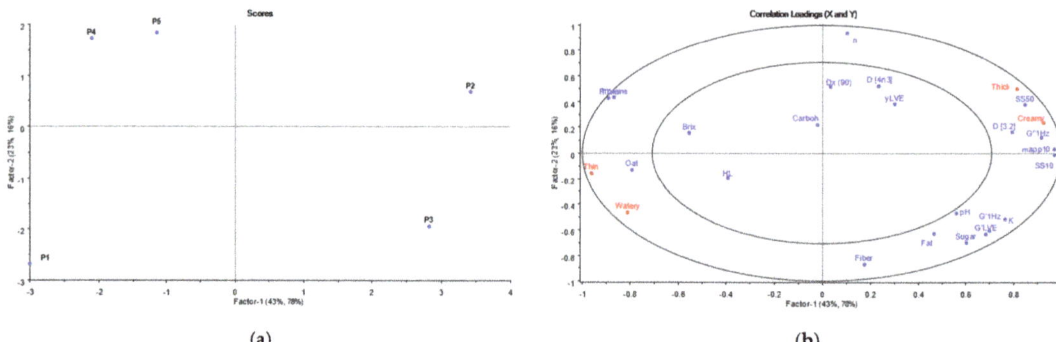

Figure 5. PLS regression bi-plots for scores (**a**) and for loadings (**b**) of sensory and physicochemical parameters for five plant-based yogurts. The abbreviations of physicochemical parameters are in accordance with Table 2.

4. Discussion

4.1. Rheological Measurements

The flow behavior index (n) values were consistent with those in previous literature, with almond yogurt showing the most similarities with oat yogurts [21], while the consistency coefficient (K) values of the studied plant-based yogurts were different to those of soy, coconut, cashew, almond, and hemp yogurts [21,22]. The greater area within the hysteresis loops reported with dairy yogurts (D1-D2) suggests stronger thixotropic behavior, which can be interpreted as a more permanent structure breakdown. This is a well-known behavior in dairy yogurts [30]. Plant-based products, however, showed visible yet significantly smaller hysteresis loop areas compared to dairy yogurts, indicating a faster structure recovery over time. This is most likely due to the difference in the gelling agents and the differences in their interactions. In plant-based yogurts, the closely packed polysaccharides can partly reform the structural network by means of noncovalent interactions. Another study showed similar results after the shearing of fermented oat-based gels, which were able to partially recover their initial structure [2]. They suggested that the higher total solid content contributes to more junction zones present between the particles, which results in the faster rearrangement of the microstructure. Our results support the same with soluble solids samples with the highest °brix (P1 and P4) had the smallest hysteresis loops, and samples with the lowest °brix had the largest hysteresis loops.

Furthermore, it has been demonstrated with oat-starch gels that the rate of the disentanglement of the macromolecules was higher than their re-entanglement during the shearing [31]. This results in a visible hysteresis loop and applies to the dairy and the majority of the plant-based yogurts in our study. In addition to the overlapping curves in hysteresis loops for sample P2, it also showed a significantly higher shear stress than the other products, reflecting its higher resistance to shear forces. The counterclockwise loop could be explained by the higher amount of remaining beta-glucan in the sample, thus contributing to the thickening behavior more effectively compared to other oat-based products. However, this interaction is not possible to discuss further, as the beta-glucan content of the samples was not analyzed. Sample P2 also contained modified starch, which could be a contributor to the different hysteresis loops. A similar pattern has been shown with solutions containing amylodextrin and beta-glucan [32].

Sample P3, however, had the lowest storage modulus, also the lowest fat content, at 0.8 g/100 mL. This is supported by another study which suggests that for a mixed food system such as fermented oat-based gel, it is likely that swollen starch granules, protein aggregates, and residual small fat droplets act as the fillers and are able to increase the rigidity (G′) of the system [2]. In addition, it has been demonstrated with polysaccharide gels that the gel-like behavior is related to molecular and physical interactions and thus the formation of the network structure [33]. It is, therefore, likely that added hydrocolloids contributed to the viscoelastic properties. Taking this into account, samples P2 and P3 were the only plant-based samples containing pectin in addition to starch. The loss moduli were also the highest for the thickest plant-based samples, P2 and P3, indicating stronger viscous behavior. At the endpoint of a linear viscoelastic area, stress (G'LVE) discriminated the samples more than the strain (γLVE) did.

4.2. The Physicochemical Dividers between Dairy and Plant-Based Yogurts

The most salient difference between plant-based and dairy yogurts is in the macromolecules that form their structures. The PCA biplot demonstrates how the macromolecules divide the samples according to their protein, fat, and carbohydrate contents. The lower protein content in plant-based yogurts (0.8 to 2.2 w-%) compared to dairy yogurts (4.0–4.2 w-%) may be seen in the lower viscosities at the beginning of the steady shear measurement, indicating a weaker initial structure. Even if bovine β-lactoglobulin has been shown to have a critical concentration for the sol-gel transition at 1%, as suggested by [34], the protein concentration in spoonable yogurts is typically greater than 3%. In the dairy yogurts of this study, the protein content was enough to build structures comparable to those obtained by various thickeners in the plant-based yogurts. All the plant-based products contained added thickeners, namely, potato, corn, tapioca starch, pectin, xanthan, or locust bean gum, which are the main contributors to the viscosities in plant-based gels. Furthermore, starch and cell wall polysaccharides are present in different amounts depending on the oat ingredient used [2–4].

4.3. Relationship between Physicochemical and Mouthfeel Properties among Plant-Based Yogurts

The PLS regression demonstrates that thickness and creaminess are associated with each other, consistent with prior studies indicating that creaminess results from a thick mouthfeel [35,36]. Moreover, increased viscosity has been linked with creaminess in dairy yogurts [10,35]. Our results indicate that of all the physicochemical parameters, rheological parameters showed the strongest connections with thickness and creaminess, particularly in large deformation tests in plant-based yogurts. Pearson correlations also support this; all the mouthfeel sensations are correlated with both steady shear rates (SS10 and SS50) and apparent viscosity either positively (thick and creamy) or negatively (thin and watery). Previously, the a shear rate of 50 s^{-1} has been regularly adopted by many authors with semisolids [27,36,37]. Positive correlations have been found with both 50 s^{-1} and 100 s^{-1} [16], as well as with mouthfeel thickness and 100 s^{-1} in dairy yogurts [14]. According to Pearson correlations, both shear rates are connected with all four mouthfeel sensations.

According to the PLS regression visualization, among small deformation tests, only the loss modulus (G″) was connected with thickness and creaminess. This indicates that the viscous properties are more strongly connected to the thickness and creaminess than the elastic properties. By contrast, the elastic properties (G′) have been associated with a viscous and fatty mouthfeel in dairy yogurts [16]. Our results indicate that fat content is associated with G′ in plant-based yogurts (Figure 5); however, the fat content or G′ do not describe the thickness or creaminess in the studied plant-based yogurts as much as they describe the thickness and creaminess in dairy yogurts (Figures 4 and 5). It has also been demonstrated that fat content is connected with a creamy and thick mouthfeel in dairy yogurts [38–40]. These differences between dairy- and plant-based yogurts could be due to the differences in fat content between the yogurts and the milk fat crystals melting in the mouth, which may contribute to the creamy mouthfeel, whereas, in plant-based yogurts, the canola and rapeseed oils are in liquid form. It has also been suggested that a creamy mouthfeel in dairy yogurts is strongly associated with the coalescence of emulsion droplets in the mouth and with the spreading of released fat at oral surfaces [41,42]. There is, however, previous evidence on the creamy mouthfeel in the following plant-based gels: an oat gel with a higher total solids content was perceived as creamier compared to a gel with a lower total solids content [2]. Furthermore, it has been suggested that structural components such as starch and protein aggregates create a smooth and thick mouthfeel in the absence of milk fat [36]. Another study with dairy and plant-based yogurts demonstrated that a high protein content provided a better gel firmness and a higher consistency coefficient (K) [21]. In addition, added starch in dairy yogurts has been shown to increase consistency, creaminess, and overall liking [43]. According to the PLS, the particle size parameter d[3.2] was associated with thick and creamy mouthfeel, whereas the other particle size parameters were not associated with any of the mouthfeel sensations. Previous literature suggests that a small particle size explains creaminess in dairy yogurts [14,16,24,26,36]. However, our results indicate that with a particle size d[3.2] of ≥ 20 μm, there is a connection to thickness and creaminess.

Both the PLS regression and the PCA graphs demonstrate that thin and watery are similar properties in plant-based yogurts (Figures 4 and 5). One explanation could be that the panelists were not able to distinguish wateriness from thinness. Another explanation could be that a watery mouthfeel is a consequence of the hydrolysis of starch by α-amylase, which present in saliva [44]. Our previous results support this: watery was perceived mainly after thinness [6]. The PLS graph and Pearson correlation indicate that thin and watery correlated negatively with fat content. A similar correlation has been demonstrated within emulsion-filled gels [20]. They showed that low-fat content relates to wateriness and that watery is the opposite of creamy. Additionally, as more saliva is added to the bolus, the perceived attributes have been found to relate to consistency (e.g., creaminess and wateriness) [8]. Furthermore, similar results were found with a descriptive analysis [45]. They concluded that the watery mouthfeel in semisolid gels is a chew-down property, whereas, sugar reduction in yogurts has been shown to result in a watery mouthfeel [19]. This should be investigated further. Interestingly, our results indicate that the samples with the lowest sugar content (P4 and P5) were perceived as watery, whereas samples with the highest sugar content (P2 and P3) were perceived as creamy.

5. Conclusions

There is a growing requirement for plant-based yogurts that meet consumer demands in terms of texture. Extensive previous literature demonstrates the relationship between physicochemical and mouthfeel properties in conventional dairy yogurts. However, more research is required on plant-based yogurts. The aim of the study was to determine the physicochemical properties of different commercial plant-based yogurts. The results were compared to those for dairy yogurts and previously studied mouthfeel sensations.

Plant-based yogurts exhibited a wide range of viscoelastic properties, which was a result of the fact that different hydrocolloids at different levels were incorporated in the samples at different levels. Our results also revealed some structural differences in the following two product groups: for example, a significantly stronger structure recovery was found in plant-based yogurts than in dairy yogurts, resulting from the differences in the gelling agents and their interactions. This study demonstrates that dairy and plant-based yogurts with a similar mouthfeel profile may have different viscoelastic properties. The considerable physicochemical differences between the two product groups are likely to also be valid with other similar yogurts as the selected samples in this study represent the typical dairy- and plant-based yogurts in the market. Further investigation is necessary to demonstrate this.

This study highlights the importance of rheological large deformation tests and their ability to explain essential mouthfeel sensations in plant-based yogurts. Thick and creamy mouthfeel sensations were positively correlated with steady shear rates and apparent viscosity. The results also suggest that oil content does not significantly affect creaminess in plant-based yogurts. The results emphasize that instrumental and sensory methods should not be considered substitutive but complementary methods when developing plant-based yogurts in a cost-effective and timely manner.

Limitations and Future Challenges

The presented relationships between the physicochemical parameters and mouthfeel are only valid within the studied plant-based yogurts. The results highlight that further investigation is necessary to demonstrate the impact of different macromolecules and hydrocolloids on the physicochemical and sensory properties in plant-based yogurts.

Author Contributions: Conceptualization and methodology: M.G., T.S., L.S., R.P. and K.K.; software, M.G. and T.S.; validation: L.S., R.P., K.K., A.J.K. and A.A.N.; formal analysis, investigation, data curation, and visualization: M.G. and T.S; writing—original draft preparation, M.G. and T.S.; writing—review and editing, R.P., L.S, K.K., A.J.K. and A.A.N.; resources, supervision, project administration, and funding acquisition: R.P., L.S. and K.K. All authors have read and agreed to the published version of the manuscript.

Funding: Maija Greis is financially supported by Valio Ltd. through a PhD position at the University of Helsinki.

Institutional Review Board Statement: The research procedure followed the ethical principles of sensory evaluation laboratory, approved by the Ethical Committee of the University of Helsinki.

Informed Consent Statement: Informed consent was obtained from all subjects involved in the study.

Data Availability Statement: Data is contained within the article.

Acknowledgments: Special thanks to Sanna Ylisjunttila-Huusko for providing valuable advice with selecting the appropriate rheological parameters in the preliminary experiments, in addition to Schlich (INRAE) for advice on using attribute durations in the PLS correlations. We also wish to thank Astrid D'Andrea for proofreading the manuscript.

Conflicts of Interest: Authors declare no conflict of interest. All the authors have adhered to conduct independent research. M.G. is financially supported by Valio Ltd. R.P. works for Valio Ltd., and T.S. was a student of University of Helsinki during the writing process, but currently she is an employee of Arla Foods Ltd. The funders had no role in the analysis, interpretation of data, or in the decision to publish the results.

Appendix A

Table A1. Pearson correlations between physicochemical and mouthfeel parameters in plant-based yogurts ($n = 5$). Correlation coefficients in bold are significant at $p < 0.05$ (*) and at $p < 0.01$ (**).

	Creamy	**Thick**	**Thin**	**Watery**
HL	−0.188	−0.541	0.260	0.373
n	0.295	0.572	−0.219	−0.466
K	0.634	0.391	−0.716	−0.493
ηapp10	**0.893 ***	0.878	**−0.955 ***	**−0.884 ***
SS10	**0.908 ***	0.846	**−0.966 ****	−0.873
SS50	0.813	**0.964 ****	−0.846	**−0.894 ***
G′LVE	0.562	0.215	−0.619	−0.318
γLVE	0.582	0.396	−0.527	−0.557
G′1 Hz	0.584	0.242	−0.640	−0.346
G″1 Hz	0.812	0.857	−0.847	−0.771
D [3.2]	0.668	0.767	−0.692	−0.633
D [4.3]	0.430	0.297	−0.281	−0.199
Dx (90)	0.283	0.108	−0.116	−0.043
°Brix	−0.684	−0.272	0.646	0.418
Fat	0.153	0.205	−0.314	−0.228
Carboh.	−0.153	0.322	0.021	−0.299
Sugar	0.272	0.181	−0.376	−0.128
Fiber	−0.257	−0.217	0.142	0.328
Protein	−0.625	−0.546	0.717	0.516
Oat	**−0.893 ***	−0.594	0.818	0.592
pH	0.477	0.064	−0.428	−0.028
TTA	−0.670	−0.559	0.757	0.544

References

1. Banovic, M.; Lähteenmäki, L.; Arvola, A.; Pennanen, K.; Duta, D.E.; Brückner-Gühmann, M.; Grunert, K.G. Foods with Increased Protein Content: A Qualitative Study on European Consumer Preferences and Perceptions. *Appetite* **2018**, *125*, 233–243. [CrossRef]
2. Brückner-Gühmann, M.; Banovic, M.; Drusch, S. Towards an Increased Plant Protein Intake: Rheological Properties, Sensory Perception and Consumer Acceptability of Lactic Acid Fermented, Oat-Based Gels. *Food Hydrocoll.* **2019**, *96*, 201–208. [CrossRef]
3. Ercili-Cura, D.; Miyamoto, A.; Paananen, A.; Yoshii, H.; Poutanen, K.; Partanen, R. Adsorption of Oat Proteins to Air-Water Interface in Relation to Their Colloidal State. *Food Hydrocoll.* **2015**, *44*, 183–190. [CrossRef]
4. Jeske, S.; Zannini, E.; Arendt, E.K. Past, Present and Future: The Strength of Plant-Based Dairy Substitutes Based on Gluten-Free Raw Materials. *Food Res. Int.* **2018**, *110*, 42–51. [CrossRef] [PubMed]
5. Mårtensson, O.; Andersson, C.; Andersson, K.; Öste, R.; Holst, O. Formulation of an Oat-Based Fermented Product and Its Comparison with Yoghurt. *J. Sci. Food Agric.* **2001**, *81*, 1314–1321. [CrossRef]
6. Greis, M.; Sainio, T.; Katina, K.; Kinchla, A.J.; Nolden, A.; Partanen, R.; Seppä, L. Dynamic Texture Perception in Plant-Based Yogurt Alternatives: Identifying Temporal Drivers of Liking by TDS. *Food Qual. Pref.* **2020**, *86*, 104019. [CrossRef]
7. Cutler, A.N.; Morris, E.R.; Taylor, L.J. Oral perception of viscosity in fluid foods and model systems. *J. Texture Stud.* **1983**, *14*, 377–395. [CrossRef]
8. De Wijk, R.A.; van Gemert, L.J.; Terpstra, M.E.J.; Wilkinson, C.L. Texture of Semi-Solids; Sensory and Instrumental Measurements on Vanilla Custard Desserts. *Food Qual. Pref.* **2003**, *14*, 305–317. [CrossRef]
9. Harte, F.; Clark, S.; Barbosa-Cánovas, G.V. Yield Stress for Initial Firmness Determination on Yogurt. *J. Food Eng.* **2007**, *80*, 990–995. [CrossRef]
10. Janhøj, T.; Petersen, C.B.; Frøst, M.B.; Ipsen, R. Sensory and Rheological Characterization of Low-Fat Stirred Yogurt. *J. Texture Stud.* **2006**, *37*, 276–299. [CrossRef]
11. Janssen, A.M.; Terpstra, M.E.J.; de Wijk, R.A.; Prinz, J.F. Relations between Rheological Properties, Saliva-Induced Structure Breakdown and Sensory Texture Attributes of Custards. *J. Texture Stud.* **2007**, *38*, 42–69. [CrossRef]
12. Jørgensen, C.E.; Abrahamsen, R.K.; Rukke, E.O.; Johansen, A.G.; Schüller, R.B.; Skeie, S.B. Improving the Structure and Rheology of High Protein, Low Fat Yoghurt with Undenatured Whey Proteins. *Int. Dairy J.* **2015**, *47*, 6–18. [CrossRef]
13. Nguyen, P.T.M.; Kravchuk, O.; Bhandari, B.; Prakash, S. Effect of Different Hydrocolloids on Texture, Rheology, Tribology and Sensory Perception of Texture and Mouthfeel of Low-Fat Pot-Set Yoghurt. *Food Hydrocoll.* **2017**, *72*, 90–104. [CrossRef]
14. Sonne, A.; Busch-Stockfisch, M.; Weiss, J.; Hinrichs, J. Improved Mapping of In-Mouth Creaminess of Semi-Solid Dairy Products by Combining Rheology, Particle Size, and Tribology Data. *LWT Food Sci. Technol.* **2014**, *59*, 342–347. [CrossRef]
15. Kilcast, D.; Clegg, S. Sensory Perception of Creaminess and Its Relationship with Food Structure. *Food Qual. Pref.* **2002**, *13*, 609–623. [CrossRef]
16. Krzeminski, A.; Tomaschunas, M.; Köhn, E.; Busch-Stockfisch, M.; Weiss, J.; Hinrichs, J. Relating Creamy Perception of Whey Protein Enriched Yogurt Systems to Instrumental Data by Means of Multivariate Data Analysis. *J. Food Sci.* **2013**, *78*, S314–S319. [CrossRef]
17. Laiho, S.; Williams, R.P.W.; Poelman, A.; Appelqvist, I.; Logan, A. Effect of Whey Protein Phase Volume on the Tribology, Rheology and Sensory Properties of Fat-Free Stirred Yoghurts. *Food Hydrocoll.* **2017**, *67*, 166–177. [CrossRef]
18. Lett, A.M.; Yeomans, M.R.; Norton, I.T.; Norton, J.E. Enhancing Expected Food Intake Behaviour, Hedonics and Sensory Characteristics of Oil-in-Water Emulsion Systems through Microstructural Properties, Oil Droplet Size and Flavour. *Food Qual. Pref.* **2016**, *47*, 148–155. [CrossRef]
19. Sodini, I.; Remeuf, F.; Haddad, C.; Corrieu, G. The Relative Effect of Milk Base, Starter, and Process on Yogurt Texture: A Review. *Crit. Rev. Food Sci. Nutr.* **2004**, *44*, 113–137. [CrossRef]
20. Sala, G.; de Wijk, R.A.; van de Velde, F.; van Aken, G.A. Matrix Properties Affect the Sensory Perception of Emulsion-Filled Gels. *Food Hydrocoll.* **2008**, *22*, 353–363. [CrossRef]
21. Grasso, N.; Alonso-Miravalles, L.; O'Mahony, J.A. Composition, Physicochemical and Sensorial Properties of Commercial Plant-Based Yogurts. *Foods* **2020**, *9*, 252. [CrossRef]
22. Gupta, M.K.; Torrico, D.D.; Ong, L.; Gras, S.L.; Dunshea, F.R.; Cottrell, J.J. Plant and Dairy-Based Yogurts: A Comparison of Consumer Sensory Acceptability Linked to Textural Analysis. *Foods* **2022**, *11*, 463. [CrossRef]
23. Wang, Y.; Sorvali, P.; Laitila, A.; Maina, N.H.; Coda, R.; Katina, K. Dextran Produced in Situ as a Tool to Improve the Quality of Wheat-Faba Bean Composite Bread. *Food Hydrocoll.* **2018**, *84*, 396–405. [CrossRef]
24. Cayot, P.; Schenker, F.; Houzé, G.; Sulmont-Rossé, C.; Colas, B. Creaminess in Relation to Consistency and Particle Size in Stirred Fat-Free Yogurt. *Int. Dairy J.* **2008**, *18*, 303–311. [CrossRef]
25. De Wijk, R.A.; Prinz, J.F.; Janssen, A.M. Explaining Perceived Oral Texture of Starch-Based Custard Desserts from Standard and Novel Instrumental Tests. *Food Hydrocoll.* **2006**, *20*, 24–34. [CrossRef]
26. Krzeminski, A.; Großhable, K.; Hinrichs, J. Structural Properties of Stirred Yoghurt as Influenced by Whey Proteins. *LWT Food Sci. Technol.* **2011**, *44*, 2134–2140. [CrossRef]
27. Shama, F.; Sherman, P. Identification of stimuli controlling the sensory evaluation of viscosity II. Oral Methods. *J. Texture Stud.* **1973**, *4*, 111–118. [CrossRef]
28. Schlich, P. Temporal Dominance of Sensations (TDS): A New Deal for Temporal Sensory Analysis. *Curr. Opin. Food Sci.* **2017**, *15*, 38–42. [CrossRef]

29. Abdi, H. *Partial Least Square Regression PLS-Regression. Encyclopedia of Measurement and Statistics*; SAGE Publications, Inc.: Thousand Oaks, CA, USA, 2007.
30. Lee, W.J.; Lucey, J.A. Formation and Physical Properties of Yogurt. *Asian-Australas. J. Anim. Sci.* **2010**, *23*, 1127–1136. [CrossRef]
31. Sikora, M.; Kowalski, S.; Tomasik, P. Binary Hydrocolloids from Starches and Xanthan Gum. *Food Hydrocoll.* **2008**, *22*, 943–952. [CrossRef]
32. Carriere, C.J.; Inglett, G.E. Nonlinear Viscoelastic Solution Properties of Oat-Based β-Glucan/Amylodextrin Blends. *Carbohydr. Polym.* **1999**, *40*, 9–16. [CrossRef]
33. Bozzi, L.; Milas, M.; Rinaudo, M. Solution and Gel Rheology of a New Polysaccharide Excreted by the Bacterium Alteromonas Sp. Strain 1644. *Int. J. Biol. Macromol.* **1996**, *18*, 83–91. [CrossRef] [PubMed]
34. Renard, D.; Lefebvre, J. Gelation of Globular Proteins: Effect of PH and Ionic Strength on the Critical Concentration for Gel Formation. A Simple Model and Its Application to β-Lactoglobulin Heat-Induced Gelation. *Int. J. Biol. Macromol.* **1992**, *14*, 287–291. [CrossRef] [PubMed]
35. Akhtar, M.; Stenzel, J.; Murray, B.S.; Dickinson, E. Factors Affecting the Perception of Creaminess of Oil-in-Water Emulsions. *Food Hydrocoll.* **2005**, *19*, 521–526. [CrossRef]
36. Dickinson, E. On the Road to Understanding and Control of Creaminess Perception in Food Colloids. *Food Hydrocoll.* **2018**, *77*, 372–385. [CrossRef]
37. Stanley, N.L.; Taylor, L.J. Rheological Basis of Oral Characteristics of Fluid and Semi-Solid Foods: A Review. *Acta Psychol.* **1993**, *84*, 79–92. [CrossRef]
38. Mosca, A.C.; Rocha, J.A.; Sala, G.; van de Velde, F.; Stieger, M. Inhomogeneous Distribution of Fat Enhances the Perception of Fat-Related Sensory Attributes in Gelled Foods. *Food Hydrocoll.* **2012**, *27*, 448–455. [CrossRef]
39. Tomaschunas, M.; Hinrichs, J.; Köhn, E.; Busch-Stockfisch, M. Effects of Casein-to-Whey Protein Ratio, Fat and Protein Content on Sensory Properties of Stirred Yoghurt. *Int. Dairy J.* **2012**, *26*, 31–35. [CrossRef]
40. Vingerhoeds, M.H.; de Wijk, R.A.; Zoet, F.D.; Nixdorf, R.R.; van Aken, G.A. How Emulsion Composition and Structure Affect Sensory Perception of Low-Viscosity Model Emulsions. *Food Hydrocoll.* **2008**, *22*, 631–646. [CrossRef]
41. Dresselhuis, D.M.; de Hoog, E.H.A.; Cohen Stuart, M.A.; Vingerhoeds, M.H.; van Aken, G.A. The Occurrence of In-Mouth Coalescence of Emulsion Droplets in Relation to Perception of Fat. *Food Hydrocoll.* **2008**, *22*, 1170–1183. [CrossRef]
42. van Aken, G.A.; Vingerhoeds, M.H.; de Wijk, R.A. Textural Perception of Liquid Emulsions: Role of Oil Content, Oil Viscosity and Emulsion Viscosity. *Food Hydrocoll.* **2011**, *25*, 789–796. [CrossRef]
43. Morell, P.; Hernando, I.; Llorca, E.; Fiszman, S. Yogurts with an increased protein content and physically modified starch: Rheological, structural, oral digestion and sensory properties related to enhanced satiating capacity. *Food Res. Int.* **2015**, *70*, 64–73. [CrossRef]
44. De Wijk, R.A.; Prinz, J.F.; Engelen, L.; Weenen, H. The Role of α-Amylase in the Perception of Oral Texture and Flavour in Custards. *Physiol. Behav.* **2004**, *83*, 81–91. [CrossRef]
45. Devezeaux de Lavergne, M.; van Delft, M.; van de Velde, F.; van Boekel, M.A.J.S.; Stieger, M. Dynamic Texture Perception and Oral Processing of Semi-Solid Food Gels: Part 1: Comparison between QDA, Progressive Profiling and TDS. *Food Hydrocoll.* **2015**, *43*, 207–217. [CrossRef]

Article

Defining Amaranth, Buckwheat and Quinoa Flour Levels in Gluten-Free Bread: A Simultaneous Improvement on Physical Properties, Acceptability and Nutrient Composition through Mixture Design

Etiene Valéria Aguiar, Fernanda Garcia Santos, Ana Carolina Ladeia Solera Centeno and Vanessa Dias Capriles *

Laboratory of Food Technology and Nutrition, Department of Biosciences, Institute of Health and Society, Campus Baixada Santista, Federal University of São Paulo (UNIFESP), Santos 11015-020, Brazil; etiene.aguiar@unifesp.br (E.V.A.); fg.santos@unifesp.br (F.G.S.); anacladeia@gmail.com (A.C.L.S.C.)
* Correspondence: vanessa.capriles@unifesp.br

Abstract: The study aimed to define the ideal proportions of pseudocereal flours (PF) in sensory-accepted gluten-free bread (GFB) formulations. The characteristics of GFB developed with PF (amaranth, buckwheat, and quinoa) were verified through a mixture design and response surface methodology. Three simplex-centroid designs were studied to analyze the effects of each PF and their interactions with potato starch (PS), and rice flour (RF) on GFB's physical and sensory characteristics, each design producing three single, three binary and six ternary GFB formulations. Results showed that using PF alone resulted in unacceptable GFB. However, the interactions between PF and RF improved the loaf specific volume and the crumb softness and also enhanced appearance, color, odor, texture, flavor, and overall liking. Moreover, the composite formulations prepared with 50% PF and 50% RF (flour basis) presented physical properties and acceptability scores like those of white GFB, prepared with 100% RF or a 50% RF + 50% PS blend (flour basis). Maximum proportions of PF to obtain well-accepted GFB (scores ≥7 for all evaluated attributes on a 10-cm hybrid hedonic scale) were defined at 60% for amaranth flour (AF), 85% for buckwheat flour (BF), and 82% for quinoa flour (QF) in blends with RF.

Keywords: gluten-free; pseudocereals; whole flour; bread quality; response surface methodology; multiple factor analysis

Citation: Aguiar, E.V.; Santos, F.G.; Centeno, A.C.L.S.; Capriles, V.D. Defining Amaranth, Buckwheat and Quinoa Flour Levels in Gluten-Free Bread: A Simultaneous Improvement on Physical Properties, Acceptability and Nutrient Composition through Mixture Design. *Foods* **2022**, *11*, 848. https://doi.org/10.3390/foods11060848

Academic Editors: Luca Serventi, Charles Brennan and Rana Mustafa

Received: 12 February 2022
Accepted: 14 March 2022
Published: 16 March 2022

Publisher's Note: MDPI stays neutral with regard to jurisdictional claims in published maps and institutional affiliations.

1. Introduction

There is constantly growing demand for gluten-free (GF) products, projected to achieve an approximate global market of USD 24 billion by 2027 [1]. Despite the increase in GF food available on the market, individuals with restrictions on gluten consumption still report difficult access to these products, since they usually have high shelf prices, restricted variety and availability, and poor palatability [2,3].

Among all GF products, bread has been the most investigated by researchers in several countries and it is also the most requested by consumers with celiac disease [4]. However, gluten-free bread (GFB) is still considered to be a product with unsatisfactory texture and flavor, lacking in nutritional content and having a short shelf life [3,5].

GFB often presented poor nutritional composition because it is mostly made using refined raw materials like white rice flour blended with corn, potato and/or cassava starches. Although these raw materials are readily available, made with cheap ingredients, and have neutral color, flavor, and odor, they lack dietary fiber, vitamins, and minerals. They have high levels of available carbohydrates, resulting in products with high glycemic response and poor nutritional quality, since they are neither enriched nor fortified [4,5]. Thus, improvement in the nutritional composition of GFB is an important objective for food research and development, a challenge that is concomitant with the improvement of the technological and sensory characteristics of these products [6].

To improve GFB formulation, the use of alternative ingredients with a rich nutrient and bioactive compounds content, such as wholemeal pseudocereal flours obtained from naturally GF grains, has been recommended [7].

The pseudocereals amaranth, buckwheat, and quinoa present high protein content, notably essential amino acids, mainly lysine (limited in cereals) and sulphur amino acids (limited in legumes). In addition, the considerable fiber, vitamin, and mineral content, and their potential as functional food are factors that increase the interest in the use of these grains for human consumption [8]. Thus, the incorporation of pseudocereals in the formulation has the potential to enhance the nutritional profile of GFB, which can benefit the health of individuals with diseases related to gluten consumption [8].

Several studies have been conducted showing that it is possible to use pseudocereal flours (PF) in GFB formulations [9–14]. However, to date, no publications report the effects of different pseudo cereal flour levels on physical characteristics, degree of liking, and nutritional profile of GFB. Therefore, this study aimed to evaluate the maximum limits and the ideal proportions of PF (amaranth (AF), buckwheat (BF) and quinoa (QF)) in combination with rice flour (RF) and potato starch (PS) using a mixture design to obtain GFB with improved technological, sensory, and nutritional properties.

2. Materials and Methods

2.1. Ingredients

The grains of amaranth (*Amaranthus caudatus*) and quinoa (*Chenopodium quinoa*), originally from Peru, were obtained from RS Blumos Industrial e Comercial Ltd.—Cotia-SP, Brazil, while the grains of buckwheat (*Fagopyrum esculentum*), originally from Bolivia, were obtained from Estação dos Grãos Ltd.—São Paulo, Brazil. The pseudocereal grains were transported to the Food Technology and Nutrition Laboratory (LAbTAN, UNIFESP) and milled using a mill (Laboratory Mill 3303, Perten Instruments, Stockholm, Sweden) at level 0, obtaining flours with the smallest particle size possible, in order to not confer or minimize the sensation of grit, which is often mentioned by consumers when tasting GFB developed with wholemeal flours [15]. Among the PF, the AF presents the largest particle size (83% $\geq$ 250 µm), followed by the QF (73% $\geq$ 250 µm), and the BF with the smallest particle size (60% < 180 µm) [16].

The xanthan gum (Ziboxan F80, Deosen Biochemical Ltd.—Mongolia, China) was donated by the company Vogler Ingredients Ltd. (São Bernardo do Campo-SP, Brazil) while carboxymethylcellulose (Denvercel FG-2504A, Denver Especialidades Químicas Ltd.—Cotia-SP, Brazil) was donated by its manufacturer. Other ingredients used for GFB preparation were obtained at the local market.

2.2. Methods

2.2.1. Formulation and Production of Gluten-Free Breads

The GFB formulation were elaborated according to Aguiar et al. (2021b) [16] and consisted of the following ingredients on a flour basis (f.b.): 100% blend of one PF with RF and/or PS, according to a mixture design, 25% egg, 10.5% whole milk powder, 6% sugar, 6% soybean oil, 2% salt, 0.8% dry yeast, 0.3% xanthan gum, 0.3% carboxymethylcellulose and 100% water.

The straight dough method was used as reported by Aguiar et al. (2021b) [16]. The analyses were conducted within up to 3 h after production. Twelve loaves of each GFB experimental formulation were produced, in two batches. Six loaves were used for the analysis of physical properties and the other six were used in the sensory analysis.

2.2.2. Experimental Design

Three simplex centroid experimental designs were conducted, combining each PF with the RF and PS: different percentage mixes of (A) AF with RF + PS; (B) BF with RF + PS; (C)

QF with RF + PS. For each of the three designs, there were twelve experimental formulations, three constituted of single components (100%), three of binary blends prepared with 50% of each of two components, one formulation of ternary blend consisting of the combination of 33.3% of each component, representing the central point of the model, which was made in three repetitions, and three formulations corresponding to the ternary blend consisting of the combination of 66% of one component and 17% of each of the others, corresponding to the axial points (Figure S1, on Supplementary Material). The sequence of execution of the experiments was randomized by a prior draw. The highest content level of each component in the blend of flours and starches (proportion = 1) represented 35.8 g of the dough (Figure S1, on Supplementary Material).

2.2.3. Bread Quality Evaluation

Physical properties were analyzed as described by Aguiar et al. (2021b) [16]. The analyses of specific loaf volume, crumb firmness and moisture content were made, respectively, according to method 10-05.01, 74-09 and 44-15.02 of AACC (2010) [17], while the crumb cell structure was analyzed according to Santos et al. (2020) [18].

The sensory acceptance of samples was conducted in ten sensory analysis sessions randomized for each design, offering in each session up to three samples of the same design, with balanced order of presentation.

In each sensory analysis session, 50 bread consumers, recruited from students and staff from the university campus, aged 18–59 years, assessed the acceptability of the attributes: appearance, color, odor, texture, flavor and overall liking of the breads, on a semi-structured 10 cm hybrid hedonic scale (0 = disliked very much, 5 = neither liked/nor disliked, 10 = liked very much) [19].

The evaluators received the samples of bread (slices of 12 mm in thickness) monadically, packaged in polypropylene bags and coded with three random digits. The participants assessed the GFB formulations in individual booths in the Sensory Analysis Laboratory, being instructed to drink water between samples to minimize residual effects.

2.2.4. Selection of Samples and Quality Verification

The GFB physical properties and acceptability served as response variables for the mixture design regression models, applying the Scheffé canonical polynomial models as explained by Aguiar et al. (2021b) [16].

Principal component analysis (PCA) also contributed to determining the content levels of these flours that do not alter the physical properties and acceptability compared with the control GFB formulation.

2.2.5. Characterization of Selected Samples

The selected samples, containing both optimum and maximum levels of PF, had their dough thermomechanical characteristics evaluated using the Chopin + 90 protocol in Mixolab® (Chopin Technologies, Villeneuve-la-Garenne, France), in which all ingredients (except yeast) were mixed in the proportions utilized in the bread preparation, using a total of 90 g of dough. Adaptations were made to method 54-60.01 of AACC (2010) [17] to allow knowing the effect of ingredients on the dough characteristics, subject to mixing and temperature variation, simulating the breadmaking process. The same parameters as those reported by of Santos et al. (2021) [20] were observed here: initial consistency (C1), weakening of protein network (C2), maximum (C3) and minimum (C4) peak during the heating phase and the value obtained after cooling (C5). Two repetitions were performed for each sample.

In addition, the selected formulations were prepared and analyzed experimentally to verify the physical properties and sensory acceptance using, respectively, the methods cited on Section 2.2.3, these results being compared statistically with the expected values of the fitted models.

The selected GFB formulations had the proximate composition analyzed. Moisture, ash, protein and lipid contents were analyzed following the respective methods 950.46, 923.03, 960.52, 920.39 of AOAC (2005) [21]. Dietary fiber (soluble and insoluble) content was verified utilizing the enzymatic-gravimetric method 991.43 of AOAC (2005) [21] and analytical kit K-ACHDF (Megazyme International Ireland Ltd., Bray, Ireland). Available carbohydrates were calculated by difference. Data were means of three repetitions and expressed as g/100 g of GFB.

2.2.6. Ethical Considerations

This study was approved by the Research Ethics Committee of UNIFESP (protocol number 1.814.143) and all the participants signed an informed consent form before enrollment in the research.

2.2.7. Statistical Analysis

The adequacy of the mixture regression model was verified through variance analysis (F test), R^2 values, lack-of-fit test, and diagnostic plots such as normal and residual plots. One-way analysis of variance (ANOVA) and Tukey's test were used to verify the differences in treatment means, comparing physical properties, sensory acceptance scores and centesimal composition of the selected GFB formulations. The Statistica 12.0 statistical software (StatSoft Inc., Tulsa, OK, USA, 2013) was used for data processing.

Multiple factor analysis (MFA) was utilized to investigate the relationships of the studied variables (physical, sensory and Mixolab parameters) using the XLSTAT 2021.2 software (Addinsoft, New York, NY, USA), with the significance level established at 0.05 for all analyses.

3. Results

3.1. Mixture Design and Response Surface Analysis

Table 1 presents the mixture regression models obtained for the physical properties and for the acceptability of the GFB. The models obtained with designs A and C were significant for the physical properties. For the acceptability scores, only models obtained with designs B and C were significant for appearance, color, and odor; only design A presented a significant model for texture, while models from designs A and B were fitted for overall liking and flavor acceptance. Linear and quadratic models were obtained but no ternary interaction was significant for the variables studied. The significant models, without lack of fit and with high coefficient of determination (R^2_{adj}), with 70 to 98% of the experimental variability being explained by the models, were used to generate the contour curves (Figures 1 and 2).

The loaf specific volume and crumb firmness are related to the sensory attributes of the bread [22,23] and were therefore evaluated more thoroughly in this study. The objective was to obtain bread with higher expansion and lower crumb firmness, indicating a softer loaf.

For specific volume, as displayed in Table 1 and Figure 1, RF showed the higher coefficient values in the regression models, therefore being responsible for higher specific volume of loaves, while PS (lower coefficient value) promotes lower specific volume. The GFB formulation prepared only with BF presented the greatest value for specific volume, followed by the formulations prepared with RF, AF, QF, and PS. The binary blends of AF with PS, similarly to QF with RF or PS showed a synergistic effect, increasing the specific volume of the bread.

No significant interactions were found for the other blends. Samples with the highest values for specific volume (~1.8 cm^3/g) are indicated in the experimental region in dark red (Figure 1), comprising different blends of two or three components containing 30–75% AF combined with RF or PS (Figure 1 Y1a), and containing 10–65% QF (Figure 1 Y1c) combined with RF or PS. The results show that it is possible to use up to 75% AF and up to 80% QF combined with RF and obtain GFB in the region of highest specific volume.

Table 1. Predicted model equations for the three mixture designs indicating the effect of each mixture component [a] and their interactions on the physical properties and acceptability scores of the gluten-free bread.

Design [b]	Predicted Model Equations [c]	R^2_{adj} (%) [d]	Model (p) [e]	Lack of Fit (p) [e]
	Loaf specific volume cm^3/g (Y_1)			
A	$Y_{1a} = 1.74RF + 1.43PS + 1.57AF + 0.65RF \times PS + 1.42PS \times AF$	86.79	0.002	0.281
B	$Y_{1b} = 1.83RF + 1.59PS + 2.01BF$	32.61	0.069	0.017
C	$Y_{1c} = 1.72RF + 1.43PS + 1.46QF + 1.11RF \times QF + 1.30PS \times QF$	80.18	0.007	0.063
	Crumb firmness N (Y_2)			
A	$Y_{2a} = 9.88RF + 9.31PS + 4.31AF$	73.62	0.001	0.749
B	$Y_{2b} = 10.19RF + 11.13PS + 22.69BF$	70.82	0.002	0.004
C	$Y_{2c} = 8.84RF + 8.97PS + 20.99QF$	83.75	0.000	0.277
	Crumb moisture % (Y_3)			
A	$Y_{3a} = 52.28RF + 55.43PS + 53.75AF$	98.19	0.000	0.137
B	$Y_{3b} = 51.59RF + 55.00PS + 52.11BF$	78.63	0.000	0.014
C	$Y_{3c} = 52.56RF + 55.63PS + 54.02QF$	83.17	0.000	0.461
	Appearance acceptability score (Y_4)			
A	$Y_{4a} = 8.73RF + 7.71PS + 7.25AF + 2.64RF \times PS+ 3.46PS \times AF$	8.55	0.001	0.587
B	$Y_{4b} = 9.13RF + 8.13PS + 7.26BF$	78.03	0.000	0.255
C	$Y_{4c} = 8.75RF + 7.73PS + 7.58QF + 1.83RF \times PS + 3.67PS \times QF$	87.83	0.002	0.269
	Color acceptability score (Y_5)			
A	$Y_{5a} = 8.61RF + 7.89PS + 7.77AF + 2.91RF \times AF$	66.25	0.033	0.571
B	$Y_{5b} = 8.98RF + 8.07PS + 6.97BF$	79.72	0.000	0.416
C	$Y_{5c} = 8.57RF + 7.91PS + 7.43QF + 3.04RF \times QF$	73.85	0.016	0.237
	Odor acceptability score (Y_6)			
A	$Y_{6a} = 8.59RF + 8.62PS + 7.62AF$	65.42	0.003	0.251
B	$Y_{6b} = 8.50RF + 8.34PS + 7.30BF$	77.10	0.001	0.607
C	$Y_{6c} = 8.41RF + 8.33PS + 6.74QF$	84.23	0.000	0.815
	Texture acceptability score (Y_7)			
A	$Y_{7a} = 8.09RF + 7.07PS + 4.87AF + 3.58RF \times PS + 3.55RF \times AF$	92.02	0.001	0.229
B	$Y_{7b} = 8.63RF + 7.30PS + 6.99BF$	59.59	0.007	0.473
C	$Y_{7c} = 8.37RF + 7.78PS + 7.46QF$	05.71	0.311	0.043
	Flavor acceptability score (Y_8)			
A	$Y_{8a} = 7.87RF + 7.88PS + 5.62AF + 3.24RF + AF + 3.19PS + AF$	91.63	0.001	0.878
B	$Y_{8b} = 7.89RF + 7.84PS + 6.51BF + 2.04RF \times PS + 2.35RF \times BF$	93.63	0.000	0.459
C	$Y_{8c} = 8.30RF + 8.29PS + 6.72QF$	58.00	0.008	0.642
	Overall liking (Y_9)			
A	$Y_{9a} = 8.14RF + 7.80PS + 5.78AF + 3.51RF \times AF + 2.78 PS \times AF$	90.34	0.001	0.440
B	$Y_{9b} = 8.20RF + 7.67PS + 7.88BF + 2.45RF \times PS - 1.99PS \times BF$	84.07	0.003	0.230
C	$Y_{9c} = 8.40RF + 8.27PS + 6.87QF$	61.38	0.005	0.411

[a] Mixture components: RF = rice flour, PS = potato starch, AF = amaranth flour, BF = buckwheat flour, QF = quinoa flour. [b] Design: amaranth (A), buckwheat (B) and quinoa (C). [c] Only the coefficients significant at a $p < 0.05$ level were selected for the predicted model construction. [d] R^2_{adj} adjusted coefficient of determination. [e] Significance of the Model and Lack of fit. p = probability level.

In general, formulations containing equivalent amounts of AF and QF show close values for specific volume, while formulations containing BF showed higher values (Figure S2, on Supplementary Material). Alvarez-Jubete, Arendt & Gallagher (2009) [9] verified a difference in pasting properties, observing that, between the PFs, BF presented the highest peak viscosity, which was associated with GFB with improved specific volume when compared with GFB containing AF or QF. BF presents a higher amount of amylose in the starch composition (>45%), which may contribute to a higher dough viscosity in formulations prepared with BF, increasing their capacity to retain gases, resulting in breads with improved volume [9]. For crumb firmness, AF showed the lowest coefficient in relation to the other single components. However, it is worth mentioning that the formulation prepared with 100% AF showed to be inadequate, as it presented very gummy texture, preventing the appropriate formation of a bread, as reported previously by Alvarez-Jubete, Arendt & Gallagher (2009) [9] in other conditions of formulation and processing. GFB developed with 100% AF showed inadequate characteristics. This could be due to the high starch content of the flour (65–75%), which has direct influence on the higher viscosity of the dough due to the gelatinization of this starch, in addition to the high protein content of the flour. The proteins present in AF have the capacity to form gels and the high concentration of these gels can affect the capacity of adequate development of alveoli in the dough [24]. The combination of AF with RF or PS causes dilution of this gel, enabling adequate development of crumb and contributing to an increase in specific volume and a decrease in crumb firmness.

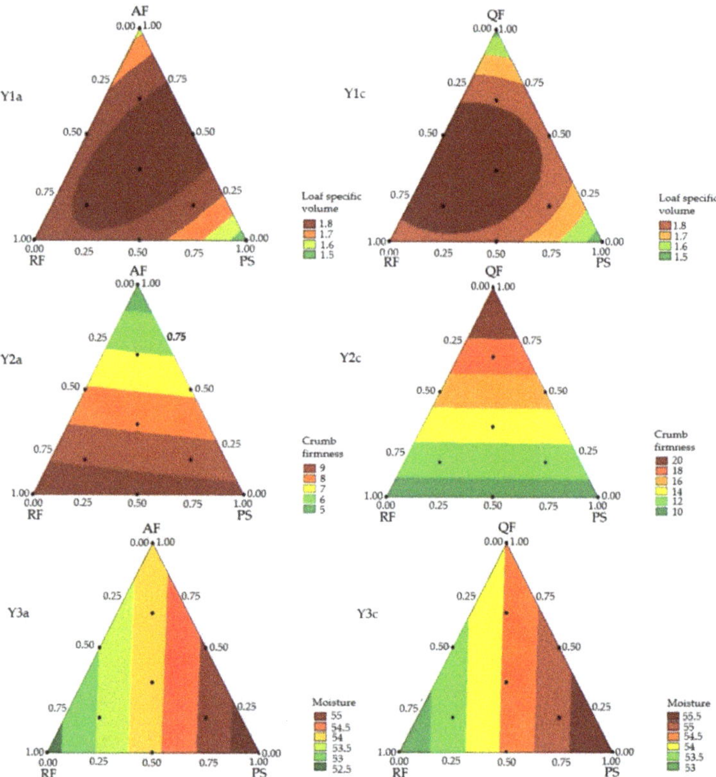

Figure 1. Contour plots from the predict model equations for physical properties of the gluten-free bread based on a mixture design. Formulation ID: AF—amaranth flour; BF—buckwheat flour; QF—quinoa flour; RF—rice flour; PS—potato starch.

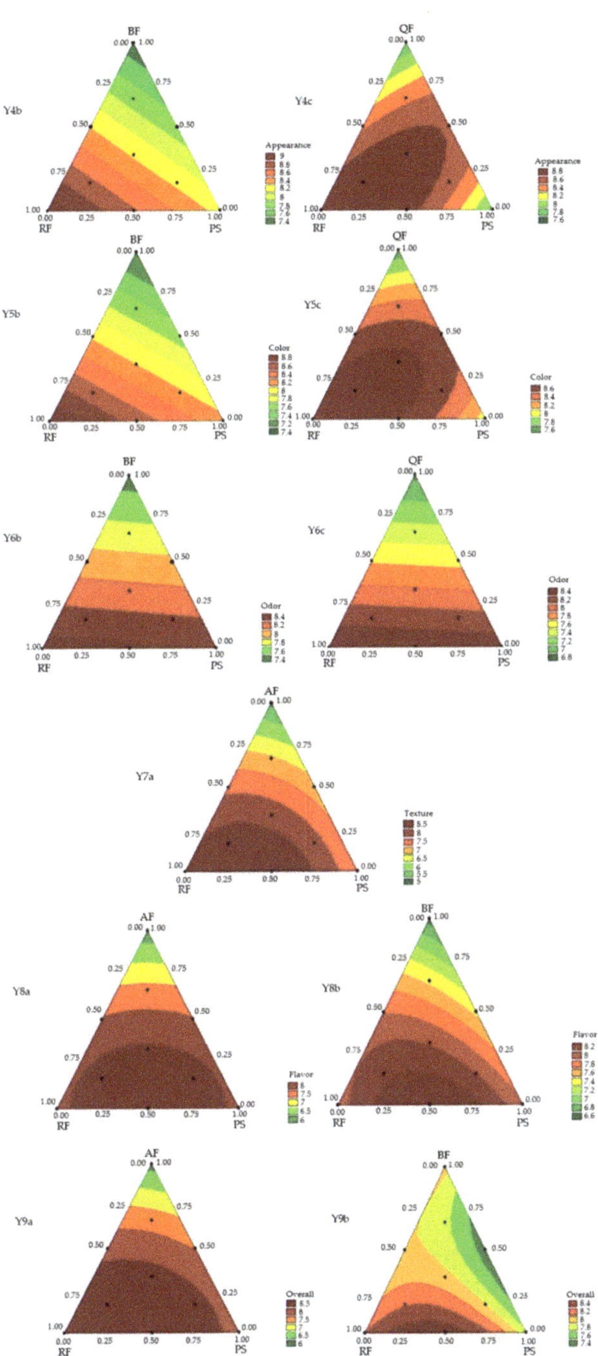

Figure 2. Contour plots from the predict model equations for the sensory acceptability scores (10-cm hybrid hedonic scale) of the gluten-free breads based on a mixture design. Formulation ID: AF—amaranth flour; BF—buckwheat flour; QF—quinoa flour; RF- rice flour; PS—potato starch.

QF showed the highest coefficient values in the regression models, therefore being responsible for the highest crumb firmness. The GFB formulation prepared only with BF showed the highest value for crumb firmness, followed by the formulations prepared with QF, RF, and PS. However, the blends of BF or QF with RF and/or PS had no significant effect on crumb firmness.

The results found in this study are similar to those reported by Föste et al. (2014) [25], as they also associate the presence of QF in GFB with increased crumb firmness, resulting in breads with lower softness. These effects may be related to the high amount of fibers present in QF, due to decreased starch gelatinization caused by the competition for water between fibers and starch [26].

Concerning crumb moisture, PS showed the highest coefficient values in the regression models, being responsible for the highest crumb moisture, while RF (lowest coefficient value) promoted the lowest crumb moisture, and AF, BF, and QF promoted intermediate values. Little variation was seen in crumb moisture between the 36 experimental formulations (variation from 51.6 to 55.6%), but no significant interactions were observed between the components of these blends in the crumb moisture of the GFB.

The variations in the crumb moisture can be associated with the differences in the properties of each flour, as they vary in the content and in the composition of the starch and protein fractions, as well as with the higher water absorption capacity of the PF and of PS compared with RF [27,28].

In general, for the GFB developed with equal PF proportions, BF enabled better expansion of the breads, providing, at all content levels, the formation of a more homogeneous crumb, with the highest number of small alveoli and a more even distribution. The use of QF enabled the formation of more uniform crumb in relation to those obtained with AF. The use of high proportions (66% and 100%) of AF resulted in bread with compact structure and crumb with few alveoli.

The contour charts for the attributes of acceptability of the GFB (Figure 2) show the possibility of using different blends of BF or QF with both RF and PS to obtain GFB with good acceptability of appearance, color, and odor (scores > 7). The results also show the synergistic effect of the blend of AF with RF or PS, which enables GFB with accepted texture. Blends of AF or BF with RF and PS enable increased flavor acceptability and overall liking, since they provide a less bitter flavor, besides improving the texture, which contributes to a higher overall liking.

Figure 2 and the equations in Table 1 obtained for each model for acceptability show that well-accepted GFB is possible (acceptability scores $\geq$ 7) for all attributes when using up to 60% AF, 85% BF, and 82% QF in blends with RF in the composition.

3.2. Optimal Gluten-Free Bread Formulations

The results of the desirability function showed that the formulation prepared with 100% RF corresponds to the optimal GFB formulation, with the highest scores for acceptability (Figure S3, in the Supplementary Material).

PCA was used to determine the relations between the physical properties (specific volume and crumb firmness) and the degree of liking of the GFB, presenting the formulations according to similarity. This enabled the finding of the most accepted formulations that contain the highest amount of PF in the composition and present similar characteristics to the optimal formulation (OF) containing 100% RF and to the control formulation (CF) prepared with 50% RF + 50% PS. The results are presented in the supplementary material (Figure S4). The two principal components explain 84.29–93.49% of the experimental variation, indicating strong correlation between the variables investigated, showing that, among the formulations with PF, those prepared with 50% AF, BF, or QF in combination with RF presented physical properties and acceptability closest to OF and CF.

3.2.1. Characterization of Selected Samples

Dough Thermomechanical Properties

Figure 3 shows the dough curves and parameters from Mixolab® for the selected GFB formulations compared to the OF and CF.

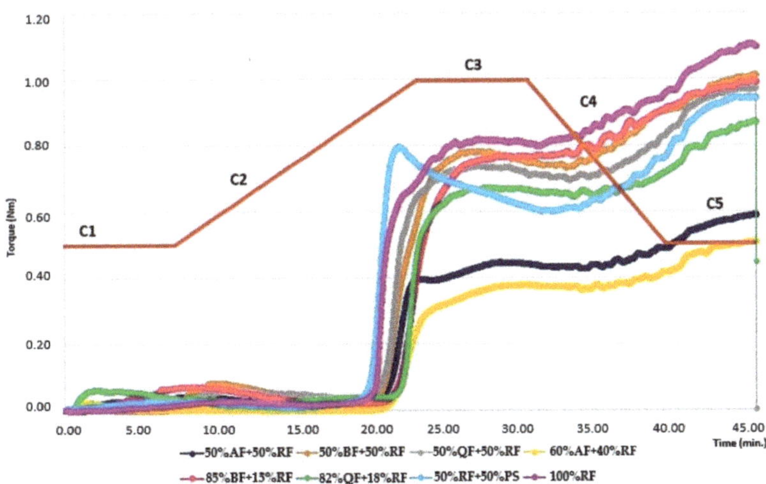

Figure 3. Dough curves of selected gluten-free bread formulations and parameters obtained by Mixolab®. Dough mixtures according to the levels of rice flour (RF), potato starch (PS) and pseudocereal flours: amaranth (AF), buckwheat (BF) and quinoa (QF).

No dough shows a significant torque for C1 and C2, which is expected for doughs prepared with GF ingredients. Without gluten, these doughs have a low consistency in the initial stage and no significant protein weakening [29].

GFB doughs containing 100% RF or those prepared with BF showed higher values of C3, C4, and C5 than the CF and doughs containing AF or QF. The doughs with 60% AF + 40% RF and 50% AF + 50% RF presented the lowest C3, C4 and C5 torques.

Alvarez-Jubete et al. (2009) [10] analyzed the peak viscosity of PF compared to RF and found that the amylose content and the particle size are the main causes of these differences. Amaranth presents the lowest content of amylose (<8%), which explains the low gelatinization of the dough containing this PF, while the doughs with higher quantities of RF and BF presented a better gelatinization due to the higher amylose content of these flours [9].

Regarding the C5 parameter, the OF prepared with 100% RF presented the higher values. Santos et al. (2020) [18] evidenced the relation between C5 values with the storage time of the GFB, so based on the results here, doughs with QF or BF show values of C5 near to the CF, except for the dough containing 50% QF + 50% RF, which presented a lower value. As observed in Figure 3, the dough parameters are dependent on the characteristics of the starch sources, which may influence the quality parameters of the final product.

Physical Properties and Acceptability Evaluation

The selected samples had the appearance, the physical properties and the acceptability evaluated and compared with the 100% RF and 50% RF + 50% PS formulations, as shown in Figure 4 and Table 2.

Table 2. Predicted and measured values to physical properties and sensory analysis of the selected gluten-free bread formulations.

Parameters		Gluten-Free Bread Formulations [a]							
		50% AF + 50% RF	50% BF + 50% RF	50% QF + 50% RF	60% AF + 40% RF	85% BF + 15% RF	82% QF + 18% RF	100% RF	50% RF + 50% PS
		Physical properties							
Specific volume (cm^3/g) [b]	PV *	1.80 (1.67–1.94)	SV	1.81 (1.65–1.97)	1.80 (1.65–1.94)	SV	1.75 (1.54–1.96)	1.77 (1.61–1.93)	1.91 (1.64–2.18)
	MV *	1.71 [e] (1.67–1.72)	2.10 [a] (1.99–2.23)	1.86 [d] (1.81–1.91)	1,94 [bc] (1.91–1.99)	2.00 [b] (1.97–2.04)	1.88 [cd] (1.87–1.89)	1.69 [e] (1.58–1.79)	1.72 [e] (0.13–1.67)
Crumb firmness (N) [c]	PV *	7.09 (6.17–8.02)	SV	14.91 (13.34–16.49)	6.54 (5. 58–7.50)	SV	18.80 (16.75–20.85)	8.90 (7.33–10.48)	8.84 (6.24–11.43)
	MV *	7.69 [e] (7.13–8.24)	12.04 [c] (11.36–12.71)	14.73 [b] (13.96–15.50)	7.24 [e] (6.85–7.61)	16.44 [a] (15.83–17.05)	14.03 [b] (13.00–5.07)	10.63 [d] (9.84–11.42)	13.48 [b] (12.73–14.23)
		Sensory acceptance [d]							
Appearance	PV *	SV	8.20 (7.94–8.45)	8.53 (8.17–8.89)	SV	7.54 (7.20–7.88)	8.22 (7.75–8.69)	8.63 (8.38–8.88)	9.13 (8.71–9.54)
	MV *	8.76 [a] (8.42–9.09)	8.39 [ab] (7.96–8.82)	8.82 [a] (8.49–9.15)	8.87 [a] (8.53–9.21)	7.83 [b] (7.28–8.38)	8.14 [ab] (7.66–8.62)	8.74 [a] (8.29–9.19)	8.75 [a] (8.34–9.16)
Color	PV *	SV	7.97 (7.71–8.23)	8.41 (8.06–8.77)	SV	7.27 (6.92–7.62)	8.10 (7.64–8.57)	8.52 (8.26–8.78)	8.98 (8.55–9.40)
	MV *	8.80 [a] (8.03- 8.98)	7.88 [bc] (7.38–8.38)	8.83 [a] (8.49–9.18)	8.96 [a] (8.69–9.24)	7.43 [c] (6.91–7.95)	8.16 [abc] (7.72–8.60)	8.50 [ab] (8.03–8.78)	8.59 [ab] (8.12–9.06)
Odor	PV *	SV	7.90 (7.71–8.08)	7.58 (7.37–7.79)	SV	7.48 (7.23–7.73)	7.04 (6.77–7.31)	8.45 (8.26–8.63)	8.50 (8.19–8.80)
	MV *	7.94 [a] (7.40–8.50)	7.62 [a] (7.02–8.21)	8.11 [a] (7.60–8.62)	7.70 [a] (7.10–8.31)	7.60 [a] (7,05–8,14)	7.90 [a] (7.37–8.42)	8.41 [a] (7.92–8.89)	8.44 [a] (7.97–8.91)
Texture	PV *	7.19 (6.65–7.72)	SV	SV	7.01 (6.30–7.41)	SV	SV	7.96 (7.61–8.32)	8.63 (8.05–9.21)
	MV *	7.67 [a] (7.97–8.85)	8.06 [a] (7.50–8.61)	7. 72 [a] (7.19–8.25)	7.36 [a] (6.84–7.84)	7.96 [a] (7.48–8.44)	7.44 [a] (6.95–7.93)	8.41 [a] (7.96–8.85)	8.18 [a] (7.71–8.64)
Flavor	PV *	7.36 (6.91–7.81)	7.59 (7.31–7.87)	SV	7.14 (6.68–7.61)	7.06 (6.68–7.44)	SV	8.23 (7.96–8.52)	8.35 (7.89–8.82)
	MV *	7.69 [a] (6.67–7.99)	8.15 [a] (7.52–8.78)	8.23 [a] (7.81–8.65)	7.33 [a] (6.78–7.91)	7.35 [a] (7.15–8.23)	7.32 [a] (6.74–7.89)	8.44 [a] (8.02–8.86)	7.83 [a] (7.32–8.34)
Overall liling	PV *	7.54 (7.13–8.04)	7.98 (7.71–8.24)	SV	7.36 (6.89–7.82)	7.69 (7.33–8.05)	SV	8.05 (7.79–8.20)	8.39 (7.94–8.83)
	MV *	8.00 [a] (7.55–8.44)	8.13 [a] (7.70–8.55)	8.33 [a] (7.99–8.67)	7.56 [a] (7.00–8.13)	7.58 [a] (7.11–8.06)	7.61 [a] (7.17–8.05)	8.43 [a] (8.04–8.82)	8.11 [a] (7.68–8.55)

[a] Bread IDs: AF—amaranth flour; BF—buckwheat flour; QF—quinoa flour; RF—rice flour; PS—potato starch. The numbers indicate the ingredient proportions in the flour weight basis (g/100 g). Values are means ± standard deviations [b] ($n = 3$), [c] ($n = 6$), [d] ($n = 54$). * PV: predicted values, MV: measured values. Values followed by different letters in each line are significantly different ($p < 0.05$).

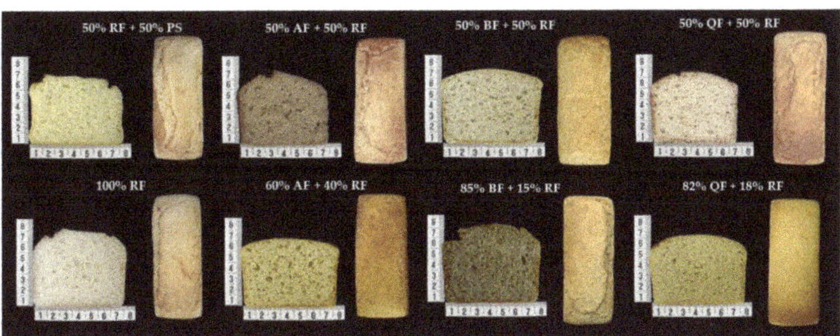

Figure 4. Photographs of central slices and crust of the selected gluten-free bread formulations according to the levels of AF—amaranth flour; BF—buckwheat flour; QF—quinoa flour; RF—rice flour; PS—potato starch.

Overall, the results in Table 2 are consistent with those expected, indicating the good quality of the fitted models.

The group of consumers of the acceptability evaluation were composed of 70% female and 30% male, presenting an average age of 27 ± 10.9 years.

Despite the difference in color between the formulations (Figure 4), high acceptance scores were obtained for this attribute (Table 2), which can indicate a higher custom and acceptance of the consumers to wholemeal products.

The formulations containing BF, QF, and 60% AF showed higher values for specific volume than the OF and CF. However, it should be noted that the crumb firmness obtained in the formulations with BF and QF are higher than those of the OF and CF. Despite the differences in physical properties between these formulations, it was possible to obtain bread with acceptability scores comparable to those of the OF and CF (Table 2).

Table S1, on supplementary material, presents the porosity data of the crumbs of the GFB of the selected formulations. The images of the center of the crumb of the selected formulations are presented in the supplementary material (Figure S5).

The results in Table S1 show that the GFB prepared with 50% QF + 50% RF, 82% QF + 18% RF, and with 85% BF + 15% RF showed the highest values for number of alveoli and the lowest values for mean size, with similar values to the formulation with 100% RF. While the GFB with 50% AF + 50% RF, 60% AF + 40% RF and with 50% BF + 50% RF showed lower number of alveoli and higher values for mean size, being similar to the CF with 50% RF + 50% PS.

Concerning total area, the GFB containing BF showed the highest values; however, they showed no significant difference in relation to the other formulations.

Burešová et al. (2017) [12], comparing the effect of different flours on the characteristics of breads, observed better porosity in breads prepared with BF and QF. While for AF the authors observed low viscosity of dough, relating the difference presented between the PF with the variation of the size of starch granules and gelatinization process of each PF [9]. In the present study, among the PF studied, BF and QF also promoted the greatest positive impact on porosity of breads prepared than AF.

Proximate Composition Evaluation

The selected samples had the proximate composition evaluated and compared with the 100% RF and 50% RF + 50% PS formulations, as shown in Table 3.

Table 3. Proximate composition of the selected gluten-free bread (GFB) formulations, compared with white GFB developed with rice flour (RF) and potato starch (PS) (g/100 g of food as eaten).

	Gluten-Free Bread Formulations [a]							
	50% AF + 50% RF	50% BF + 50% RF	50% QF + 50% RF	60% AF + 40% RF	85% BF + 15% RF	82% QF + 18% RF	100% RF	50% RF + 50% PS
Moisture	47.03 + 0.02 [bcd]	47.05 + 0.13 [bcd]	46.22 + 0.05 [cd]	47.47 + 0.27 [abc]	47.41 + 0.06 [abcd]	47.58 + 0.03 [ab]	46.16 + 0.26 [d]	48.59 + 1.24 [a]
Ash	1.71 + 0.03 [c]	1.66 + 0.00 [d]	1.80 + 0.00 [b]	1.80 + 0.02 [b]	1.00 + 0.01 [g]	2.03 + 0.01 [a]	1.40 + 0.00 [e]	1.32 + 0.00 [f]
Protein	6.90 + 0.03 [d]	6.72 + 0.05 [e]	7.67 + 0.02 [b]	6.92 + 0.07 [d]	7.29 + 0.06 [c]	8.02 + 0.08 [a]	5.12 + 0.00 [f]	4.10 + 0.09 [g]
Fat	5.13 + 0.47 [ab]	4.78 + 0.38 [abc]	4.05 + 0.07 [de]	4.71 + 0.14 [bcd]	4.17 + 0.17 [cd]	3.42 + 0.13 [e]	5.39 + 0.11 [a]	4.17 + 0.03 [cd]
Total dietary fiber	10.39 + 0.47 [c]	12.74 + 0.51 [b]	10.78 + 0.20 [c]	11.41 + 0.07 [c]	13.40 + 0.47 [ab]	14.37 + 0.57 [a]	7.66 + 0.18 [d]	4.89 + 0.12 [e]
Insoluble fiber	7.82 + 0.17 [d]	10.24 + 0.33 [b]	8.05 + 0.14 [cd]	8.61 + 0.29 [c]	10.31 + 0.12 [b]	11.47 + 0.36 [a]	5.50 + 0.22 [e]	3.10 + 0.10 [f]
Soluble fiber	2.56 + 0.48 [abc]	2.50 + 0.19 [abc]	2.74 + 0.34 [abc]	2.80 + 0.26 [ab]	3.09 + 0.37 [a]	2.91 + 0.22 [ab]	2.15 + 0.04 [bc]	1.78 + 0.13 [c]
Available carbohydrate	28.83 + 0.84 [cd]	27.05 + 0.59 [de]	29.46 + 0.10 [c]	27.69 + 0.16 [cde]	26.74 + 0.66 [e]	24.57 + 0.55 [f]	34.27 + 0.43 [b]	36.93 + 1.18 [a]

[a] Bread IDs: AF—amaranth flour; BF—buckwheat flour; QF—quinoa flour; RF—rice flour; PS—potato starch. The numbers indicate the ingredient proportions in the flour weight basis (g/100 g). Values are means ± standard deviations ($n = 3$). Values followed by different letters in each line are significantly different ($p < 0.05$).

Based on the composition of the formulations, GFB had higher values of protein, fat and a lower carbohydrate content. Regarding the content of total dietary fiber, the use of PF to develop GFB can contribute to an improvement in the nutritional profile, mainly in the amount of insoluble fiber in the formulations. Alvarez-Jubete, Arendt & Gallagher (2010) [11] reported the nutritional potential of the PF used to develop GFB. The authors noted that the partial replacement of RF with 50% PF resulted in increased content levels of proteins, lipids (high levels of unsaturated fatty acids), dietary fibers and minerals, such as calcium, magnesium, zinc and iron. According to a recent review made by Aguiar et al. (2021a) [5], GFB, available in the market, is mainly classified as low (<3 g/100 g) or source of (>3 g/100 g) fiber content, while the selected GFB formulations can be classified as a product with high fiber content (>6 g/100 g) [30], which evidence the nutritional improvement from the use of whole flours like PF. Therefore, the use of these whole flours in the development of GF products can contribute to a better quality of the diets of CD patients, contributing to a higher consumption of fibers, which can improve the deficient intake of this group and, also, the general population.

3.3. Relationships between Dough Properties and Instrumental and Sensory Parameters of GFB

Figure 5 shows the relation between variables studied in MFA, having factors sum explained 87.41% of the data total variation.

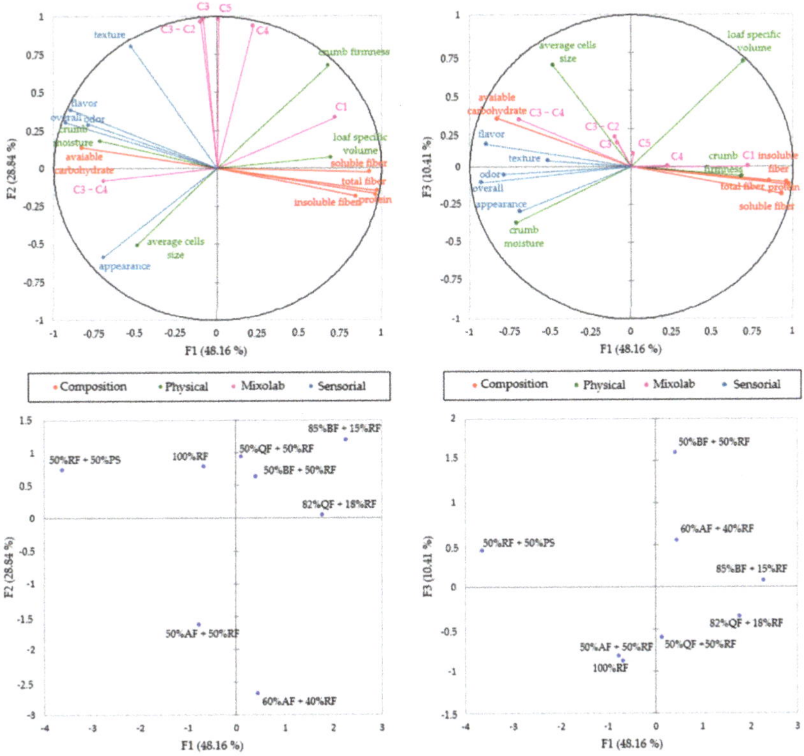

Figure 5. Multiple factor analysis correlating the physical (in green), sensorial (in blue), proximate composition (in red) and Mixolab parameters (in purple) of the selected gluten-free bread formulations. Bread ID: rice flour (RF), potato starch (PS) and pseudocereal flours: amaranth (AF), buckwheat (BF) and quinoa (QF).

Figure 5 shows that F1 explained 48.16% of the data variation and positively discriminates all the rheological parameter C1, and the composition parameters protein, insoluble fiber, soluble fiber and the total dietary fiber describing the GFB sample containing 85% BF + 15% RF and 82% QF + 18% RF, induced by the higher amounts of PF in those formulations. Still on F1, the vectors negatively discriminate the variables C3–C4, appearance, odor, flavor, overall, crumb moisture, crumb firmness and available carbohydrate, related to CF (50% RF + 50% PS).

F2 explained 28.84% of the data variation and positively describes the dough parameters C3, C4, C5, C3–C2, and the physical property parameter, texture, related to OF (100% RF) and 50% QF + 50% RF. It negatively describes the same parameters to samples developed with 50% AF + 50% RF and 60% AF + 40% RF, both having higher amounts of AF in the dough.

F3, on the other hand, explained 10.41% of the data variation and was positively discriminated with loaf-specific volume and the average cell size, related to the sample containing 50% BF + 50% RF.

The AFM sorted the selected GFB formulations into four distinct groups (Figure S6, in the Supplementary Material): the first group included samples containing AF (50% AF + 50% RF and 60% AF + 40% RF). The second group was composed of the samples containing the higher amounts of whole PF (85% BF + 15% RF and 82% QF + 18% RF). The third group contained the OF (100% RF) and the formulations with a lower quantity of PF (50% BF + 50% RF and 50% QF + 50% RF). The fourth group was composed of the CF (50% RF + 50% PS).

Based on the data relationship, the combination of PF and RF provides a better dough than those that use PS, resulting in GFB with improved technological, nutritional, and sensory properties, contributing to a better food profile for people who choose or need to follow a GF diet.

4. Conclusions

The mixture design showed that the use of PF alone resulted in GFB with low acceptability, due to changes in odor and flavor of the product which the consumers are not so used to.

The results indicate that PF needs to be blended with RF to get possible positive effects, contributing to improved physical properties and better acceptability of the GFB. This shows that adding high PF levels to develop high-quality GFB enriched with protein, fat and dietary fiber is possible.

Blends of 50% AF, BF, or QF with 50% RF (flour basis) to obtain GFB with high acceptance, being similar to GFB formulated with 100% RF and also with the control formulation developed with 50% RF + 50% PS on flour basis.

The mixture design allowed for the determination of the maximum PF proportions that can be used to obtain well-accepted formulations (scores ≥ 7) for appearance, color, odor, texture, and overall liking: 60% AF, 85% BF, and 82% QF in combination with RF.

The promising results of this study indicate an alternative for simultaneous improvement of physical properties, acceptability and nutritional content of GFB, which is very important for the nutrition and health of individuals with restrictions for gluten consumption.

Supplementary Materials: The following supporting information can be downloaded at: https://www.mdpi.com/article/10.3390/foods11060848/s1, Figure S1: Flow chart of the experimental design. * Formulation ID: rice flour (RF), potato starch (PS) and pseudocereals flours—amaranth flour (AF), buckwheat flour (BF) or quinoa flour (QF). Figure S2: Scanned images of the gluten-free bread (GFB) formulations obtained from the experimental mixture designs. * Bread ID: rice flour (RF), potato starch (PS) and pseudocereals flours—amaranth flour (AF), buckwheat flour (BF) or quinoa flour (QF). Figure S3: Profiles for predicted mixture experimental design of rice flour (RF), potato starch (PS) and pseudocereal flour (amaranth (AF), buckwheat (BF) or quinoa (QF)) and the desirability level for acceptability factor for optimum gluten-free bread. Figure S4: Principal component analysis of mixture design to evaluate the effect of rice flour—RF, potato starch—PS and pseudocereal flour (amaranth flour—AF (A), or buckwheat flour—BF (B) or quinoa flour—QF (C)) on physical properties and acceptability of gluten-free breads. Figure S5: Crumb porosity of gluten-free bread formulations selected from the mixture designs. Figure S6. Dendrogram obtained by hierarchical cluster analysis for data of selected gluten-free bread formulations. Table S1: Crumb porosity of gluten-free bread formulations selected from a mixture design to study the effects of pseudocereal flour: amaranth (AF), buckwheat (BF) and quinoa (QF) based gluten-free breads, comparing with white formulations developed with rice flour (RF) and potato starch (PS).

Author Contributions: Conceptualization and methodology, V.D.C.; validation and formal analysis, investigation, E.V.A., F.G.S. and A.C.L.S.C.; data curation, E.V.A. and V.D.C.; writing—original draft preparation, E.V.A.; writing—review and editing, F.G.S. and V.D.C.; supervision, project administration, funding acquisition, V.D.C. All authors have read and agreed to the published version of the manuscript.

Funding: This research was funded by São Paulo Research Foundation (FAPESP), grant number 2012/17838-4, 2016/01505-7 and 2017/10843-6.

Institutional Review Board Statement: The study was conducted in accordance with the Declaration of Helsinki and approved by the Ethics Committee of Federal University of São Paulo (protocol code 1.814.143 and date of approval 09/12/2016).

Informed Consent Statement: Informed consent was obtained from all subjects involved in the study.

Data Availability Statement: Not applicable.

Acknowledgments: The authors gratefully acknowledge Vogler Ingredients Ltd. and Denvercel FG-2504A, Denver Especialidades Químicas Ltd. who donated, respectively, xantham gum and carboxymethylcellulose samples and the volunteers who kindly participated in this study.

Conflicts of Interest: The authors declare no conflict of interest. The funders had no role in the design of the study; in the collection, analyses, or interpretation of data; in the writing of the manuscript, or in the decision to publish the results.

References

1. Grand View Research. *Gluten-Free Products Market Size, Share and Trends Analysis Report by Product (Bakery Products, Dairy/Dairy Alternatives), by Distribution Channel (Grocery Stores, Mass Merchandiser), by Region, and Segment Forecasts, 2020–2027*; Grand View Research: San Francisco, CA, USA, 2020.
2. Do Nascimento, A.B.; Fiates, G.M.R.; Dos Anjos, A.; Teixeira, E. Gluten-free is not enough—Perception and suggestions of celiac consumers. *Int. J. Food Sci. Nutr.* **2014**, *65*, 394–398. [CrossRef] [PubMed]
3. Alencar, N.M.M.; de Araújo, V.A.; Faggian, L.; Araújo, M.B.D.S.; Capriles, V.D. What about gluten-free products? An insight on celiac consumers' opinions and expectations. *J. Sens. Stud.* **2021**, *36*, e12664. [CrossRef]
4. Capriles, V.D.; Santos, F.G.; Aguiar, E.V. Innovative gluten-free breadmaking. In *Trends in Wheat and Bread Making*, 30th ed.; Galanakis, C., Ed.; Academic Press: Cambridge, MA, USA; Elsevier: Amsterdam, The Netherlands, 2021; pp. 371–404.
5. Aguiar, E.V.; Santos, F.G.; Krupa-Kozak, U.; Capriles, V.D. Nutritional facts regarding commercially available gluten-free bread worldwide: Recent advances and future challenges. *Crit. Rev. Food Sci. Nutr.* **2021**, 1–13. [CrossRef] [PubMed]
6. Capriles, V.; Santos, F.G.; Arêas, J.A.G. Gluten-free breadmaking: Improving nutritional and bioactive compounds. *J. Cereal Sci.* **2016**, *67*, 83–91. [CrossRef]
7. Woomer, J.S.; Adedeji, A.A. Current applications of gluten-free grains—A review. *Crit. Rev. Food Sci. Nutr.* **2020**, *61*, 14–24. [CrossRef]
8. Martínez-Villaluenga, C.; Peñas, E.; Hernández-Ledesma, B. Pseudocereal grains: Nutritional value, health benefits and current applications for the development of gluten-free foods. *Food Chem. Toxicol.* **2020**, *137*, 111178. [CrossRef]
9. Alvarez-Jubete, L.; Arendt, E.K.; Gallagher, E. Nutritive value and chemical composition of pseudocereals as gluten-free ingredients. *Int. J. Food Sci. Nutr.* **2009**, *60*, 240–257. [CrossRef]
10. Alvarez-Jubete, L.; Auty, M.; Arendt, E.K.; Gallagher, E. Baking properties and microstructure of pseudocereal flours in gluten-free bread formulations. *Eur. Food Res. Technol.* **2009**, *230*, 437–445. [CrossRef]
11. Alvarez-Jubete, L.; Arendt, E.K.; Gallagher, E. Nutritive value of pseudocereals and their increasing use as functional gluten-free ingredients. *Trends Food Sci. Technol.* **2010**, *21*, 106–113. [CrossRef]
12. Burešová, I.; Tokár, M.; Mareček, J.; Hřivna, L.; Faměra, O.; Šottníková, V. The comparison of the effect of added amaranth, buckwheat, chickpea, corn, millet and quinoa flour on rice dough rheological characteristics, textural and sensory quality of bread. *J. Cereal Sci.* **2017**, *75*, 158–164. [CrossRef]
13. Alencar, N.M.M.; Steel, C.J.; Alvim, I.D.; de Morais, E.C.; Bolini, H. Addition of quinoa and amaranth flour in gluten-free breads: Temporal profile and instrumental analysis. *LWT Food Sci. Technol.* **2015**, *62*, 1011–1018. [CrossRef]

14. Drub, T.F.; dos Santos, F.G.; Centeno, A.C.L.S.; Capriles, V.D. Sorghum, millet and pseudocereals as ingredients for gluten-free whole-grain yeast rolls. *Int. J. Gastron. Food Sci.* **2020**, *23*, 100293. [CrossRef]
15. Capriles, V.D.; Arêas, J.A.G. Novel Approaches in Gluten-Free Breadmaking: Interface between Food Science, Nutrition, and Health. *Compr. Rev. Food Sci. Food Saf.* **2014**, *13*, 871–890. [CrossRef]
16. Aguiar, E.V.; Santos, F.G.; Centeno, A.C.L.S.; Capriles, V.D. Influence of pseudocereals on gluten-free bread quality: A study integrating dough rheology, bread physical properties and acceptability. *Food Res. Int.* **2021**, *150*, 110762. [CrossRef]
17. AACC. *Approved Methods of Analysis*; AACC: Washington, DC, USA, 2010.
18. Santos, F.G.; Aguiar, E.V.; Centeno, A.C.L.; Rosell, C.M.; Capriles, V.D. Effect of added psyllium and food enzymes on quality attributes and shelf life of chickpea-based gluten-free bread. *LWT* **2020**, *134*, 110025. [CrossRef]
19. Villanueva, N.D.; Petenate, A.J.; Da Silva, M.A. Performance of three affective methods and diagnosis of the ANOVA model. *Food Qual. Prefer.* **2000**, *11*, 363–370. [CrossRef]
20. Santos, F.G.; Aguiar, E.V.; Rosell, C.M.; Capriles, V.D. Potential of chickpea and psyllium in gluten-free breadmaking: Assessing bread's quality, sensory acceptability, and glycemic and satiety indexes. *Food Hydrocoll.* **2021**, *113*, 106487. [CrossRef]
21. AOAC—Association of Official Analytical Chemists. *Official Methods of the AOAC International*, 18th ed.; Association of Official Analytical Chemists: Rockville, MD, USA, 2005.
22. Fratelli, C.; Muniz, D.G.; Santos, F.G.; Capriles, V.D. Modelling the effects of psyllium and water in gluten-free bread: An approach to improve the bread quality and glycemic response. *J. Funct. Foods* **2018**, *42*, 339–345. [CrossRef]
23. Fratelli, C.; Santos, F.; Muniz, D.; Habu, S.; Braga, A.; Capriles, V. Psyllium Improves the Quality and Shelf Life of Gluten-Free Bread. *Foods* **2021**, *10*, 954. [CrossRef]
24. Avanza, M.; Puppo, M.; Añón, M. Rheological characterization of amaranth protein gels. *Food Hydrocoll.* **2005**, *19*, 889–898. [CrossRef]
25. Föste, M.; Nordlohne, S.D.; Elgeti, D.; Linden, M.H.; Heinz, V.; Jekle, M.; Becker, T. Impact of quinoa bran on gluten-free dough and bread characteristics. *Eur. Food Res. Technol.* **2014**, *239*, 767–775. [CrossRef]
26. Collar, C.; Santos, E.; Rosell, C.M. Significance of Dietary Fiber on the Viscometric Pattern of Pasted and Gelled Flour-Fiber Blends. *Cereal Chem. J.* **2006**, *83*, 370–376. [CrossRef]
27. Wronkowska, M.; Haros, M.; Soral-Śmietana, M. Effect of Starch Substitution by Buckwheat Flour on Gluten-Free Bread Quality. *Food Bioprocess Technol.* **2012**, *6*, 1820–1827. [CrossRef]
28. De la Hera, E.; Rosell, C.M.; Gomez, M. Effect of water content and flour particle size on gluten-free bread quality and digestibility. *Food Chem.* **2014**, *151*, 526–531. [CrossRef]
29. Matos, M.E.; Rosell, C.M. Quality Indicators of Rice-Based Gluten-Free Bread-Like Products: Relationships Between Dough Rheology and Quality Characteristics. *Food Bioprocess Technol.* **2012**, *6*, 2331–2341. [CrossRef]
30. Codex Alimentarius Comission. *Guidelines for Use of Nutritionand Health Claims—CAC/GL 23-1997*; Codex Alimentarius Comission: Rome, Italy, 2013.

Article

Textural, Color and Sensory Features of Spelt Wholegrain Snack Enriched with Betaine

Jovana Kojić [1,*], Miona Belović [1], Jelena Krulj [1], Lato Pezo [2], Nemanja Teslić [1], Predrag Kojić [3], Lidija Peić Tukuljac [1], Vanja Šeregelj [3] and Nebojša Ilić [1]

1   Institute of Food Technology, University of Novi Sad, Bulevar Cara Lazara 1, 21000 Novi Sad, Serbia; miona.belovic@fins.uns.ac.rs (M.B.); jelena.krulj@fins.uns.ac.rs (J.K.); nemanja.teslic@fins.uns.ac.rs (N.T.); lidija.peictukuljac@fins.uns.ac.rs (L.P.T.); nebojsa.ilic@fins.uns.ac.rs (N.I.)
2   Institute of General and Physical Chemistry, University of Belgrade, Studentski Trg 12-16, 11000 Beograd, Serbia; latopezo@yahoo.co.uk
3   Faculty of Technology, University of Novi Sad, Bulevar Cara Lazara 1, 21000 Novi Sad, Serbia; kojicpredrag@uns.ac.rs (P.K.); vanjaseregelj@tf.uns.ac.rs (V.Š.)
*   Correspondence: jovana.kojic@fins.uns.ac.rs

Abstract: The influence of different extrusion parameters, including screw speed (250–750 rpm), feed rate (15–25 kg/h) and feed moisture content (15–25%), on the textural and color properties of spelt wholegrain snack products produced on a co-rotating twin-screw extruder with added betaine was investigated. In order to determine the relative influence of input variables in the artificial neural network (ANN) model, Yoon's interpretation method was used, and it was concluded that feed moisture content has the greatest influence on L* values, while screw speed has the greatest influence on a* and b* values. The softest samples were obtained at the lowest moisture content. Sensory analysis was carried out on selected samples, and it showed that betaine addition did not intensify the bitter taste. The sample with the largest expansion exhibited the lowest hardness and chewiness before and after immersion in milk, and this sample is the most suitable for enrichment with betaine.

Keywords: extrusion; snack; betaine; functional foods

Citation: Kojić, J.; Belović, M.; Krulj, J.; Pezo, L.; Teslić, N.; Kojić, P.; Tukuljac, L.P.; Šeregelj, V.; Ilić, N. Textural, Color and Sensory Features of Spelt Wholegrain Snack Enriched with Betaine. *Foods* **2022**, *11*, 475. https://doi.org/10.3390/foods11030475

Academic Editors: Luca Serventi, Charles Brennan and Rana Mustafa

Received: 31 December 2021
Accepted: 1 February 2022
Published: 6 February 2022

Publisher's Note: MDPI stays neutral with regard to jurisdictional claims in published maps and institutional affiliations.

1. Introduction

Recently, there has been an increasing amount of interest in the replacement of extruded products based on corn grits, which are the most common ones in the market, with nutritional, rich, cereals-based extrudates. Numerous cereals (such as wheat, amaranth and quinoa) have been used to improve the nutritional value and textural properties of extruded snack without reducing product quality in terms of organoleptic properties and consumers acceptability [1–4]. Thus far, the influence of spelt flour addition on the physical and rheological properties of extruded products based on corn grits has been investigated [5]. In comparison with common wheat, spelt flour has a higher content of protein (especially prolamin) and some amino acids (proline, glutamic acid, tyrosine and aspartic acid), as well as vitamin B, fiber, lipids and mineral elements, and it also has higher bioavailability [6–9]. The high content of nutritionally valuable components makes spelt flour suitable for the production of a wide range of food products. Although described as poorer in technological quality compared to common wheat, spelt flour is used for the production of pasta, bread, snacks and other food products [8,10,11]. In the last few years, cultivation of spelt flour has increased in Serbia, and a large number of studies have been conducted related to the protective role of the spelt husk of grain [12], for its use in bakeries [13], pasta production [14,15], as well as in solving byproduct issues by pelleting spelt grain husks [16]. Foods based on cereals have been presented as the largest source of betaine in the Western diet [17].

Betaine as a bioactive compound provides many health benefits. The main role of betaine in the human organism is to supply methyl groups for essential physiological processes [18]. The requirements of the organism cannot be satisfied with the endogenous synthesis of betaine, and therefore, its intake is necessary through diet. In the 2017 study of Kojić et al., the following order was determined among cereals in terms of the highest betaine content, with spelt flour at the top: buckwheat < millet < wheat < oats < rye < barley < amaranth < spelt [19]. Functional snack products from spelt wholegrain flour with the addition of betaine have been produced to satisfy the need for the recommended daily intake of betaine of 1500 mg in accordance with Commission Regulation (EU) No 432/2012. In previous work, it was shown that enriched spelt-flour-based extrudates satisfy the recommended daily intake of betaine [20]. Many critical parameters during extrusion, such as the feed rate of the mixture, screw speed and temperature in the barrel and die, affect the sensory properties of extrudate products which are primarily related to taste, texture and color.

Qualitative evaluation of snack products includes sensory, instrumental and microstructural characterization, which represent the final evaluation to determine consumer acceptability. In order to obtain a sensory profile of snack product samples, an objective sensory evaluation needs to be performed using a panel of trained evaluators. Human perception of a product is often closely related to the instrumental analysis of texture. Instrumental texture determination is an objective, fast and relatively inexpensive analysis of the characteristics of final products [21]. Color is one of the most important attributes of food products, providing information regarding the degree of cooking of the product and appearance and freshness of food [22]. It is a very important quantitative characteristic of extrudate quality that is directly related to consumer acceptance [23]. In addition, a change in food color may be a qualitative indicator of the extent of deterioration in food quality due to heat treatment [24]. Since human perception of color is subjective and individual, instrumental techniques for defining the color of a product provide more reliable results. The effects of extrusion parameters and the application of different raw materials on the color of extruded products have been the subject of numerous studies [25–30]. The main aim of this research was to evaluate the influence of extrusion cooking parameters (moisture, feed rate and screw speed) on the texture, color and sensory characteristics of snack products based on wholegrain spelt flour with added betaine.

2. Materials and Methods

2.1. Extrusion Processing—Experimental Design

Spelt flour enriched with betaine (9% w/w addition) was extruded using a co-rotating twin-screw extruder (Bühler BTSK 30/28D, 7 sections, length/diameter ratio = 28:1, Bühler, Uzwil, Switzerland). The extruder contains two temperature control units (the first unit-controlled temperature in sections set at 60 °C and the second set temperature at 120 °C). Screw configuration specially designed for the production of directly expanded snack products was used (the die opening diameter was 4 mm). A betaine addition of 9% was chosen in accordance with our preliminary trials with wholegrain spelt flour enriched with 1% of betaine, taking into account betaine loss during the extrusion cooking process. Before the extrusion process, the blends were mixed in a twin-shaft paddle mixer that is part of the laboratory vacuum coater (model F-6-RVC, Forberg International AS, Oslo, Norway). Total creation of snack products from spelt wholegrain flour with added 9% w/w betaine that can be beneficial to the human health and contribute to the recommended daily betaine intake was successfully carried out [23]. The content of betaine was measured by the developed and validated HPLC-ELSD method, and it was in the range from 1248.0 to 1543.1 mg/40 g [19].

The effects of the three extrusion factors, i.e., moisture (M; 15–25%), feed rate (FR; 15–25 kg/h) and screw speed (SS; 250–750 rpm), on hardness and color coordinates L*, a* and b* during the extrusion process of snacks was studied. The experimental data used for the analysis were fully determined using a central composite rotatable design (CCRD; $\alpha = 1.682$), explained with eight cube points, six axial points and three central points (Table 1). The CCRD experimental design was applied to limit the number of samples to a value of 17 that was sufficient for the calculation of the second-order polynomial coefficients in the model and to develop the artificial neural network (ANN). The RSM model describes the effects of process variables on the observed responses, determines interrelationships between process variables and represents the combined effect of all process variables to responses. The developed ANN consisted of three layers (input, hidden and output) with hyperbolic tangent function as the activation function. The Broyden–Fletcher–Gol–dfarb–Shanno (BFGS) calculation showed better model criteria than other training algorithms, such as Levenberg–Marquardt, Bayesian regularization, etc. Having in mind that the ANN results, including weight values, depend on the initial assumptions of parameters and number of hidden neurons, each topology was run several times to avoid overfitting. The coefficient of determination was higher than 0.9 for all ANN runs. In the extrusion process of input-toward-outputs, the ANN was implemented in Yoon's interpretation method to determine the relative influence of input process variables. The following equation was used:

$$RI_{ij} = \frac{\sum_{k=0}^{n}\left(w_{ik}w_{kj}\right)}{\sum_{i=0}^{m} abs \sum_{k=0}^{n}\left(w_{ik}w_{kj}\right)} \cdot 100\%$$

where RI_{ij} is the relative importance of the ith input variable on the jth output, w_{ik} is the weight between the ith input and the kth hidden neuron and w_{kj} is the weight between the kth hidden neuron and the jth output [31].

Table 1. Experimental EI, BD and hardness values of snack according to the adopted central composite rotatable design (CCRD) experimental plan.

	Variables			Product Response		
CCRD Runs	M (%)	FR (kg/h)	SS (o/min)	EI	BD (g/L)	Hardness (N)
1	20	20	500	1.82 ± 0.18 [a]	328.7 ± 34.5 [ab]	302.4 ± 21.3 [bcd]
2	20	20	250	1.54 ± 0.16 [a]	478.0 ± 51.2 [c]	372.2 ± 14.9 [e]
3	17	17	350	1.60 ± 0.17 [a]	367.0 ± 34.5 [b]	268.1 ± 16.1 [ab]
4	20	20	750	1.93 ± 0.18 [a]	244.4 ± 24.6 [a]	262.7 ± 20.5 [ab]
5	20	20	500	1.81 ± 0.20 [a]	333.1 ± 30.4 [ab]	302.4 ± 12.8 [bcd]
6	23	17	350	1.69 ± 0.18 [a]	359.1 ± 37.2 [b]	351.4 ± 22.0 [de]
7	17	17	650	1.90 ± 0.18 [a]	367.2 ± 38.7 [b]	300.6 ± 16.8 [bcd]
8	25	20	500	1.72 ± 0.16 [a]	380.0 ± 40.0 [bc]	366.9 ± 34.6 [e]
9	20	25	500	1.78 ± 0.19 [a]	359.1 ± 34.8 [b]	342.0 ± 21.3 [cde]
10	23	17	650	1.88 ± 0.19 [a]	299.9 ± 28.4 [ab]	276.7 ± 14.8 [b]
11	20	15	500	1.83 ± 0.17 [a]	321.8 ± 31.0 [ab]	282.8 ± 10.4 [bc]
12	23	23	650	1.86 ± 0.19 [a]	298.7 ± 27.3 [ab]	343.9 ± 15.9 [de]
13	20	20	500	1.81 ± 0.20 [a]	341.6 ± 34.9 [ab]	302.4 ± 27.7 [bcd]
14	15	20	500	1.92 ± 0.20 [a]	324.7 ± 29.5 [ab]	261.8 ± 26.8 [ab]
15	17	23	650	1.89 ± 0.19 [a]	249.8 ± 27.1 [a]	214.3 ± 20.2 [a]
16	23	23	350	1.62 ± 0.16 [a]	385.7 ± 35.6 [bc]	377.0 ± 36.6 [e]
17	17	23	350	1.82 ± 0.19 [a]	380.0 ± 41.1 [bc]	357.1 ± 25.4 [de]

M (%)—feed moisture, FR (kg/h)—feed rate, SS (rpm)—screw speed, CCRD—central composite rotatable design. Means in the same column with different superscript are statistically different ($p \leq 0.05$); EI—expansion index, BD—bulk density.

2.2. Characterization of Extrudates

2.2.1. Textural Properties

Snack hardness was determined by diametric compression on a TA-XT.2, Texture Analyzer (Stable Micro Systems, Godalming, Surrey, UK) in accordance with the method described by Svihus et al. (2004) [32]. The hardness of the whole snack product (13.5–15.8 mm height, 7.08–7.72 mm diameter) was determined in accordance with a modified method (dry catfood_CTF1_P35). In all, 15 extrudates were taken from each sample, and 3 individual extrudates were placed horizontally on a flat surface of the device and then compressed with a cylindrical probe made of stainless steel with a diameter of 45 mm, load cell of 50 kg and trigger force of 100 g. The hardness of the sample is expressed as the mean value of 15 measurements and is expressed in kilograms. The parameters of the instrument adjustment during the test were as follows: pre-test speed: 2.0 mm/s; test speed: 1 mm/s; post-test speed: 10 mm/s, probe path: 2.5 mm.

2.2.2. Color Measurement

The color of wholegrain spelt flour with the addition of betaine (9% *w/w*) and grounded snack products was determined in ten replicates using Chroma Meter CR-400 (Konica Minolta Co., Ltd., Osaka, Japan) and a suitable extension (CR-A50), adapted for measurements of this type of sample in the CIE L* a* b* color space. Total color change (ΔE) between flour and betaine blend and spelt wholegrain flour was calculated based on the following formula:

$$\Delta E = [(L - L_0)^2 + (a - a_0) + (b - b_0)^2]^{1/2} \tag{1}$$

where subscript zero indicates the color parameters of the raw material blend.

2.2.3. Sensory Evaluation of Snack Products Using a Panel of Trained Evaluators

Eight trained panelists, between 25 and 50 years old, from the Institute of Food Technology in Novi Sad, participated in the examination of the sensory properties of snack products. The panelists had more than 4 years of experience in working with commercial products and products developed within scientific research projects. Their training included exercises in identifying, developing terminology and evaluating the intensity of sensory attributes. The panelists had additional training on snack products for the purposes of this study. Two-hour sessions were held to establish the sensory terminology for the tested snack products. Initially, panelists used descriptors from previously published papers with similar topics [33,34] as a starting point, and they could keep, delete or add any descriptor.

A consensus approach was used to determine the final descriptors for snack products. The panel leader led a discussion of each descriptor in order to determine the appropriateness of the terms, definitions and assessment techniques. A final list of descriptors with definitions is given in Supplementary Table S1, which were used by the panel to evaluate all samples in terms of intensity. Intensity assessment was performed using an unstructured linear scale with points 0—imperceptible and 100—very intense. Since it is predicted that the created snack products will be consumed after immersion in milk, the sensory evaluation consisted of two parts.

In the first part, 6 selected attributes of snack products before immersion in milk were evaluated (color, hardness, chewiness, sweet taste, bitter taste). In the second part, 5 g of flips product was immersed in 50 mL of milk (1.5% milk fat) at room temperature. After 5 min, the attributes describing taste and texture were re-evaluated (hardness, chewiness, sweet taste, bitter taste). Distilled water was used to clean the mouth between samples during the evaluation. The assessment was performed in a sensory testing laboratory with appropriate control of environmental conditions [35].

2.3. Statistical Analysis

Statistical analysis was obtained by analysis of variance (ANOVA) followed by Tukey's Test. The results, expressed as mean $\pm$ standard deviation, were considered statistically significant with $p \leq 0.05$. Different letters indicate significant differences in the results ($p \leq 0.05$).

In order to obtain a better insight into the relationship between sensory properties, instrumentally measured quality parameters (diameter, color and textural properties) and betaine content in snack products, principal component analysis (PCA) was performed using the PanelCheck software (version 1.4.0, Nofima Mat, Norway, Norway, 2010, https://www.panelcheck.com/, accessed on 27 December 2021).

3. Results and Discussion

3.1. Impact of Process Conditions on Texture Properties of Extrudates

Texture is an important sensory indicator for the quality of snack products. In snack products, expansion is desirable, and texture plays an important role in terms of consumer acceptance [36]. The most commonly used tests to measure the texture of snack are cutting or shear tests, compression and puncture tests. There is no single term that describes the texture of extruded snack products, and the most common terms are hardness, brittleness and crunchiness [34]. In this study, the hardness of snack products from wholegrain spelt flour with the addition of betaine was determined through diametrical compression. Table 1 presents the experimental hardness values for the obtained snack products.

Figure 1 shows the influence of process parameters (M, FR and SS) on the expansion index, bulk density and hardness of the snack products. A tendency of increasing hardness with increasing feed moisture content (M, %) can be observed. This result is in agreement with the results for bulk density and the expansion index, which has been previously published in our snack from wholegrain spelt flour with the addition of 9% of betaine (Table 1) [37,38]. Namely, a smaller expansion index occurs with an increase in moisture content, and an increase in bulk density with a decrease in the expansion index, which is confirmed in our study. A negative correlation between expansion index and bulk density was observed in our study (r = −0.785; p = 0.000 (p < 0.001)). Liu et al. (2011) also link the results for bulk density and the expansion index with hardness, which is confirmed in our results through positive correlation between hardness and bulk density (=+0.736; p = 0.001 (p < 0.01)) (Supplementary Figure S1) [39].

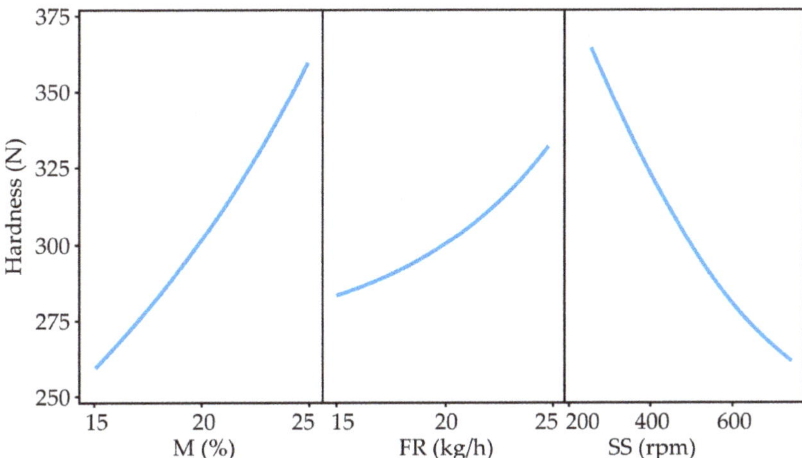

Figure 1. Influence of process parameters moisture (M), feed rate (FR) and screw speed (SS) on hardness.

Numerous studies have confirmed that the hardness of the extrudate increased with increasing feed moisture content [40,41]. Increasing the feed moisture content leads to plasticization of the sample, forming a protective layer and compressing the sample, resulting in a high density and hardness of rice [42] and wheat extrudates [2]. The results showed that low hardness was associated with low feed moisture and high screw speed (Figure 1). As screw speed increases, viscosity decreases, which results in lower density and less hardness of extrudate. By contrast, with an increase in feed rate, viscosity increases, giving extrudates with high density and hardness. As screw speed increases, the sample expands and thus becomes softer, while with an increase in feed rate, the barrel of the extruder is filled, and therefore, pressure increases, which leads to the material being compressed and firm. Ding et al. (2006) concluded that feed rate and screw speed have a significant effect on the hardness of wheat extrudate [2]. Diaz et.al (2013) showed that changes in the hardness of extrudates containing kañiwa were caused by screw speed more than changes in feed moisture content [4].

An analysis of operating parameters on hardness is presented in Table 2. The most influential in the second-order polynomial approximation SOP model for hardness evaluation was the linear term of SS statistically significant at $p < 0.01$ level, as well as the linear term of M (statistically significant at $p < 0.05$ level). The coefficient of determination value (R^2) for the SOP model was 0.823, which can be considered satisfactory for predicting hardness (Table 2).

Table 2. Analysis of variance for second-order polynomial for hardness calculation.

Terms	df	Hardness
M	1	113.128 *
M^2	1	1.194
FR	1	28.863
FR^2	1	0.729
SS	1	122.982 [+]
SS^2	1	2.158
M × FR	1	10.557
M × SS	1	0.009
SS × SS	1	23.222
Error	7	64.989
r^2		0.823

df—degrees of freedom; [+] statistically significant at $p < 0.01$ level, * statistically significant at $p < 0.05$ level.

Moreover, from Figure 2, which presents the relative influence of process parameters on the hardness of extrudates obtained by Yoon's model, it is clear that feed moisture content and screw speed are the ones that most significantly affect the hardness of the extrudate. In fact, hardness shows a high positive correlation with feed moisture content (increasing with increasing moisture content) and a negative correlation with screw speed (decreasing with increasing screw speed). Feed rate had the smallest effect on the hardness of the extrudate, and the hardness of the extrudate increased with increasing feed rate (Figure 2). These results are in accordance with the results obtained by Brnčić et al. (2006), who concluded that feed moisture content has the greatest positive effect on hardness, while screw speed and temperature have a significant negative effect on hardness [43].

3.2. Impact of Process Conditions on Color Properties of Extrudates

The color of extruded products can vary depending on the combination of established parameters such as raw material moisture content, temperature and chemical components of each raw material and their ratio in the mixture. Therefore, it is important to control the color of the ingredients, as well as to monitor the product throughout the production process to obtain and maintain the desired color [44]. The values of lightness (L*) of ground snack products from wholegrain spelt flour with added betaine ranged from 65.04 to 73.50, and the redness value (a*) and yellowness value (b*) of the same samples was in the range of 3.20–4.97 and 15. 71–17.25, respectively (Table 3). The value of L* for the control whole grain flour with 9% betaine was 81.232, while the values of a* and b* were 1.084 and 11.614, respectively. ΔE values calculated between snack products and blend ranged from 10.15 to 17.06, indicating a very pronounced color change. These results are in agreement with the results obtained in the study by Wani and Kumar (2015), who examined the effect of the addition of different vegetable raw materials on the color change of corn, rice and barley extrudates and recorded values of 56.3–71.3 for L*, 4, 44–6.47 for a* and 11.89–19.88 for b* [45].

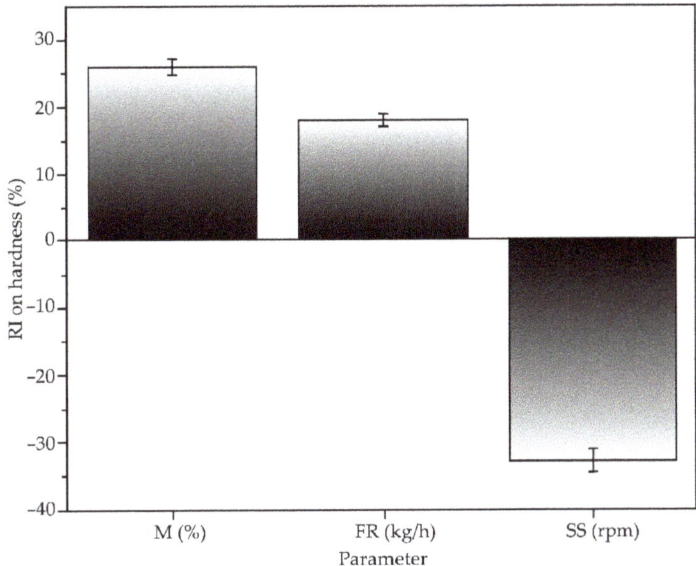

Figure 2. The relative importance of extrusion parameters on hardness using Yoon's interpretation method.

Table 3. Experimental color values of snack according to the adopted CCRD experimental plan.

	Variable			Product Response			
CCRD Runs	**M (%)**	**FR (kg/h)**	**SS (o/min)**	**L***	**a***	**b***	**ΔE**
1	20	20	500	69.00 ± 5.65 [a]	4.18 ± 0.25 [bc]	16.89 ± 1.35 [a]	13.68 ± 1.45 [abcd]
2	20	20	250	66.07 ± 5.20 [a]	4.01 ± 0.23 [bc]	15.74 ± 1.30 [a]	15.98 ± 1.46 [cd]
3	17	17	350	67.80 ± 4.08 [a]	4.97 ± 0.30 [d]	17.00 ± 1.52 [a]	14.98 ± 1.53 [bcd]
4	20	20	750	69.10 ± 4.57 [a]	3.20 ± 0.19 [a]	16.94 ± 1.27 [a]	13.42 ± 1.27 [abc]
5	20	20	500	69.00 ± 5.70 [a]	4.25 ± 0.26 [bcd]	16.50 ± 1.02 [a]	13.55 ± 1.40 [abcd]
6	23	17	350	69.12 ± 4.46 [a]	4.25 ± 0.35 [bcd]	16.41 ± 1.32 [a]	13.41 ± 1.28 [abc]
7	17	17	650	68.79 ± 3.97 [a]	4.30 ± 0.25 [bcd]	15.76 ± 0.83 [a]	13.50 ± 1.28 [abcd]
8	25	20	500	73.50 ± 4.75 [a]	4.66 ± 0.36 [cd]	17.14 ± 1.31 [a]	10.15 ± 1.10 [a]
9	20	25	500	68.38 ± 5.91 [a]	3.91 ± 0.22 [ab]	16.74 ± 1.34 [a]	14.12 ± 1.28 [bcd]
10	23	17	650	69.44 ± 3.65 [a]	3.88 ± 0.30 [ab]	17.25 ± 1.04 [a]	13.37 ± 1.39 [abc]
11	20	15	500	68.28 ± 3.81 [a]	3.80 ± 0.27 [ab]	17.07 ± 1.04 [a]	14.31 ± 1.44 [bcd]
12	23	23	650	68.75 ± 5.91 [a]	4.22 ± 0.35 [bc]	16.53 ± 0.87 [a]	13.78 ± 1.29 [bcd]
13	20	20	500	69.00 ± 3.89 [a]	4.35 ± 0.26 [bcd]	16.98 ± 1.46 [a]	13.75 ± 1.50 [bcd]
14	15	20	500	65.04 ± 5.79 [a]	4.26 ± 0.21 [bcd]	15.93 ± 0.98 [a]	17.06 ± 1.62 [d]
15	17	23	650	70.38 ± 6.01 [a]	3.86 ± 0.27 [ab]	16.31 ± 1.23 [a]	12.15 ± 1.29 [ab]
16	23	23	350	69.41 ± 5.47 [a]	4.23 ± 0.27 [bc]	15.71 ± 0.99 [a]	12.90 ± 1.31 [abc]
17	17	23	350	68.86 ± 5.36 [a]	4.34 ± 0.26 [bcd]	15.73 ± 1.09 [a]	13.44 ± 1.37 [abc]

M (%)—feed moisture, FR (kg/h)—feed rate, SS (rpm)—screw speed, CCRD—central composite rotatable design. Means in the same column with different superscript are statistically different ($p \leq 0.05$).

The presented results indicate that the values of L* after extrusion are reduced, while the values of a* and b* are increased (Table 3). These results are in agreement with the results of Menegassi et al. (2011) and Durge et al. (2013) [44,46]. Changes in the color of extrudates may be related to the potential role of betaine as an amino acid in Maillard reactions and may be due to a reaction between betaine and sugar that contributes to the formation of colored compounds (products of Maillard reactions) that reduce the lightness of extrudates.

From Figure 3, it is clear that L* and b* color values increased with increasing feed moisture, while parameter a* decreased with increasing feed moisture up to 20% and then started to increase.

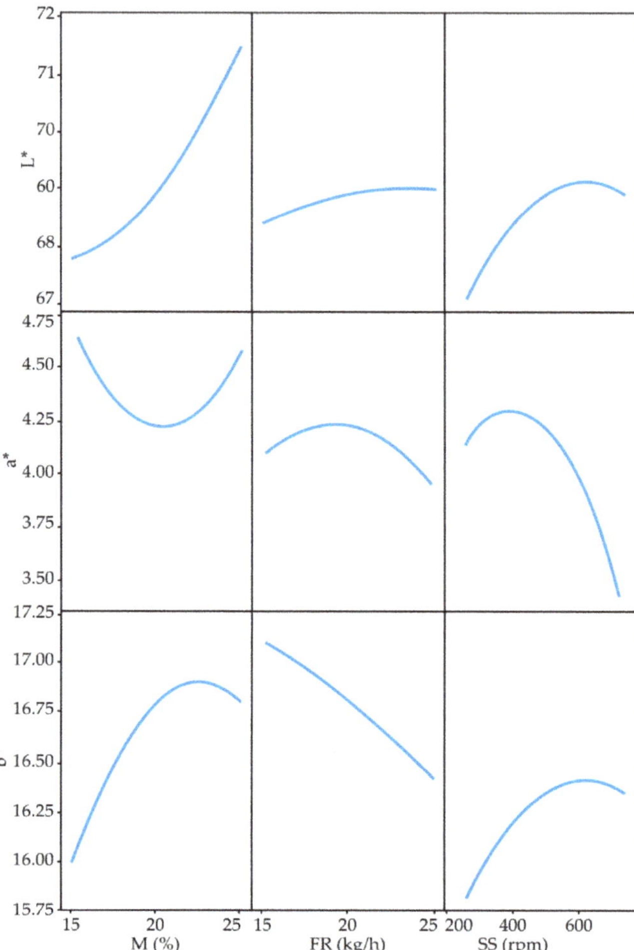

Figure 3. Influence of process parameters moisture (M), feed rate (FR) and screw speed (SS) on L*, a* and b*.

Feed moisture is an important factor, and its increase gives a lighter product, i.e., it prevents its darkening and has a protective role in the extrusion process. It is considered that increased feed moisture lowers the temperature of the extrusion process, which in turn reduces the potential for darkening of the product through Maillard reactions between reducing sugars and free amino groups [47].

Increasing screw speed in the extrusion cooking process increased the values of L* and b*, while the value of a* was decreased (Figure 3). As concluded by Gulati et al. (2016), an increasing value of b* with an increase in screw speed may be associated with a lower retention time of the material in the extruder barrel, thus achieving less sample cooking [48]. The obtained results are in accordance with extruded rice flour [49]. Additionally, increasing the feed rate increased the L* values, while a* and b* decreased (Figure 3).

Yoon's model (Figure 4) has shown that feed moisture content has the greatest influence on the L* values, while screw speed has the greatest influence on the a* and b* values.

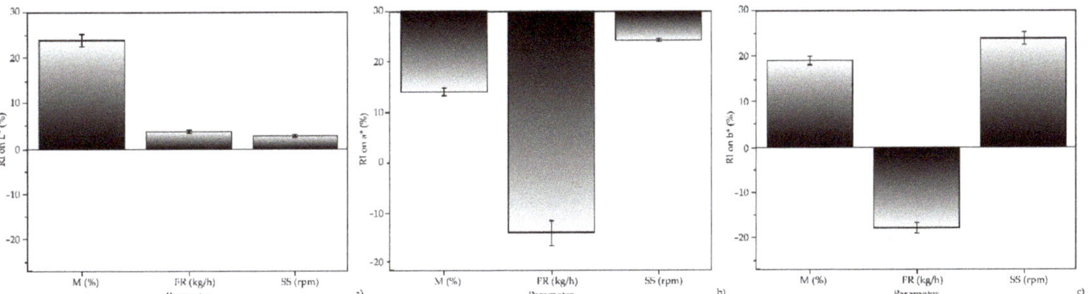

Figure 4. The relative importance of the extrusion parameters on the color parameters (**a**) L*—lightness; (**b**) a*—red/green color; (**c**) b*—yellow/blue color, using Yoon's interpretation method.

Gulati et al. (2016) showed that feed moisture content is the main factor influencing the values of L* and a*, while feed moisture in interaction with temperature was the main factor influencing the value of b* [48].

For L* value evaluation, the most influential was the linear term of M in the SOP model (statistically significant at $p < 0.1$ level), as well as the linear term of SS for the evaluation of value a* (statistically significant at $p < 0.05$ level). For calculating the b* value, the most influential were the linear terms M and SS, as well as the combined effect of these two variables (statistically significant at the level of $p < 0.10$), Table 4.

Table 4. Analysis of variance for second-order polynomial for color parameter calculation.

Terms	df	L*	a*	b*
M	1	16.576 **	0.004	0.717 **
M^2	1	0.769	0.225	0.239
FR	1	0.431	0.024	0.534
FR^2	1	0.060	0.062	0.002
SS	1	3.846	0.612 *	0.664 **
SS^2	1	1.291	0.300 **	0.521
M × FR	1	1.163	0.242	0.061
M × SS	1	1.015	0.074	0.673 **
FR × SS	1	0.025	0.038	0.405
Error	7	22.756	0.546	1.174
r^2		0.533	0.766	0.760

df—degrees of freedom; * statistically significant at $p < 0.05$ level; ** statistically significant at $p < 0.10$ level; L*—lightness; a*—red/green color; b*—yellow/blue color.

The coefficients of determination for the calculation of L*, a* and b* had values of 0.533, 0.766 and 0.760, respectively, which can be considered relatively satisfactory for predicting the stated color coordinates.

3.3. Sensory Evaluation of Snack Products

The relationship between the diameter, instrumentally measured color and hardness, betaine content and sensory descriptors of the evaluated snack samples was visually presented by linear combinations of variables identified by PCA (Loading Plot) and the position of samples in the factor space (Score plot) together in a Bi-plot (Figure 5). The first two principal components (F1 and F2) explained 90.73% of the total variability, which can be explained by a good selection of sensory variables and a relatively small number of tested samples. If the relationship between variables is considered, three groups can be observed, and it can be concluded that all variables within one group are in a significant positive correlation with each other (r close to +1). The first group of variables consisted of diameter and sweet taste. These two parameters could be related since a higher expansion index occurs when there is a higher degree of starch gelatinization, and starch hydrolysis into fragments with smaller molecular weights and higher sweetness could also occur at the same time [50].

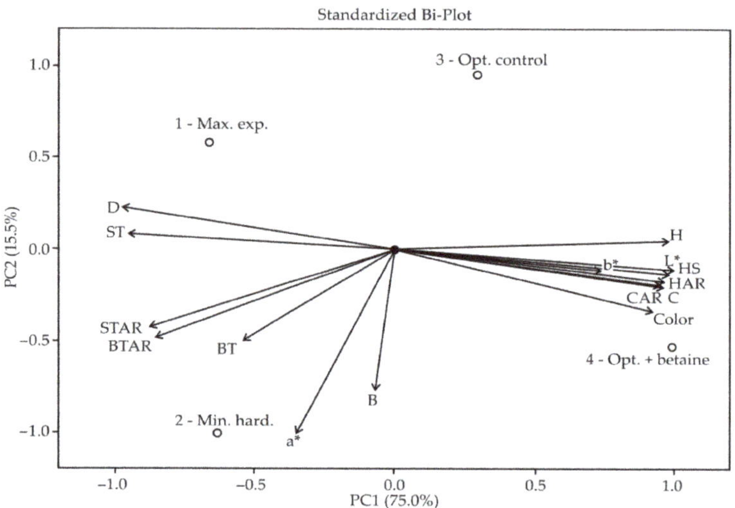

Figure 5. Linear combinations of variables identified via PCA analysis and the position of samples in the factor space (Bi-plot).

PCA—principal component analysis; B—betaine; BT—bitter taste; BTAR—bitter taste after rehydration; C—chewiness; CAR—chewiness after rehydration; D—diameter; H—hardness; HAR—hardness after rehydration; HS—hardness (sensory); ST—sweet taste; STAR—sweet taste after rehydration.

The second group of variables contained L* and b* color parameters, instrumentally measured hardness and sensory determined color, hardness and chewiness before and after immersion in milk and were opposite to the variables in the first group.

This indicates that the variables in the first and second groups are mutually negatively correlated (r close to −1). Hardness determined by a sensory panel was highly correlated with hardness determined instrumentally, and color intensity perceived by a sensory panel was highly correlated with lightness (L*) and yellow tone intensity (b*), indicating that these instrumental parameters could be used successfully for fast determination of the sensory quality of snack samples. Additionally, this grouping indicated that smaller diameter snack samples (samples 3 and 4) are at the same time harder, tougher (require more time to be masticated) and darker, and that after immersion in milk, the relationship between their mechanical properties remains the same.

The variables in the third group were a* color parameter, betaine content, bitter taste before immersion, as well as bitter and sweet taste after immersion in milk. These correlations suggest that betaine content in snack samples could be related to the perceived bitter taste, and a more pronounced red tone (a* values) could be a consequence of the aforementioned Maillard reactions.

According to the Score plot, the selected sensory descriptors enabled a clear distinction between the obtained snack samples. The sample with the maximum expansion (sample 1) is distinguished, as expected, with the largest diameter, but also with the most pronounced sweet taste. The sample with the lowest hardness (sample 2) is distinguished through bitter taste before and after immersion in milk, as well as through sweet taste after immersion in milk and betaine content. Since this sample had the best rehydration properties (the weakest mechanical properties) and therefore absorbed the largest amounts of milk, both flavors present in the sample came to the fore due to the dissolution of substances that give a sweet and bitter taste in milk.

Snack extruded products that are obtained via an optimized extrusion process (samples 3 and 4) are characterized by higher hardness and chewiness both before and after immersion in milk, as well as by a darker color.

Since sample 1 has the best mechanical properties (largest diameter, the lowest hardness before and after immersion, the lowest chewiness before and after immersion) and a distinctly sweet taste, it can be considered the most suitable for consumers. It is important to note that betaine slightly influenced bitter taste (sample number 4) but reduced the mechanical properties (samples 3 and 4). Additionally, it is assumed that sample 1 would be the most suitable for enrichment with betaine.

4. Conclusions

This study confirmed that the operational parameters of the extrusion cooking process (moisture content (M, %), feed rate (FR, (kg/h)) and screw speed (SS, rpm)) affect the success of the sensory experience of the snack product, which is related to texture and color. The softest spelt wholegrain snack was produced at the lowest level of feed moisture content. The results for hardness obtained by Yoon's model showed that feed moisture content and screw speed are the most influential parameters during the production of spelt wholegrain snacks with added betaine. Decrease in the lightness of the extrudate may be associated with a reaction between betaine and sugar that contributes to the formation of colored compounds. L* color values were decreased after extrusion, while a* and b* values were increased. In addition to satisfying the nutritional recommended daily intake of betaine, it is important to obtain a product that is acceptable in terms of sensory properties. Products obtained via the optimized extrusion process are not rated as the most acceptable, which emphasizes the importance of sensory analysis, which represents the final assessment carried out by consumers. The sample with the largest expansion, lowest hardness before and after immersion and lowest chewiness before and after immersion can be considered the most appropriate for supplementation with betaine and for consumers.

Supplementary Materials: The following supporting information can be downloaded at: https://www.mdpi.com/article/10.3390/foods11030475/s1; Table S1: Sensory descriptors and definitions used in the sensory analysis of snack product samples; Figure S1: Correlation between (**a**) the expansion index and bulk density and between (**b**) hardness and bulk density.

Author Contributions: Conceptualization, J.K. (Jovana Kojić) and M.B.; methodology, N.I.; software, L.P. and P.K.; validation, V.Š. and J.K. (Jelena Krulj); formal analysis, L.P.T., J.K. (Jelena Krulj) and V.Š.; investigation, N.T. and L.P.T.; writing—original draft preparation, J.K. (Jovana Kojić) and M.B.; writing—review and editing, P.K.; visualization, L.P. and N.T.; supervision, N.I. All authors have read and agreed to the published version of the manuscript.

Funding: This research received no external funding.

Institutional Review Board Statement: The study did not require ethical approval; it was the part of PhD study.

Informed Consent Statement: Not applicable.

Data Availability Statement: The data that support the findings of this study are available from the corresponding author upon reasonable request.

Acknowledgments: The Ministry of Education, Science and Technological Development of the Republic of Serbia. (451-03-9/2021-14/200222).

Conflicts of Interest: The authors declare no conflict of interest.

References

1. O'Shea, N.; Gallagher, E. Evaluation of novel-extruding ingredients to improve the physicochemical and expansion characteristics of a corn-puffed snack-containing pearled barley. *Eur. Food Res. Technol.* **2019**, *245*, 1293–1305. [CrossRef]
2. Ding, Q.B.; Ainsworth, P.; Plunkett, A.; Tucker, G.; Marson, H. The effect of extrusion conditions on the functional and physical properties of wheat-based expanded snacks. *J. Food Eng.* **2006**, *73*, 142–148. [CrossRef]
3. Chávez-Jáuregui, R.N.; Silva, M.E.M.P.; Arêas, J.A.G. Extrusion Cooking Process for Amaranth (Amaranthus caudatus L.). *J. Food Sci.* **2000**, *65*, 1009–1015. [CrossRef]
4. Diaz, J.M.R.; Kirjoranta, S.; Tenitz, S.; Penttilä, P.A.; Serimaa, R.; Lampi, A.-M.; Jouppila, K. Use of amaranth, quinoa and kañiwa in extruded corn-based snacks. *J. Cereal Sci.* **2013**, *58*, 59–67. [CrossRef]
5. Jozinović, A.; Subari, D.; Ačkar, Đ.; Babić, J.; Miličević, B. Influence of spelt flour addition on properties of extruded products based on corn grits. *J. Food Eng.* **2016**, *172*, 31–37. [CrossRef]
6. Ruibal-Mendieta, N.L.; Delacroix, D.L.; Mignolet, E.; Pycke, J.M.; Marques, C.; Rozenberg, R.; Larondelle, Y. Spelt (Triticum aestivum ssp. spelta) as a source of breadmaking flours and bran naturally enriched in oleic acid and minerals but not phytic acid. *J. Agric. Food Chem.* **2005**, *53*, 2751–2759. [CrossRef]
7. Pruska- Kędzior, A.; Kędzior, Z.; Klockiewicz-Kaminska, E. Comparison of viscoelastic properties of gluten from spelt and common wheat. *Eur. Food Res. Technol.* **2008**, *227*, 199–207. [CrossRef]
8. Kohajdová, Z.; Karovičová, J. Nutritional value and baking applications of spelt wheat. *Acta Sci. Pol. Technol. Aliment.* **2008**, *7*, 5–14.
9. Bodroža Solarov, M.; Vujić, Đ.; Ačanski, M.; Pezo, L.; Filipčev, B.; Mladenov, N. Characterization of the liposoluble fraction of common wheat (Triticum aestivum) and spelt (T. aestivum ssp. spelta) flours using multivariate analysis. *J. Sci. Food Agric.* **2014**, *94*, 2613–2617. [CrossRef]
10. Pasqualone, A.; Piergiovanni, A.R.; Caponio, F.; Paradiso, V.M.; Summo, C.; Simeone, R. Evaluation of the technological characteristics and bread-making quality of alternative wheat cereals in comparison with common and durum wheat. *Food Sci. Technol. Int.* **2011**, *17*, 135–142. [CrossRef]
11. Filipčev, B.; Šimurina, O.; Bodroža Solarov, M.; Brkljača, J. Dough rheological properties in relation to cracker making performance of organically grown spelt cultivars. *Int. J. Food Sci. Technol.* **2013**, *48*, 2356–2362. [CrossRef]
12. Vučković, J.; Bodroža-Solarov, M.; Vujić, Đ.; Bočarov-Stančić, A.; Bagi, F. The protective effect of hulls on the occurrence of Alternaria mycotoxins in spelt wheat. *J. Sci. Food Agric.* **2013**, *93*, 1996–2001. [CrossRef]
13. Filipčev, B.; Šimurina, O.; Bodroža-Solarov, M.; Obreht, D. Comparison of the bread-making performance of spelt varieties grown under organic conditions in the environment of norther Serbia and their responses to dough strengthening improvers. *Hem. Ind.* **2013**, *67*, 443–453. [CrossRef]
14. Filipović, J.; Pezo, L.; Filipović, N.; Filipović, V.; Bodroža-Solarov, M.; Plančak, M. Mathematical approach to assessing spelt cultivars (Triticum aestivum subsp. spelt) for pasta making. Inter. *J. Food Sci. Technol.* **2013**, *48*, 195–203. [CrossRef]
15. Filipović, J.; Pezo, L.; Filipović, V.; Brkljača, J.; Krulj, J. The effects of omega-3 fatty acids and inulin addition to spelt pasta quality. *LWT-Food Sci.Technol.* **2015**, *63*, 43–51. [CrossRef]
16. Brlek, T.; Bodroža-Solarov, M.; Vukmirović, Dj.; Čolovic, R.; Vučković, J.; Lević, J. Utilization of Spelt Wheat Hull as a Renewable Energy Source by Pelleting. *Bulg. J. Agric. Sci.* **2012**, *18*, 752–758.
17. Filipčev, B.; Kojić, J.; Krulj, J.; Bodroža-Solarov, M.; Ilić, N. Betaine in cereal grains and grain-based products. *Foods* **2018**, *7*, 49. [CrossRef]
18. Craig, S.A. Betaine in human nutrition. *Am. J. Clin. Nutr.* **2004**, *80*, 539–549. [CrossRef]
19. Kojić, J.; Krulj, J.; Ilić, N.; Lončar, E.; Pezo, L.; Mandić, A.; Bodroža-Solarov, M. Analysis of betaine levels in cereals, pseudocereals and their products. *J. Funct. Foods* **2017**, *37*, 157–163. [CrossRef]
20. Kojić, J.; Ilić, N.; Kojić, P.; Pezo, L.; Banjac, V.; Krulj, J.; Bodroža-Solarov, M. Multiobjective process optimization for betaine enriched spelt flour based extrudates. *J. Food Process Eng.* **2019**, *42*, e12942. [CrossRef]
21. Chen, L.; Opara, U.L. Texture measurement approaches in fresh and processed foods—A review. *Food Res. Int.* **2013**, *51*, 823–835. [CrossRef]
22. Valadez-Blanco, R.; Virdi, A.I.S.; Balke, S.T.; Diosady, L.L. In-line colour monitoring during food extrusion: Sensitivity and correlation with product colour. *Food Res. Int.* **2007**, *40*, 1129–1139. [CrossRef]
23. Do Nascimento, E.M.D.G.C.; Carvalho, C.W.P.; Takeiti, C.Y.; Freitas, D.D.G.C.; Ascheri, J.L.R. Use of sesame oil cake (Sesamum indicum L.) on corn expanded extrudates. *Food Res. Int.* **2012**, *45*, 434–443. [CrossRef]
24. Lei, H.; Fulcher, R.G.; Ruan, R.; van Lengerich, B. Assessment of color development due to twin-screw extrusion of rice-glucose-lysine blend using image analysis. *LWT Food Sci. Technol.* **2007**, *40*, 1224–1231. [CrossRef]
25. Ondo, S.E.; Singkhornart, S.; Ryu, G.H. Effects of die temperature alkalized cocoa powder content and CO_2 gas injection on physical properties of extruded cornmeal. *J. Food Eng.* **2013**, *117*, 173–182. [CrossRef]
26. Singkhornart, S.; Edou-Ondo, S.; Ryu, G.H. Influence of germination and extrusion with CO_2 injection on physicochemical properties of wheat extrudates. *Food Chem.* **2014**, *143*, 122–131. [CrossRef]
27. Wang, Y.Y.; Ryu, G.H. Physical properties of extruded corn grits with corn fibre by CO_2 injection extrusion. *J. Food Eng.* **2013**, *116*, 14–20. [CrossRef]

28. Poliszko, N.; Kowalczewski, P.Ł.; Rybicka, I.; Kubiak, P.; Poliszko, S. The effect of pumpkin flour on quality and acoustic properties of extruded corn snacks. *J. fur Verbraucherschutz und Leb.* **2019**, *14*, 121–129. [CrossRef]

29. Blandino, M.; Bresciani, A.; Loscalzo, M.; Vanara, F.; Marti, A. Extruded snacks from pigmented rice: Phenolic profile and physical properties. *J. Cereal Sci.* **2022**, *103*, 103347. [CrossRef]

30. Grasso, S. Extruded snacks from industrial by-products: A review. *Trends Food Sci.Technol.* **2020**, *99*, 284–294. [CrossRef]

31. Yoon, Y.; Swales, G., Jr.; Margavio, T.M. A comparison of discriminant analysis versus artificial neural networks. *J. Oper. Res. Soc.* **1993**, *44*, 51–60. [CrossRef]

32. Svihus, B.; Kløvstad, K.H.; Perez, V.; Zimonja, O.; Sahlström, S.; Schuller, R.B.; Jeksrud, W.K.; Prestløkken, E. Physical and nutritional effects of pelleting of broiler chicken diets made from wheat ground to different coarsenesses by the use of roller mill and hammer mill. *Anim. Feed Sci. Technol.* **2004**, *117*, 281–293. [CrossRef]

33. Lazou, A.; Krokida, M. Structural and textural characterization of corn–lentil extruded snacks. *J. Food Eng.* **2010**, *100*, 392–408. [CrossRef]

34. Paula, A.M.; Conti-Silva, A.C. Texture profile and correlation between sensory and instrumental analyses on extruded snacks. *J. Food Eng.* **2014**, *121*, 9–14. [CrossRef]

35. *ISO 8589*; Sensory Analysis—General Guidance for the Design of Test Rooms, Amendment 1, 2014; International Organization for Standardization: Geneva, Switzerland, 2007.

36. Anton, A.A.; Luciano, F.B. Instrumental texture evaluation of extruded snack foods: A review. *Cienc. Tecnol. Aliment.* **2007**, *5*, 245–251. [CrossRef]

37. Kojić, J.; Krulj, J.; Peić Tukuljac, L.; Jevtić Mučibabić, R.; Cvetković, B.; Kojić, P.; Ilić, N. The effect of extrusion conditions on the bulk density of spelt wholegrain snack product. In Proceedings of the Poster Session Presentation in the Book of Abstracs of the VII International Congress "Engineering, Environment and Materials in Process Industry" EEM2021, Jahorina, Bosnia and Herzegovina, 17–19 March 2021.

38. Kojić, J.; Krulj, J.; Šimurina, O.; Cvetković, B.; Pezo, L.; Đermanović, B.; Ilić, N. The effect of extrusion conditions on the expansion of spelt wholegrain snack product. In Proceedings of the Poster Session Presentation in the Book of Abstracts of the VII International Conference Sustainable Postharvest and Food Technologies, Vršac, Serbia, 18–23 April 2021.

39. Liu, C.; Zhang, Y.; Liu, W.; Wan, J.; Wang, W.; Wu, L.; Zuo, N.; Zhou, Y.; Yin, Z. Preparation, physicochemical and texture properties of texturized rice produce by Improved Extrusion Cooking Technology. *J. Cereal Sci.* **2011**, *54*, 473–480. [CrossRef]

40. Tsokolar-Tsikopoulos, K.C.; Katsavou, I.D.; Krokida, M.K. The effect of inulin addition on structural and textural properties of extruded products under several extrusion conditions. *J. Food Sci. Technol.* **2015**, *52*, 6170–6181. [CrossRef]

41. Liu, Y.; Hsieh, F.; Heymann, H.; Huff, H.E. Effect of process conditions on the physical and sensory properties of extruded oat–corn puff. *J. Food Sci.* **2000**, *65*, 1253–1259. [CrossRef]

42. Ding, Q.B.; Ainsworth, P.; Tucker, G.; Marson, H. The effect of extrusion conditions on the physicochemical properties and sensory characteristics of rice-based expanded snacks. *J. Food Eng.* **2005**, *66*, 283–289. [CrossRef]

43. Brnčić, M.; Tripalo, B.; Ježek, D.; Semenski, D.; Drvar, N.; Ukrainczyk, M. Effect of twin screw extrusion parameters on mechanical hardness of direct-expanded extrudates. *Sādhanā* **2006**, *31*, 527–536. [CrossRef]

44. Menegassi, B.; Pilosof, A.M.; Areas, J.A. Comparison of properties of native and extruded amaranth (Amaranthus cruentus L.–BRS Alegria) flour. *LWT-Food Sci. Technol.* **2011**, *44*, 1915–1921. [CrossRef]

45. Wani, S.A.; Kumar, P. Characterization of extrudates enriched with health promoting ingredients. *Food Measure.* **2015**, *9*, 592–598. [CrossRef]

46. Durge, A.V.; Sarkar, S.; Survase, S.A.; Singhal, R.S. Impact of extrusion on red beetroot colour used as pre-extrusion colouring of rice flour. *Food Bioprocess Technol.* **2013**, *6*, 570–575. [CrossRef]

47. Sun, Y.; Muthukumarappan, K. Changes in functionality of soy-based extrudates during single-screw extrusion processing. *Int. J. Food Prop.* **2002**, *5*, 379–389. [CrossRef]

48. Gulati, P.; Weier, S.A.; Santra, D.; Subbiah, J.; Rose, D. Effects of feed moisture and extruder screw speed and temperature on physical characteristics and antioxidant activity of extruded proso millet (Panicum miliaceum) flour. *Int. J. Food Sci. Technol.* **2016**, *51*, 114–122. [CrossRef]

49. Hagenimana, A.; Ding, X.L.; Fang, T. Evaluation of rice flour modified by extrusion cooking. *J. Cereal Sci.* **2006**, *43*, 38–46. [CrossRef]

50. Torbica, A.; Belović, M.; Popović, L.; Čakarević, J. Heat and hydrothermal treatments of non-wheat flours. *Food Chem.* **2021**, *334*, 127523. [CrossRef]

foods

MDPI

Article

Physicochemical, Microbial, and Volatile Compound Characteristics of *Gochujang*, Fermented Red Pepper Paste, Produced by Traditional Cottage Industries

Srinivasan Ramalingam [1,†], Ashutosh Bahuguna [1,†], SeMi Lim [1], Ah-Ryeong Joe [1], Jong-Suk Lee [2], So-Young Kim [3] and Myunghee Kim [1,*]

1   Department of Food Science and Technology, Yeungnam University, Gyeongsan 38541, Korea; sribt27@gmail.com (S.R.); ashubahuguna@gmail.com (A.B.); thfvkalfpeh7@naver.com (S.L.); whdkfud12@naver.com (A.-R.J.)
2   Division of Food & Nutrition and Cook, Taegu Science University, Daegu 41453, Korea; jslee1213@ynu.ac.kr
3   Department of Agrofood Resources, National Institute of Agricultural Sciences, Rural Development Administration, Wanju 55365, Korea; foodksy@korea.kr
*   Correspondence: foodtech@ynu.ac.kr; Tel.: +82-53-810-2958
†   These authors contributed equally to this work.

Abstract: *Gochujang*, fermented red pepper paste, is a grain-based Korean traditional food. The quality of *gochujang* produced by cottage industries is not well-documented. Thus, the present study aimed to analyze the quality of *gochujang* from 35 traditional cottage industries for physicochemical and microbial characteristics, along with volatile compound contents. In addition to microbial characteristics, salinity, pH, free amino nitrogen, and alcohol content were evaluated. Ethanol was detected as the predominant alcohol and 57% of tested *gochujang* products harbored >1% of total alcohol content, which was above the recommended level for halal products. *Gochujang* products contained hexadecanoic and linoleic acids predominantly and several volatile compounds belonging to the classes of alcohols, aldehydes, alkanes, nitrogen-containing compounds, and terpenes. A wide range of aerobic mesophilic bacteria (2.79–8.73 log CFU/g) and yeast counts (1.56–7.15 log CFU/g) was observed. Five distinct yeast species were identified, including *Zygosaccharomyces rouxii*. Eight *gochujang* products were found to be contaminated with *Bacillus cereus* (>4 log CFU/g). This study suggests that there is a need to limit *B. cereus* contamination in cottage industry products and reduce alcohol content to comply with halal food guidelines.

Keywords: alcohol; *gochujang*; *Bacillus cereus*; free amino nitrogen; *Zygosaccharomyces rouxii*

Citation: Ramalingam, S.; Bahuguna, A.; Lim, S.; Joe, A.-R.; Lee, J.-S.; Kim, S.-Y.; Kim, M. Physicochemical, Microbial, and Volatile Compound Characteristics of *Gochujang*, Fermented Red Pepper Paste, Produced by Traditional Cottage Industries. *Foods* **2022**, *11*, 375. https://doi.org/10.3390/foods11030375

Academic Editors: Luca Serventi, Charles Brennan and Rana Mustafa

Received: 14 December 2021
Accepted: 21 January 2022
Published: 27 January 2022

Publisher's Note: MDPI stays neutral with regard to jurisdictional claims in published maps and institutional affiliations.

1. Introduction

Gochujang (fermented red pepper paste) is one of the most important grain-based traditional Korean fermented foods and is generally used as a sauce in Korean cuisines and as a seasoning in spicy foods. *Gochujang* has a distinguished flavor and savory taste [1]. In 2017, the total domestic and international retail market revenue of *gochujang* accounted for approximately USD 149.55 million and USD 31.98 million, respectively. *Gochujang* products are exported to several countries, including the US, China, Japan, and Middle Eastern countries [2]. Owing to the high amount of saccharified grain starch (from rice, wheat, or barley), and powdered red hot pepper (*Capsicum annuum* L.), *gochujang* is a red and thick paste. Furthermore, significant amounts of salt, powdered *meju*, and potable water are used in the preparation of *gochujang*. *Meju* is a naturally fermented soybean, which acts as the source of microorganisms (starter culture) in the fermentation of *gochujang*. The mixture of these ingredients starts the fermentation and aging processes [3]. Two major types of *gochujang* are available in the market: a modern large-scale industrial *gochujang* and traditional homemade *gochujang* [4,5]. The modern large-scale industrial *gochujang* is produced in a quality-controlled environment with the use of specific starter cultures (*Aspergillus* and *Bacillus* species) in a short period of fermentation [6].

The preparation of homemade and cottage industrial *gochujang* relies on traditional fermentation techniques using simple equipment. The traditional homemade *gochujang* is produced using an extensive fermentation process with natural microorganisms. The process includes saccharification via heating of glutinous rice and malt, followed by the addition of *meju*, red pepper powder, and salt, depending on the desired characteristics of taste and flavor, and, finally, a fermentation stage, which can last from 1 to 2 years [7]. The microbial composition of *meju* can affect the quality of *gochujang* [7]. The traditional *gochujang* fermentation is influenced by several elements, including local microorganisms such as the *meju* microflora, and surrounding environmental factors such as weather conditions [4,5,8]. Thus, the *gochujang* cottage industry in different provinces generates products with diversified nutritional values and organoleptic properties [9]. Raw ingredients, process methods, microorganisms involved in the fermentation, and duration of the fermentation significantly influence the organoleptic properties of *gochujang*, including its aroma, taste, and texture [6,8].

To make appropriate choices and optimize the production of traditional *gochujang*, it is essential to investigate the physicochemical and microbial characteristics. Although various laboratory and homemade unbranded *gochujang* products have been previously examined [9,10], these studies have not focused on the physicochemical and microbial properties and volatile compound characteristics of indigenous-branded, traditional cottage industrial *gochujang* products. Generally, because of the use of traditional processing technologies adopted by cottage industries, the interbatch quality of *gochujang* remains unvaried [10]. Hence, consumers are highly interested in indigenous branded traditional *gochujang* products owing to the consistently outstanding quality. These traditional cottage industries are operated with minimal capital, and thus lack a quality control department for the analyses of *gochujang* products. Moreover, the physicochemical and microbial features of these products are not monitored by any food and health organization and, therefore, not publicly available.

A previous study detected a significant amount of different alcohol types, particularly ethanol, in *gochujang* during the fermentation process [11]. In addition to the basic ethanol content, some companies supplement the product with a considerable amount of ethanol during the packaging phase to prevent microbial activity. The *gochujang* products containing more than 1% ethanol are prohibited for trade in Muslim countries (halal markets). Moreover, the risk of contamination of traditional cottage industry *gochujang* products with food pathogens, particularly *Bacillus cereus*, remains unexplored. The quality of *gochujang* products produced by cottage industries has not been sufficiently examined. This study aimed to determine the physicochemical and microbial properties, alcohol content, and volatile compounds of *gochujang* products collected from nationwide cottage industries, and to categorize such products based on the findings of biostatistical analyses.

2. Materials and Methods

2.1. Chemicals

All chemicals used were of analytical grade. Potassium chromate, 0.1 N sodium hydroxide, silver nitrate, methyl alcohol, ethyl alcohol, and sodium chloride were obtained from Duksan Pure Chemicals (Ansan, Gyeonggi-do, Korea). Sodium hydroxide, sodium hydrogen carbonate, and ammonium hydroxide were purchased from Junsei Chemicals (Tokyo, Japan). Formalin solution, standard methanol, ethanol, pentanol, propanol, and butanol were purchased from Sigma-Aldrich (St. Louis, MO, USA). Plate count agar, nutrient agar, potato dextrose agar (PDA), and potato dextrose broth were purchased from Difco (Becton, Dickinson and Company, Sparks, MD, USA). Mannitol egg yolk polymyxin agar (MYP), egg yolk emulsion, and polymyxin B supplement were purchased from Oxoid LTD (Basingstoke, Hampshire, UK). 3M Yeast and Mold Petrifilm was purchased from 3M

Health Care (St. Paul, MN, USA). API 50CHB and API 20E were obtained from bioMerieux (Marcy l'Etoile, France).

Instruments and Apparatus

A pH meter (Orion Star A211, Thermo Fisher Scientific, Beverly, MA, USA) and Konica Minolta Chromameter, equipped with a CR-400 model chromameter measuring head and DP-400 model data processor, were used to measure the pH and color values, respectively. GC-MS-QP2010 SE (Shimadzu Co., Kyoto, Japan) gas chromatography–mass selective detection (GC-MSD) system with SH-Stabilwax column (30 m × 0.32 mm × 0.25 μm) and Agilent 7890B and 5977B GC-MS system (Agilent, Santa Clara, CA, USA), which includes an Agilent DB-WAX 122-7062 column (60 m × 250 μm × 0.25 μm), were used for the detection of volatile compounds and alcohol content, respectively. Plastic Petri plates (SPL Life Sciences, Pocheon, Gyeonggi, Korea) were used for the microbiological analysis. Internal transcribed spacer (ITS) sequencing of isolated microbes was conducted using the ABI PRISM 3730XL DNA analyzer (Applied Biosystems, Foster City, CA, USA).

2.2. Sample Collection

A total of 35 *gochujang* products were purchased from various cottage industries located in different provinces of the Republic of Korea, as previously reported [1]. The major ingredients of *gochujang* products include red pepper powder, glutinous rice powder, powdered soybeans, grain syrup, malt, salt, and water. Detailed ingredients of the purchased *gochujang* products were also previously reported [1].

2.3. Physicochemical Characteristics

2.3.1. Determination of pH, Salinity, Color Values, and Free Amino Nitrogen

pH values of *gochujang* products were analyzed according to the protocol of Ramalingam et al. [12]. The salinity of *gochujang* was determined using the Korea Food and Drug Administration method [13]. Color values of *gochujang* were obtained using a chromameter. The tristimulus color analyzer was calibrated to a reference (white porcelain plate) prior to the experiment [14]. The total free amino nitrogen contents of the *gochujang* samples were determined using the titration method as described by the Korea Food and Drug Administration [15] and Cho et al. [16].

2.3.2. Determination of Total Alcohol Content

The alcohol content profiles of *gochujang* products were investigated using gas chromatography–mass spectrometry (GC-MS), according to the method described by Lee et al. [17] and Gil et al. [18]. Briefly, 0.5 g of a sample was mixed with 9.5 mL of dimethyl sulfoxide and stirred at 100 rpm at 40 °C for 1 h in a 20 mL closed container. The reaction solution settled before the supernatant was filtered using the Whatman syringe filter. Subsequently, the supernatant was used for the GC-MS analysis via a GC-MSD system. A temperature of 160 °C was maintained in the GC injector, and 20 μL of the sample was injected with a split ratio of 40:1. The oven temperature was programmed to start at 40 °C for 5 min, and increase 10 °C/min up to 240 °C, and then stop at (isothermal) 240 °C for 9 min. Mass spectrum analysis (70 eV, ion-source temperature 200 °C) was performed at 0.5 s scan intervals. Standard methanol, ethanol, pentanol, butanol, and propanol solutions (0.2%) were used to estimate each alcohol concentration in the *gochujang* samples.

2.4. Determination of Volatile Compounds

The volatile compound profiles of *gochujang* products were investigated using a solid-phase microextraction (SPME) method, followed by GC-MS [12]. Approximately 5 g of sample was heated to 70 °C for 20 min in a closed 20 mL container. A carbowax/divinylbenzene polydimethylsiloxane SPME fiber assembly was allowed to absorb the volatile compounds within the samples for 30 min at 70 °C. Temperatures of 250 °C and 230 °C were maintained in the GC injector and MS source, respectively. A split ratio of 20:1 was used to inject the

SPME fiber at a purge flow rate of 3 mL/min (with a total flow rate of 24 mL/min) at 18.5 psi. The oven temperature was programmed to start at 40 °C for 2 min, and increase at a rate of 2 °C/min up to 220 °C and 10 °C/min up to 240 °C, and then stop at 240 °C for 10 min. Mass spectrum analysis (70 eV, ion-source temperature 230 °C) was performed at 0.5 s scan intervals. Mass spectra of the unknown compounds of samples were interpreted using the data available in the National Institute of Standards and Technology MS library [19]. The molecular weights, names, and structures of volatile compounds in the samples were determined.

2.5. Microbial Profile

The standard methods of the Association of Official Analytical Chemists [20] were adopted to analyze the total number of aerobic mesophilic bacteria and *B. cereus* in *gochujang*. 3M Petrifilm, plate count agar, and MYP culturing medium were used according to the manufacturer's protocol to estimate the total yeast and mold (yeast/mold) [21], aerobic mesophilic bacteria, and *B. cereus* counts, respectively. API 50CHB and API 20E kits were used to identify *B. cereus* using the manufacturer's protocol. PDA was used to isolate yeast/mold. The isolated yeast/mold from *gochujang* was subjected to ITS sequencing analysis [12]. The analyzed sequences were aligned with the help of the sequence alignment editor software BioEdit (version 7.0.4). The data on ITS sequences of the isolated microorganisms were documented in the NCBI GenBank database using the BLAST program. Phylogenetic analysis was performed for the isolated microorganisms using the neighbor-joining method [12].

2.6. Statistical Analysis

All the experiments were performed at least in triplicate, and the values were presented as the mean ± standard deviation. Statistical analyses were performed using the SPSS software 23 (IBM, Chicago, IL, USA). One-way analysis of variance in a completely randomized design and Duncan's multiple range comparison tests were used to explore the significant differences between the samples with a 95% confidence interval at $p < 0.05$. The multivariate exploratory techniques of principal component analysis (PCA) were conducted to categorize the *gochujang* samples based on their pH, lightness, redness, yellowness, amino nitrogen content, aerobic mesophilic bacteria count, yeast/mold count, and major volatile compound profile using the XLSTAT package on Microsoft Office Excel 2016 version [1].

3. Results and Discussion

3.1. Physicochemical Analysis of Gochujang Products

3.1.1. pH

Optimal pH is one of the prerequisite physicochemical parameters of fermented foods and is the main factor influencing the occurrence of several biochemical activities [12]. All selected *gochujang* products exhibited acidic pH between the ranges of 3.57 ± 0.01–4.98 ± 0.01 (Table 1). The mean pH value of *gochujang* was 4.44 ± 0.35. Based on the pH values, all the *gochujang* products were grouped into two categories: samples with pH higher than 4.6 (low-acidic food), and samples with pH below 4.6 (acidic food) (USFDA, Code of Federal Regulations) [22]. A total of 40% of the *gochujang* products ($n = 14$) showed a pH higher than 4.6 (in the range of low-acid food), whereas 60% ($n = 21$) presented pH values below 4.6 (acidic food). The variation in pH between the different *gochujang* products is probably due to the origin of different basic raw materials and the contribution of different microorganisms. A previous report showed a range of low-acidic pH (4.59 ± 0.36–4.79 ± 0.15) measured in 80 different homemade *gochujang* products [23]. However, Lee et al. [23] did not report pH values below 4.0 for any sample. The present investigation detected a slightly acidic pH for some samples, similar to that reported by Kim et al. [4] in several laboratory-made *gochujang–meju* samples. In general, the initial pH values of the *gochujang* products ranged from 5.5–6. These values are then reduced to the level of either low-acidic food or acidic food pH values during the fermentation process [24,25]. The decrease in the pH value is dependent on the fermentation time [26], fermenting microbes [24], environmental factors [27], and raw materials [25]. The mean pH value (4.44 ± 0.35) measured in this investigation was similar to that previously reported for other *gochujang* products [23–28].

Table 1. The pH, salinity, color values, and free amino nitrogen content of *gochujang* products.

Product Code	pH [#]	Salinity (%) [#]	Color Values [#]			Free Amino Nitrogen (mg/100 g) [#]
			Lightness (L*)	Redness (a*)	Yellowness (b*)	
Go-1	4.96 ± 0.01 [b]	5.01 ± 0.30 [pq]	29.70 ± 0.58 [bcdefg]	13.36 ± 0.63 [d]	8.74 ± 0.13 [cdef]	28.03 ± 8.09 [kl]
Go-2	4.49 ± 0.01 [n]	10.59 ± 0.32 [c]	28.26 ± 0.19 [bcdefgh]	12.24 ± 0.10 [e]	8.92 ± 0.07 [cd]	65.40 ± 8.09 [defgh]
Go-3	4.78 ± 0.01 [d]	7.74 ± 0.00 [fg]	26.98 ± 0.59 [bcdefgh]	10.16 ± 0.16 [hi]	7.09 ± 0.13 [jk]	74.74 ± 14.01 [cdefg]
Go-4	4.65 ± 0.01 [i]	8.16 ± 0.00 [ef]	27.25 ± 0.29 [bcdefgh]	10.12 ± 0.13 [hi]	7.95 ± 0.08 [gh]	46.71 ± 14.01 [hijk]
Go-5	4.74 ± 0.01 [ef]	4.81 ± 0.27 [qr]	28.22 ± 0.31 [bcdefgh]	12.03 ± 0.34 [ef]	8.93 ± 0.11 [cd]	74.74 ± 14.01 [cdefg]
Go-6	4.74 ± 0.01 [f]	5.22 ± 0.26 [opq]	29.21 ± 0.87 [bcdefg]	11.35 ± 0.69 [g]	8.86 ± 0.30 [cde]	51.38 ± 8.09 [ghijk]
Go-7	4.12 ± 0.02 [u]	6.64 ± 0.33 [jk]	26.91 ± 0.51 [cdefgh]	9.33 ± 0.59 [jk]	7.24 ± 0.44 [jk]	65.40 ± 16.18 [defgh]
Go-8	4.62 ± 0.01 [j]	4.72 ± 0.00 [qr]	26.06 ± 1.16 [efghi]	5.76 ± 0.39 [n]	5.47 ± 0.30 [opq]	130.80 ± 0.00 [b]
Go-9	4.20 ± 0.01 [t]	7.03 ± 0.00 [hij]	24.53 ± 0.34 [hi]	6.88 ± 0.10 [m]	5.89 ± 0.01 [no]	51.38 ± 16.18 [ghjik]
Go-10	4.30 ± 0.03 [p]	6.0 ± 0.00 [klmn]	25.95 ± 0.68 [fghi]	7.26 ± 0.07 [m]	6.47 ± 0.06 [lm]	93.43 ± 8.09 [c]
Go-11	3.99 ± 0.00 [v]	4.98 ± 0.00 [pq]	26.01 ± 1.03 [efghi]	5.58 ± 0.78 [no]	5.80 ± 0.29 [no]	37.37 ± 8.09 [ijkl]
Go-12	4.26 ± 0.01 [q]	5.67 ± 0.67 [mnop]	28.05 ± 0.26 [bcdefgh]	10.64 ± 0.16 [h]	8.47 ± 0.04 [def]	60.73 ± 0.00 [efghi]
Go-13	4.40 ± 0.01 [o]	4.81 ± 0.25 [qr]	25.88 ± 1.21 [fghi]	6.03 ± 0.50 [n]	5.74 ± 0.15 [nop]	32.70 ± 0.00 [jkl]
Go-14	4.58 ± 0.02 [k]	7.35 ± 0.31 [ghi]	26.33 ± 0.34 [defgh]	8.91 ± 0.10 [jkl]	7.07 ± 0.01 [jk]	42.04 ± 8.09 [hijkl]
Go-15	3.84 ± 0.01 [x]	5.81 ± 0.53 [lmno]	26.72 ± 0.74 [cdefgh]	7.47 ± 0.55 [m]	7.03 ± 0.16 [jk]	37.37 ± 8.09 [ijkl]
Go-16	4.62 ± 0.01 [j]	5.95 ± 0.78 [klmn]	26.76 ± 0.34 [cdefgh]	9.50 ± 0.12 [ij]	7.36 ± 0.06 [ij]	46.71 ± 0.00 [hijk]
Go-17	4.84 ± 0.01 [c]	5.14 ± 0.89 [opq]	27.30 ± 0.68 [bcdefgh]	10.52 ± 0.09 [h]	8.10 ± 0.02 [fgh]	51.38 ± 8.09 [ghijk]
Go-18	4.26 ± 0.01 [qr]	6.46 ± 0.50 [jkl]	27.28 ± 0.43 [cdefghi]	12.46 ± 0.21 [e]	9.00 ± 0.03 [cde]	37.37 ± 8.09 [ijkl]
Go-19	4.69 ± 0.01 [g]	7.54 ± 0.30 [fgh]	28.75 ± 0.62 [bcdefgh]	11.45 ± 0.23 [fg]	7.88 ± 0.01 [gh]	56.06 ± 8.09 [fghij]
Go-20	4.67 ± 0.01 [h]	4.20 ± 0.32 [r]	26.09 ± 0.08 [efghi]	4.96 ± 0.12 [o]	4.97 ± 0.10 [q]	42.04 ± 8.09 [hijkl]
Go-21	4.98 ± 0.01 [a]	4.93 ± 0.30 [q]	27.97 ± 0.16 [bcdefgh]	9.22 ± 0.24 [jk]	6.77 ± 0.09 [kl]	42.04 ± 8.09 [hijkl]
Go-22	4.29 ± 0.02 [p]	8.52 ± 0.50 [de]	31.21 ± 0.57 [ab]	16.63 ± 0.22 [b]	11.19 ± 0.14 [a]	168.17 ± 16.18 [a]
Go-23	4.23 ± 0.01 [s]	5.01 ± 0.31 [pq]	30.28 ± 0.14 [abcde]	12.63 ± 0.12 [e]	8.38 ± 0.07 [efg]	65.4 ± 16.18 [defgh]
Go-24	3.94 ± 0.01 [w]	6.06 ± 0.33 [klm]	30.41 ± 0.17 [abcd]	8.77 ± 0.21 [kl]	6.05 ± 0.24 [mn]	126.13 ± 16.18 [b]
Go-25	4.79 ± 0.02 [d]	6.95 ± 0.00 [hij]	25.67 ± 0.10 [ghi]	6.95 ± 0.31 [m]	5.42 ± 0.10 [opq]	28.03 ± 8.09 [kl]
Go-26	4.55 ± 0.01 [l]	12.68 ± 0.33 [a]	30.90 ± 0.07 [abc]	14.76 ± 0.04 [c]	10.08 ± 0.02 [b]	79.41 ± 8.09 [cdef]
Go-27	4.56 ± 0.01 [l]	11.36 ± 0.57 [b]	28.39 ± 0.03 [bcdefgh]	8.33 ± 0.04 [l]	7.06 ± 0.02 [jk]	46.71 ± 14.01 [hijk]
Go-28	4.78 ± 0.00 [d]	6.63 ± 0.34 [jk]	29.14 ± 0.22 [i]	10.40 ± 0.15 [ij]	7.98 ± 0.09 [cde]	18.69 ± 0.00 [l]
Go-29	3.99 ± 0.01 [v]	5.68 ± 0.00 [mnop]	30.09 ± 0.04 [abcdef]	13.93 ± 0.07 [d]	9.05 ± 0.01 [c]	18.69 ± 0.00 [l]
Go-30	3.57 ± 0.01 [z]	6.82 ± 0.57 [ij]	30.57 ± 0.35 [abcd]	17.72 ± 0.13 [a]	10.52 ± 0.03 [b]	79.41 ± 21.41 [cdef]
Go-31	4.52 ± 0.01 [m]	5.30 ± 0.33 [nopq]	30.85 ± 0.17 [abc]	13.43 ± 0.17 [d]	9.13 ± 0.03 [c]	37.37 ± 8.09 [ijkl]
Go-32	3.76 ± 0.01 [y]	5.41 ± 0.33 [mnopq]	33.90 ± 0.23 [a]	6.82 ± 0.14 [m]	7.04 ± 0.07 [jk]	84.08 ± 16.18 [cde]
Go-33	4.76 ± 0.01 [e]	8.94 ± 0.28 [d]	28.24 ± 0.49 [bcdefgh]	9.20 ± 0.02 [jk]	7.07 ± 0.02 [jk]	65.40 ± 21.41 [defgh]
Go-34	4.24 ± 0.00 [rs]	3.44 ± 0.00 [s]	28.11 ± 0.08 [bcdefgh]	10.49 ± 0.23 [h]	7.76 ± 0.11 [hi]	32.70 ± 14.01 [jkl]
Go-35	4.53 ± 0.01 [m]	11.58 ± 0.00 [b]	25.77 ± 0.04 [fghi]	5.93 ± 0.06 [n]	5.24 ± 0.03 [pq]	93.43 ± 21.41 [c]
Mean ± SD	4.44 ± 0.35	6.66 ± 2.18	28.11 ± 2.04	10.04 ± 3.15	7.59 ± 1.53	60.33 ± 32.51

[#]—The values are the mean of triplicates with standard deviation. Different superscript letters (a–z) within a column indicate significant differences ($p < 0.05$) between the selected *gochujang* products when subjected to Duncan's multiple comparison test.

3.1.2. Salinity

The salinity of the tested *gochujang* products was between 3.44 ± 0.00% and 12.68 ± 0.33%, and the mean salinity value was 6.66 ± 2.18% (Table 1). The *gochujang* products were categorized based on salinity in three broad groups, group I (salinity <5%), group II (salinity, 5–10%), and group III (salinity >10%). Most of the samples (68.57%) were placed in group II, followed by groups I (20%) and III (11.43%). This was due to the initial amount of salt added during the manufacturing phase of the *gochujang* products at the cottage industry. In the present study, all the tested *gochujang* products were prepared using salt supplements between 5% and 12%, which further impacted the salinity of the final product [1]. During the *gochujang* fermentation process, salinity increase was also detected by Beak et al. [27], whereas a decrease in salinity was reported by Ryu et al. [24]. The water content of the raw materials and external environment humidity showed a significant influence on the salt concentration of *gochujang* products [4]. The salinity of *gochujang* products reported in previous reports [23,24,27] was consistent with mean salinity observed in the present study (6.66 ± 2.18%). Moreover, Lee et al. [23] reported that none of the *gochujang* samples had salinity below 5% or above 10%.

3.1.3. Free Amino Nitrogen Content

In the 35 *gochujang* products, free amino nitrogen content presented mean values of 60.33 ± 32.51 mg/100 g (Table 1). The free amino nitrogen content in all the samples ranged from 18.69 ± 0.00 mg/100 g to 168.17 ± 16.18 mg/100 g. *Gochujang* products were grouped into three categories based on the free amino nitrogen content, including group I (free amino nitrogen, 0–50 mg/100 g), group II (50–100 mg/100 g), and group III (100–200 mg/100 g). A total of 45.7% of the *gochujang* products were assigned to groups I and II, whereas only 8.6% of samples were placed in group III. It has been reported that the fermentation process increases the amino nitrogen in the *gochujang* products [24,27–30]. Similarly, the prevalence of *Bacillus* spp. and *Zygosaccharomyces* spp. has a significant correlation with an amino-type nitrogen concentration of *gochujang* products [24]. The difference in free amino nitrogen content in the tested *gochujang* products was due to the distinct initial raw material used (particularly protein-rich matter), fermentation period, and the microorganisms involved in the fermentation of *gochujang* [24,27]. Because the 35 *gochujang* products were prepared with different raw materials, including powdered soybean (a major protein substrate) [12], they had diverse free amino nitrogen content. Similarly, a previous study reported the difference in the free amino nitrogen content in various industrial *gochujang* products [31]. Accordingly, Kim et al. [4] reported the variation in free amino nitrogen content in homemade *gochujang* products prepared with four different types of *meju*.

3.1.4. Color Values

The surface color of all the *gochujang* products was measured using a chromameter and are presented in Table 1. Color is an essential food quality for consumer acceptability. The color of the fermented food is highly dependent on the raw material used and the composition of the final product [12]. The mean values of lightness (L*), redness (a*), and yellowness (b*), of *gochujang* products, were 28.11 ± 2.04, 10.04 ± 3.15, and 7.59 ± 1.53, respectively. The most influential factor responsible for the redness of the products is the red pepper powder. In the present study, Go-30 displayed the highest value for redness (17.72 ± 0.13) owing to the high percentage of red pepper (34%) during preparation, whereas Go-20 had the lowest value (4.96 ± 0.12) due to the limited amount (19%) of red pepper. A previous report revealed that the progression in the fermentation process increased the a* and L* values of *gochujang* products, whereas b* values were decreased [24]. In another investigation, a* and L* values decreased, and no significant changes were observed in the b* values during the 1-year fermentation of *gochujang* products evaluated [27]. In addition to the raw material, the variation in the color values for different *gochujang* products is associated with the microbial composition, which metabolizes the complex biomolecules and converts them into simple molecules responsible for a unique taste, aroma, and color. The present results, including the mean color values of the *gochujang* products, were consistent with those reported in previous reports [24,27,28].

3.1.5. Alcohol Content

Alcohols, particularly ethanol, are important volatile components of fermented foods, responsible for imparting a unique flavor and aroma [32]. A wide range (0–4.99%) was noticed in the alcohol content and proportions of the tested *gochujang* products. The mean total alcohol content was 1.58 ± 1.28% (Figure 1 and Supplementary Table S1). Among the tested alcohols (methanol, propanol, butanol, and pentanol), ethanol content was the highest, ranging from 0 to 4.9%. Therefore, ethanol represented the single major contributor to the total alcohol content of *gochujang* products (Figure 1 and Supplementary Table S1). The mean ethanol content of the *gochujang* products was 1.53 ± 1.23%, whereas the mean methanol content was 0.004 ± 0.005% (Figure 1).

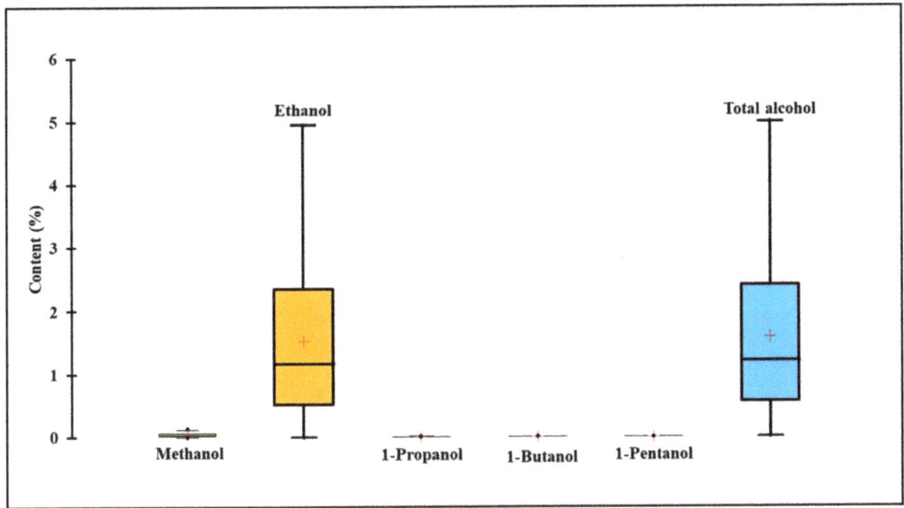

Figure 1. The content of various alcohols in 35 *gochujang* products from the traditional cottage industry.

None of the products showed an excessively high amount of propanol, butanol, and pentanol, and these alcohols were detected in the range from not detected to 0.016%. *Gochujang* products were grouped into two categories based on the total alcohol content, including group I (0–1%) and group II (>1%) [33]. Approximately 42.85% of the samples (n = 15) were placed in group I, whereas the rest of the *gochujang* products (n = 20) were assigned in group II. The alcohol content in 57.15% of the tested *gochujang* products (n = 20) was higher than the recommended amount (1%) for halal foods [32]. Moreover, six *gochujang* products contained more than 3% of ethanol. In these cases, the extra ethanol was added by the manufacturing companies during the packaging of *gochujang* products to prevent spoilage and microbial activity. Furthermore, the basic ethanol content in *gochujang* products is linked with types and populations of fungi participating in the fermentation [34]. In particular, *Zygosaccharomyces* spp. and *Saccharomyces* spp. yeast isolated from *gochujang* products produced 1.6–3.2% of the ethanol [17], thus contributing to the higher basic level of ethanol in *gochujang* products. Previous data generated using the electronic nose analysis of 25 traditional *gochujang* products revealed the presence of 0.14–2.7% of ethanol in *gochujang* products, and 44% of the products (n = 11) contained more than 1% of ethanol content, thus supporting the present findings [35]. An abnormally high amount of alcohol content in *gochujang* products leads to alteration in taste and may cause spoilage. Moreover, ethanol content higher than 1% in food restricts their consumption in Muslim countries due to halal requirements [33].

3.2. Volatile Compounds

GC–MS analysis results revealed the presence of various volatile compounds in all tested *gochujang* products. The total number of compounds identified in the *gochujang* products ranged between 53 and 104 (Supplementary Table S2). The compound names, retention times, and percentage peak area for all *gochujang* products are listed in Supplementary Table S2. The predominant compound was identified as 2,3,5,6-tetramethyl pyrazine with a peak area of 54.31% and retention time of 40.054 min in Go-19, followed by ethanol with a peak area of 49.26% and retention time of 9.138 min in Go-7.

Ethanol was found in all tested *gochujang* products, with a peak area percentage range of 1.17–49.26%. Among the 35 *gochujang* products, 17 showed ethanol as the predominant component, depicting a percentage range of the highest peak area between 7.54–49.26%. Linoleic acid ethyl ester and 2,3,5,6-tetramethyl pyrazine were detected as the predominant components in 11 *gochujang* products with a peak area range of 7.92–31.19%, and to a lesser extent, in four other *gochujang* products (Go-14, Go-16, Go-19, and Go-30) with peak area range of 2.7–54.31%. Hexadecanoic acid, an ethyl/methyl ester, was detected in all *gochujang* products and was either the second or third most abundant compound in 26 *gochujang* products with a peak area range of 6.49–25.09%. Compounds detected in the *gochujang* products included low quantities of acids, alcohols, aldehydes, alkanes, alkenes, benzene derivatives, carboxylic acids, cyclic and bicyclic ketones, cyclosiloxanes, esters, fatty acids, furans, hydrocarbons, nitrogen-containing compounds, phenolics, pyranones, pyrazines, sulfur-containing compounds, and terpenes.

Several compounds (such as ethanol, acetic acid, benzaldehyde, benzene acetaldehyde, hexadecanoic acid, ethyl ester, hexanoic acid, hexanol, hexyl ester, linoleic acid ethyl ester, methyl salicylate, nonanoic acid, ethyl ester, octadecanoic acid, ethyl ester, 1-propanol, octanoic acid, and ethyl ester) were previously reported in *gochujang* [36–42], thus supporting the present results. Similar to the present investigation, previous studies reported a high amount of ethanol in the commercial, traditional, and improved *gochujang* products generated during the fermentation process via the yeast-dependent glycolytic pathway [36,37,43–46]. Most of the ester compounds detected in the *gochujang* products were ethyl esters, which render a fruity aroma to the product; fatty acid esters are produced due to the esterification of organic acids and fatty acids with the ethanol generated during the fermentation process by the yeast [37]. Several aroma-active compounds, such as acetic acid (pungent sour), 2-furanmethanol (cooked sugar), methyl salicylate (peppermint), ethanol (sweet), linalool (flower and lavender), hexanal (grass, tallow, and fat), benzaldehyde (almond and burned sugar), benzene acetaldehyde (fruity and rosy), nonanal (soapy), tetramethylpyrazine (cocoa, mocha, and milk coffee), acetic acid (sour) and ethyl hexanoate (apple peel, and fruit) were detected in the tested *gochujang* products [37,47]. Diversified classes of compounds with different peak areas were detected in each *gochujang* product due to the difference in raw materials, method of raw material processing, microbial diversity during the fermentation process, and fermentation period and conditions.

3.3. Microbial Profile Analysis

3.3.1. Aerobic Mesophilic Bacteria and Yeast/Mold

The tested *gochujang* products harbored aerobic mesophilic bacteria in the range of 2.79 ± 0.10 to 8.73 ± 0.30 log CFU/g (Table 2). The mean value of aerobic mesophilic bacteria present in the *gochujang* products was 6.98 ± 1.42 log CFU/g. In addition to the aerobic mesophilic bacteria, the active presence of fungi and yeast in *gochujang* was previously reported [7]. The yeast/mold count in all products ranged from 1.56 ± 0.06 to 7.15 ± 0.02 log CFU/g (Table 2). The mean value of *gochujang* products' yeast/mold population was 4.47 ± 1.47 log CFU/g (Table 2). In general, during the *gochujang* fermentation phase, aerobic mesophilic bacterial populations increased from approximately 5 log CFU/g to 8 log CFU/g, and yeast/mold counts decreased [24,26,28]. The findings from the present study are in agreement with those from previous reports that showed similar bacterial and yeast count in different *gochujang* products [25,26,48]. The microbial population in the products highly depended on external environmental factors, physicochemical and microbial profiles of raw materials, and the *meju* used as a starter culture [24,27]. Even though changes occurred in the microbial composition, the total count of aerobic bacteria was nearly constant after 3 months of *gochujang* fermentation [27]. Both bacteria and yeast/mold play important roles in the final features (taste, color, and aroma) of the fermented *gochujang* products [24,27,37,38].

Table 2. Microbial profile of the *gochujang* products.

Product Code	Aerobic Bacteria (log CFU/g) *	Yeast and Mold (log CFU/g) *	Isolated and Identified Yeast	GenBank Accession Number
Go-1	6.64 ± 0.16 [o]	3.89 ± 0.04 [hi]	*Zygosaccharomyces rouxii*	OL679471
Go-2	7.20 ± 0.17 [kl]	3.71 ± 0.02 [i]	*Zygosaccharomyces rouxii*	OL679472
Go-3	8.29 ± 0.07 [b]	2.67 ± 0.05 [kl]	*Zygosaccharomyces rouxii*	OL679473
Go-4	7.97 ± 0.23 [d]	4.92 ± 0.03 [ef]	*Zygosaccharomyces rouxii*	OL679474
Go-5	7.75 ± 0.16 [fg]	3.74 ± 0.10 [hi]	*Zygosaccharomyces rouxii*	OL679475
Go-6	7.82 ± 0.08 [ef]	3.10 ± 0.08 [jk]	*Zygosaccharomyces rouxii*	OL679476
Go-7	6.04 ± 0.12 [q]	5.16 ± 0.06 [def]	*Zygosaccharomyces rouxii*	OL679477
Go-8	7.31 ± 0.15 [h]	3.66 ± 0.09 [i]	*Zygosaccharomyces rouxii*	OL679478
Go-9	7.19 ± 0.11 [i]	2.69 ± 0.01 [m]	*Zygosaccharomyces rouxii*	OL679479
Go-10	7.10 ± 0.12 [lm]	6.22 ± 0.03 [b]	*Zygosaccharomyces rouxii*	OL679480
Go-11	6.17 ± 0.52 [s]	3.15 ± 0.04 [j]	*Zygosaccharomyces rouxii*	OL679481
Go-12	7.94 ± 0.17 [efg]	6.10 ± 0.04 [b]	*Zygosaccharomyces rouxii*	OL679482
Go-13	7.86 ± 0.11 [fg]	5.96 ± 0.04 [bc]	*Starmerella lactis-condensi*	OL679483
Go-14	7.92 ± 0.22 [g]	4.94 ± 0.12 [ef]	*Starmerella lactis-condensi*	OL679484
Go-15	7.92 ± 0.17 [efg]	5.84 ± 0.10 [bc]	*Zygosaccharomyces rouxii*	OL679485
Go-16	7.93 ± 0.00 [e]	4.68 ± 0.03 [fg]	*Zygosaccharomyces rouxii*	OL679486
Go-17	8.42 ± 0.04 [a]	5.30 ± 0.15 [de]	*Zygosaccharomyces rouxii*	OL679487
Go-18	6.35 ± 0.54 [s]	2.37 ± 0.05 [l]	*Starmerella lactis-condensi*	OL679488
Go-19	8.12 ± 0.09 [c]	4.04 ± 0.03 [hi]	*Zygosaccharomyces rouxii*	OL679489
Go-20	7.01 ± 0.06 [kl]	4.02 ± 0.10 [hi]	*Zygosaccharomyces rouxii*	OL679490
Go-21	6.10 ± 0.17 [r]	4.23 ± 0.02 [gh]	*Zygosaccharomyces rouxii*	OL679491
Go-22	7.33 ± 0.20 [jk]	4.23 ± 0.04 [gh]	*Zygosaccharomyces rouxii*	OL679492
Go-23	6.39 ± 0.22 [p]	5.50 ± 0.04 [cd]	*Zygosaccharomyces rouxii*	OL679493
Go-24	8.73 ± 0.30 [a]	6.13 ± 0.03 [b]	*Zygosaccharomyces rouxii*	OL679494
Go-25	8.06 ± 0.17 [d]	2.29 ± 0.02 [l]	*Zygosaccharomyces rouxii*	OL679495
Go-26	7.86 ± 0.09 [efg]	6.12 ± 0.03 [b]	*Zygosaccharomyces rouxii*	OL679496
Go-27	4.65 ± 0.14 [n]	5.31 ± 0.01 [de]	*Zygosaccharomyces rouxii*	OL679497
Go-28	7.84 + 0.09 [efg]	5.06 ± 0.07 [def]	*Zygosaccharomyces rouxii*	OL679498
Go-29	3.48 ± 0.12 [s]	7.15 ± 0.02 [a]	*Zygosaccharomyces rouxii*	OL679499
Go-30	7.23 ± 0.19 [j]	4.82 ± 0.02 [ef]	*Zygosaccharomyces rouxii*	OL679500
Go-31	7.28 ± 0.24 [kl]	6.90 ± 0.02 [a]	*Wikerhamomyces subpelliculosus*	OL679501
Go-32	3.55 ± 0.43 [u]	1.56 ± 0.06 [m]	*Cladosporium welwitschiicola*	OL679502
Go-33	7.53 ± 0.13 [h]	2.50 ± 0.02 [l]	*Zygosaccharomyces rouxii*	OL679503
Go-34	2.79 ± 0.10 [t]	6.14 ± 0.03 [b]	*Pichia membranifaciens*	OL679504
Go-35	4.97 ± 0.50 [m]	2.37 ± 0.04 [l]	*Wikerhamomyces subpelliculosus*	OL679505
Mean ± SD	6.98 ± 1.42	4.47 ± 1.47		

*—The values are mean of triplicates with standard deviation. Different superscript letters (a–z) within a column indicate significant differences ($p < 0.05$) between the selected *gochujang* products when subjected to Duncan's multiple comparison test.

Several reasons can be identified for the variation in yeast/mold count among the *gochujang* products, among which the selection of *meju* may be the most critical. The present results are in accordance with several published reports that indicated the presence of several microorganisms in *gochujang* [7,48,49]. In *gochujang*, various bacterial species have been identified and extensively studied [7,37]. However, studies regarding the presence of yeast in *gochujang* are limited [7]. Thus, the present investigation focused on the isolation and identification of yeast from all *gochujang* products. More than 100 yeast colonies were isolated from 35 *gochujang* products. After microscopic examination and evaluation of colony characteristics, five distinct yeast colonies were analyzed using ITS sequencing and comparative phylogenetic analysis (Supplementary Figures S1–S5). These colonies were identified as *Zygosaccharomyces rouxii*, *Starmerella lactis-condensi*, *Wikerhamomyces subpelliculosus*, *Pichia membranifaciens*, and *Cladosporium welwitschiicola* (Table 2). To the best of our knowledge, *P. membranifaciens*, *C. welwitschiicola*, and *W. subpelliculosus* were reported in the traditional *gochujang* products for the first time. *Zygosaccharomyces rouxii* was detected as a predominant yeast in 82.85% of *gochujang* products ($n = 29$). It produces several aromatic secondary metabolites during fermentation, such as esters, aldehydes, and ketones, with leavening properties [50] that improve the quality of *gochujang* products [24,51,52]. *Z. rouxii* is the main yeast species found in the traditional *gochujang* products, whereas *Candida* and *Cryptococcus* species were dominant in the commercial *gochujang* products, supporting the present results [7,48,49]. Phylogenetic analysis displayed a minor variation between the identified *Z. rouxii* strains. The high occurrence of *Z. rouxii* in *gochujang* products led to its high isolation frequency in the present study (Table 2), which was in accordance with the data from Jang et al. [7].

3.3.2. Detection of *B. cereus* in *Gochujang* Products

The presence of pathogenic bacteria in food represents a major concern for food safety. *Escherichia coli*, *B. cereus*, *Salmonella* species, and *Staphylococcus aureus* are common foodborne pathogens responsible for significant health and economic losses. Although the acidic pH of *gochujang* products acts as a barrier for most of the pathogenic microbes, *B. cereus* can proliferate in *gochujang* [8]. In the present study, the presence of *B. cereus* was observed in eight *gochujang* products (22.85%) at a level higher than the safety limit (4 log CFU/g) recommended by the Korean Food and Drug Administration [53] (Table 3). Yim et al. [54] measured *B. cereus* counts below 4 log CFU/g in all the tested commercial *gochujang* products. Kim et al. [10] reported the presence of *B. cereus* in nine industrial and 23 homemade *gochujang* samples and revealed that three homemade *gochujang* samples contained *B. cereus* levels higher than the safety limit. In general, the *B. cereus* counts increase during the *gochujang* fermentation process [8,27]. The source of *B. cereus* in *gochujang* may include contaminated raw materials and cross-contamination during the fermentation process. In summary, the present investigation, supported by several other studies [8,10,52], indicated that although present in *gochujang* products, the *B. cereus* count in most products was within the safety limit, suggesting that appropriate sterilization measures were adopted during the preparation process. However, a few *gochujang* products showed higher *B cereus* counts, thus leading to concerns regarding *B. cereus* contamination and the need for necessary preventive measures against such contamination.

Table 3. *Bacillus cereus* count in *gochujang* products.

Product Code	*Bacillus cereus* (Log CFU/g) *
Go-13	4.26
Go-16	5.30
Go-17	4.60
Go-19	4.60
Go-22	5.90
Go-24	6.26
Go-26	6.94
Go-31	5.26

* Safe limit of *Bacillus cereus* is 4 log CFU/g (Korea Food and Drug Administration, 2010).

3.4. Principal Component Analysis and Hierarchical Clustering of Gochujang Products

The PCA and agglomerative hierarchical clustering analysis were performed based on the physicochemical characteristics, microbial count, alcohol content, and the distribution of major volatile components of different *gochujang* products (Figure 2). PC1 grouped Go-1, Go-6, Go-7, Go-12, Go-17, Go-23, Go-31, and Go-34, in the positive plane from the other samples (Figure 2A). The *gochujang* samples located in the positive values of PCA1 were influenced by yeast population and alcohol content. The PC2 showed the variance and grouped Go-2, Go-5, Go-19, Go-22, Go-26, Go-28, Go-29, and Go-30 (in positive values) (Figure 2A). PCA separated and grouped the different clusters of 35 *gochujang* samples based on their pH, salinity, free amino nitrogen, lightness, yellowness, redness, aerobic bacterial count, yeast and mold count, methanol, ethanol, propanol, pentanol, and butanol content, and major components from GC-MS analysis (ethanol, linoleic acid, and hexadecanoic acid). The *gochujang* products grouped in the positive region of PCA2 were influenced by free amino nitrogen content and color values. The *gochujang* samples in the negative plane of PCA1 and 2 varied from other samples owing to the differences in salinity, aerobic bacterial count, methanol content, and linoleic acid composition (one of the major components detected in GC-MS analysis). The pH and hexadecanoic acid content displayed significant variance in *gochujang* products (Go-8, Go-10, Go-11, Go-13, Go-14, Go-15, Go-21, Go-27, and Go-33) (Figure 2A). Agglomerative hierarchical clustering analysis revealed dissimilarities between the *gochujang* products in two key groups (Figure 2B). The group I consisted of five closely-related clusters with 20 *gochujang* samples (cluster 1 = Go-29, Go-34, Go-22, and Go-30; cluster 2 = Go-3, Go-2, Go-4, and Go-18; cluster 3 = Go-16, Go-19, Go-20 Go-21, Go-14, and Go-15; cluster 4 = Go-12, and Go-17; and cluster 5 = Go-1, Go-28, Go-5, and Go-6) (Figure 2B). Group II also comprised five clusters with 15 *gochujang* products (cluster 1 = Go-7 and Go-23; cluster 2 = Go-26, Go-33, Go-27, and Go-31; cluster 3 = Go-24 and Go-32; cluster 4 = Go-10, Go-8, and Go-13; and cluster 5 = Go-25, Go-35, Go-9, and Go-11) (Figure 2B). The *gochujang* products within these 10 clusters were closely related in terms of tested parameters. To the best of our knowledge, no study has categorized *gochujang* products based on their physicochemical and microbial features using multivariate PCA. Only a limited number of previous studies have employed PCA to represent the profiling of microbes and biogenic amines in *gochujang* products [1,8,27,49].

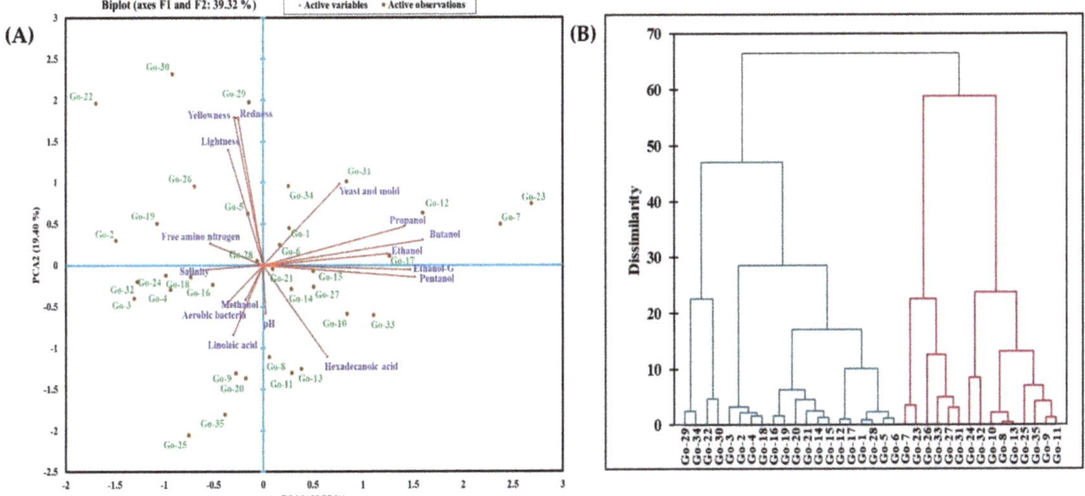

Figure 2. Principal component analysis (**A**) and agglomerative hierarchical clustering analysis (**B**) of 35 different *gochujang* products from the traditional cottage industry.

4. Conclusions

Diversified physiochemical and microbial profiles were detected in *gochujang* products collected from different provinces of the Republic of Korea. This study highlighted the presence of a variety of alcohols with a predominance of ethanol in the gochujang products. A few samples had ethanol content higher than the recommended limit for halal foods. Similarly, 22% of *gochujang* products were contaminated with *B. cereus*. The study demonstrated the variations in physicochemical, microbiological, and volatile compound characteristics of *gochujang* products, which may be due to the influence of raw material and fermentation conditions. The variation in the microbial profile also influenced physicochemical constituents and volatile compounds of the *gochujang* products. Nonetheless, an exact correlation needs to be established in future studies. The outcome of the study indicates that most of the *gochujang* products were free from toxicogenic microorganisms, though a few *gochujang* products had high ethanol content and *B.* cereus contamination, which needs to be addressed to satisfy the guidelines of food safety and marketability. This study also recommends regular analysis of the *gochujang* products prepared by the cottage industry to ensure their safety toward consumers.

Supplementary Materials: The following supporting information can be downloaded at: https://www.mdpi.com/article/10.3390/foods11030375/s1, Figure S1: Phylogenetic tree analyses of the ITS sequences of the *Zygosaccharomyces rouxii* isolated from *gochujang* products constructed in MEGA 6 software by employing UPGMA method; Figure S2. Phylogenetic tree analyses of the ITS sequences of the *Starmerella lactis-condensi* isolated from *gochujang* products constructed in MEGA 6 software by employing UPGMA method; Figure S3. Phylogenetic tree analyses of the ITS sequences of the *Wikerhamomyces subpelliculosus* isolated from *gochujang* products constructed in MEGA 6 software by employing UPGMA method; Figure S4. Phylogenetic tree analyses of the ITS sequences of the *Pichia membranifaciens* isolated from *gochujang* products constructed in MEGA 6 software by employing UPGMA method; Figure S5. Phylogenetic tree analyses of the ITS sequences of the *Cladosporium welwitschiicola* isolated from *gochujang* products constructed in MEGA 6 software by employing UPGMA method; Table S1. Various alcohol contents in *gochujang* products; Table S2. Various volatile compounds found in *gochujang* products.

Author Contributions: Conceptualization, M.K.; Data curation, S.R., A.B., S.L., A.-R.J. and J.-S.L.; Formal analysis, S.R., A.B., S.L., A.-R.J. and J.-S.L.; Funding acquisition, M.K.; Investigation, S.R., A.B., S.L., A.-R.J. and J.-S.L.; Methodology, S.R., A.B., S.L., J.-S.L., A.-R.J. and S.-Y.K.; Project administration, S.-Y.K. and M.K.; Resources, S.-Y.K. and M.K.; Software, S.R., A.B. and S.L.; Supervision, M.K.; Validation, S.R., A.B., S.L., A.-R.J. and J.-S.L.; Writing—original draft, S.R. and A.B.; Writing—review and editing, M.K. All authors have read and agreed to the published version of the manuscript.

Funding: This research was funded by the Cooperative Research Program for Agricultural Science & Technology Development, Rural Development Administration, the Republic of Korea [Project No. PJ013833].

Institutional Review Board Statement: Not applicable.

Informed Consent Statement: Not applicable.

Data Availability Statement: Data is contained within the article.

Acknowledgments: The authors thank the Core Research Support Center for Natural Products and Medicinal Materials (CRCNM) for providing technical support related to sample preparation.

Conflicts of Interest: The authors declare no conflict of interest.

References

1. Ramalingam, S.; Bahuguna, A.; Lim, S.; Joe, A.R.; Lee, J.S.; Kim, S.Y.; Kim, M. Quantification of biogenic amines in 35 Korean cottage industry traditional *gochujang* (fermented red pepper paste) products. *Foods* **2021**, *10*, 2370. [CrossRef]
2. Information Statistics System, 2018, Food Market Newsletter Gochujang-2018. Available online: https://www.atfis.or.kr/article/M001010000/view.do?articleId=3034 (accessed on 19 February 2021).
3. CODEX Alimentarius. Regional Standard for Gochujang (Asia): CODEX STAN 294R-2009. 2009. Available online: http://www.fao.org/fao-who-codexalimentarius/sh-proxy/en/?lnk=1&url=https%253A%252F%252Fworkspace.fao.org%252Fsites%252Fcodex%252FStandards%252FCXS%2B294R-2009%252FCXS_294Re.pdf (accessed on 31 December 2018).
4. Kim, H.E.; Han, S.Y.; Kim, Y.S. Quality characteristics of *gochujang meju* prepared with different fermentation tools and inoculation time of *Aspergillus oryzae*. *Food Sci. Biotechnol.* **2010**, *19*, 1579–1585. [CrossRef]
5. Shin, D.; Jeong, D. Korean traditional fermented soybean products: Jang. *J. Ethn. Foods* **2015**, *2*, 2–7. [CrossRef]
6. Kim, D.H.; Choi, H.J. Physicochemical properties of *kochujang* prepared by *Bacillus* sp. *koji*. *Korean J. Food Sci. Technol.* **2003**, *35*, 1174–1181.
7. Jang, S.J.; Kim, Y.J.; Park, J.M.; Park, Y.S. Analysis of microflora in *gochujang*, Korean traditional fermented food. *Food Sci. Biotechnol.* **2011**, *20*, 1435–1440. [CrossRef]

8. Ryu, J.A.; Kim, E.; Kim, M.-J.; Lee, S.; Yoon, S.R.; Ryu, J.G.; Kim, H.Y. Physicochemical Characteristics and Microbial Communities in *Gochujang*, a Traditional Korean Fermented Hot Pepper Paste. *Front. Microbiol.* **2021**, *11*, 620478. [CrossRef]

9. Burges, P. Modification of a traditional Korean food product (*gochujang*) to enhance its consumer acceptability as an ethnic food. *J. Ethn. Foods* **2014**, *1*, 13–18. [CrossRef]

10. Kim, J.B.; Kim, C.W.; Cho, S.H.; No, W.S.; Kim, W.J. Proposal of statistical sampling plans for in Korean fermented soybean pastes. *Food Sci. Biotechnol.* **2015**, *24*, 765–770. [CrossRef]

11. Park, Y.K.; Lee, J.H.; Mah, J.H. Occurrence and reduction of biogenic amines in *kimchi* and Korean fermented seafood products. *Foods* **2019**, *8*, 547. [CrossRef]

12. Ramalingam, S.; Dhatchanamoorthi, I.; Arumugam, A.; Bahuguna, A.; Krishnamoorthy, M.; Lee, J.S.; Devarajan, N.; Kim, M. Functional, nutritional, antinutritional, and microbial assessment of novel fermented sugar syrup fortified with pre-mature fruits of Totapuri mango and star gooseberry. *LWT-Food Sci. Technol.* **2021**, *136*, 110276. [CrossRef]

13. Korea Food and Drug Administration (KFDA). 2019. Available online: http://www.foodsafetykorea.go.kr/foodcode/01_03.jsp?idx=308 (accessed on 27 August 2020).

14. Lee, J.S.; Ramalingam, S.; Jo, I.G.; Kwon, Y.S.; Bahuguna, A.; Oh, Y.S.; Kwon, O.-J.; Kim, M. Comparative study of the physicochemical, nutritional, and antioxidant properties of some commercial refined and non-centrifugal sugars. *Food Res. Inter.* **2018**, *109*, 614–625. [CrossRef] [PubMed]

15. Korea Food and Drug Administration (KFDA). 2014. Available online: http://www.foodsafetykorea.go.kr/foodcode/01_03.jsp?idx=11009 (accessed on 26 August 2020).

16. Cho, K.M.; Kang, J.R.; Kim, G.M.; Kang, M.J.; Hwang, C.E.; Jeong, Y.S.; Kim, J.H.; Shin, J.H. Quality characteristics of low salted garlic *Doenjang* during fermentation. *Korean J. Food Preserv.* **2014**, *21*, 627–635. [CrossRef]

17. Lee, J.S.; Choi, Y.J.; Kwon, S.J.; Yoo, J.Y.; Chung, D.H. Screening and characterization of osmotolerant and gas-producing yeasts from traditional *Doenjang* and *Kochujang*. *Food Sci. Biotechnol.* **1996**, *5*, 54–58.

18. Gil, N.Y.; Song, J.; Eom, J.S.; Park, S.Y.; Choi, H.S. Changes of physicochemical properties of *Cheonggukjang* prepared with various soybean cultivars and *Bacillus subtilis* HJ18-9. *Korean J. Food Preserv.* **2016**, *23*, 811–818. [CrossRef]

19. National Institute of Standards and Technology (NIST). Standard Reference Database. 2014. Available online: https://www.nist.gov/system/files/documents/srd/NIST1aVer22Man.pdf (accessed on 27 August 2020).

20. Association of Official Analytical Chemists (AOAC). Official methods of analysis. In *Official Methods of Analysis*, 16th ed.; AOAC: Washington, DC, USA, 1999; Available online: https://www.aoac.org/AOAC_Prod_Imis/AOAC_Member/Default.aspx?WebsiteKey=2e25ab5a-1f6d-4d78-a498-19b9763d11b4&hkey=8fc2171a-6051-4e64-a928-5c47dfa25797 (accessed on 19 February 2020).

21. Bird, P.; Flannery, J.; Crowley, E.; Agin, J.; Goins, D.; Jechorek, R. Evaluation of the 3M™ Petrifilm™ rapid yeast and mold count plate for the enumeration of yeast and mold in food: Collaborative study, first action 2014.05. *J. AOAC Inter.* **2015**, *98*, 767–783. [CrossRef]

22. United States Food and Drug Administration (USFDA). Code of Federal Regulations, revised on 2021. Available online: https://www.accessdata.fda.gov/scripts/cdrh/cfdocs/cfcfr/CFRSearch.cfm?CFRPart=114&showFR=1 (accessed on 2 November 2020).

23. Lee, S.; Yoo, S.M.; Park, B.R.; Han, H.M.; Kim, H.Y. Analysis of quality state for *gochujang* produced by regional rural families. *J. Korean Soc. Food Sci. Nutr.* **2014**, *43*, 1088–1094. [CrossRef]

24. Ryu, M.H.; Zhang, J.; Toth, T.; Khokhani, D.; Geddes, B.A.; Mus, F.; Amaya, G.C.; Peters, J.W.; Poole, J.M.; Ane, C.A. Control of nitrogen fixation in bacteria that associate with cereals. *Nat. Microbiol.* **2020**, *5*, 314–330. [CrossRef]

25. Oh, H.I.; Shon, S.H.; Kim, J.M. Changes in microflora and enzyme activities of *kochujang* prepared with *Aspergillus oryzae*, *Bacillus licheniformis* and *Saccharomyces rouxii* during fermentation. *Korean J. Food Sci. Technol.* **2000**, *32*, 410–416.

26. Kim, G.T.; Hwang, Y.I.; Lim, S.I.; Lee, D.S. Carbon dioxide production and quality changes in korean fermented soybean paste and hot pepper-soybean paste. *J. Korean Soc. Food Sci. Nutr.* **2000**, *29*, 807–813.

27. Baek, S.Y.; Gil1, N.Y.; Han, M.H.; Kang, H.Y.; Lee, H.Y.; Yoon, Y.S.; Lee, J.; Song, Y.E.; Lee, S.K.; Ryu, J.A.; et al. Study on relationship between quality characteristics and exterior environment of the Korean traditional *gochujang* produced in 2018 by 8 regions of Korea. *Korean J. Food Preserv.* **2019**, *26*, 2287–7428. [CrossRef]

28. Yang, H.J.; Lee, Y.S.; Choi, I.S. Comparison of physicochemical properties and antioxidant activities of fermented soybean-based red pepper paste, *gochujang*, prepared with five different red pepper (*Capsicum annuum* L.) varieties. *J. Food Sci. Technol.* **2018**, *55*, 792–801. [CrossRef] [PubMed]

29. Hill, A.E.; Stewart, G.G. Free amino nitrogen in brewing. *Fermentation* **2019**, *5*, 22. [CrossRef]

30. Xu, D.; Li, C.; Zhao, M.; Feng, Y.; Sun, L.; Wang, Y. Assessment on the improvement of soy sauce fermentation by *Aspergillus oryzae* HG76. *Biocatal. Agric. Biotechnol.* **2013**, *2*, 344–351. [CrossRef]

31. Jang, Y.K.; Shin, G.R.; Jung, E.S.; Lee, S.; Lee, S.; Singh, D.; Jang, E.S.; Shin, D.J.; Kim, H.J.; Shin, H.W.; et al. Process specific differential metabolomes for industrial *gochujang* types (pepper paste) manufactured using white rice, brown rice, and wheat. *Food Chem.* **2017**, *234*, 416–424. [CrossRef] [PubMed]

32. Devanthi, P.V.P.; Gkatzionis, K. Soy sauce fermentation: Microorganisms, aroma formation, and process modification. *Food Res. Inter.* **2019**, *120*, 364–374. [CrossRef]

33. Alzeer, J.; Hadeed, K.A. Ethanol and its Halal status in food industries. *Trends Food Sci. Technol.* **2016**, *58*, 14–20. [CrossRef]

34. Buratti, S.; Benedetti, S. Alcoholic fermentation using electronic nose and electronic tongue. In *Electronic Noses and Tongues in Food Science*; Academic Press: Cambridge, MA, USA, 2016; pp. 291–299.

35. Park, E.S.; Heo, J.H.; Ju, J.; Park, K.Y. Changes in quality characteristics of *gochujang* prepared with different ingredients and *meju* starters. *J. Korean Soc. Food Sci. Nutr.* **2016**, *45*, 880–888. [CrossRef]

36. Choi, J.Y.; Lee, T.S.; Noh, B.S. Characteristics of volatile flavor compounds in improved *kochujang* prepared with glutinous rice *koji* during fermentation. *Korean J. Food Sci. Technol.* **1999**, *31*, 1221–1226.

37. Kang, K.M.; Baek, H.H. Aroma quality assessment of Korean fermented red pepper paste (*gochujang*) by aroma extract dilution analysis and headspace solid-phase micro extraction–gas chromatography–olfactometry. *Food Chem.* **2014**, *145*, 488–495. [CrossRef]

38. Choi, J.Y.; Lee, T.S.; Noh, B.S. Characteristics of volatile flavor compounds in *kochujangs* with *meju* and soybean *koji* during fermentation. *Korean J. Food Sci. Technol.* **2000**, *32*, 1035–1042.

39. Choi, J.Y.; Lee, T.S. Characteristics of volatile flavor compounds in *kochujang* prepared with commercial enzyme during fermentation. *J. Korean Soc. Agric. Chem. Biotechnol.* **2003**, *46*, 207–213.

40. Park, H.K.; Kim, J.G. Character impact compounds of *kochujang*. *Korea Soybean Digest.* **2003**, *20*, 1–11.

41. Oh, J.Y.; Kim, Y.S.; Kim, Y.H.; Shin, D.H. Volatile flavor components of *kochujang* prepared with different *kojis*. *Food Sci. Biotechnol.* **2001**, *10*, 45–53.

42. Shon, S.; Kim, J.; Oh, H.; Ha, J. Volatile flavor components of *kochujang* prepared with *Aspergillus oryzae*, *Bacillus licheniformis* and *Saccharomyces rouxii*. *Food Sci. Biotechnol.* **2003**, *12*, 18–22.

43. Byun, H.O.; Park, M.J.; Park, Y.S.; Bai, H.S. Analysis of microflora and volatile flavor components in traditional *gochujang* with different concentrations of salt during fermentation. *Food Eng. Prog.* **2014**, *18*, 282–292. [CrossRef]

44. Choi, J.Y.; Lee, T.S.; Park, S.O. Characteristics of volatile flavor compounds in improved *kochujang* prepared with soybean *koji* during fermentation. *Korean J. Food Sci. Technol.* **1997**, *29*, 1144–1150.

45. Choi, J.Y.; Lee, T.S.; Park, S.O.; Noh, B.S. Changes of volatile flavor compounds in traditional *kochujang* during fermentation. *Korean J. Food Sci. Technol.* **1997**, *29*, 745–751.

46. Kim, Y.S.; Oh, H.I. Volatile flavor components of traditional and commercial *kochujang*. *Korean J. Food Sci. Technol.* **1993**, *25*, 494–501.

47. Zhang, L.; Che, Z.; Xu, W.; Yue, P.; Li, R.; Li, Y.; Pei, X.; Zeng, P. Dynamics of physicochemical factors and microbial communities during ripening fermentation of *Pixian Doubanjiang*, a typical condiment in Chinese cuisine. *Food Microbiol.* **2020**, *86*, 103342. [CrossRef]

48. Cho, S.H.; Park, H.S.; Jo, S.W.; Yim, E.J.; Yang, H.Y.; Ha, G.S.; Jeong, D.Y. Comparison of microbial community profiling on traditional fermented soybean products (*Deonjang*, *Gochujang*) produced in Jeonbuk, Jeonnam, and Jeju province area. *Korean J. Microbiol.* **2017**, *53*, 39–48. [CrossRef]

49. Nam, Y.D.; Park, S.I.; Lim, S.I. Microbial composition of the Korean traditional food "*kochujang*" analyzed by a massive sequencing technique. *J. Food Sci.* **2012**, *77*, 250–256. [CrossRef] [PubMed]

50. Zhou, N.; Schifferdecker, A.J.; Gamero, A.; Compagno, C.; Boekhout, T.; Piškur, J.; Knecht, W. *Kazachstania gamospora* and *Wickerhamomyces subpelliculosus*: Two alternative baker's yeasts in the modern bakery. *Inter. J. Food Microbiol.* **2017**, *250*, 45–58. [CrossRef] [PubMed]

51. Oh, N.S.; Shin, D.B.; In, M.J.; Chang, Y.I.; Han, M. Effects of capsaicin on the growth and ethanol production of *Zygosaccharomyces rouxii* KFY80 isolated from *gochujang* (fermented hot pepper paste). *Food Sci. Biotechnol.* **2004**, *13*, 749–753.

52. Hong, Y.J.; Son, S.H.; Kim, H.Y.; Hwang, I.G.; Yoo, S.S. Volatile components of traditional *gochujang* produced from small farms according to each cultivation region. *J. East Asian Soc. Diet. Life* **2013**, *23*, 451–460.

53. Korea Food and Drug Administration (KFDA). Food Borne Pathogen Test Methods, Seoul, Korea. 2010. Available online: http://www.kfda.go.kr (accessed on 5 April 2020).

54. Yim, J.H.; Kim, K.Y.; Chon, J.W.; Kim, D.H.; Kim, H.S.; Choi, D.S.; Choi, I.S.; Seo, K.H. Incidence, antibiotic susceptibility, and toxin profiles of *Bacillus cereus sensulato* isolated from Korean fermented soybean products. *J. Food Sci.* **2015**, *80*, M1266–M1270. [CrossRef]

Review

Vegan Egg: A Future-Proof Food Ingredient?

Fatma Boukid [1] and Mohammed Gagaoua [2,*]

[1] Food Safety and Functionality Programme, Institute of Agriculture and Food Research and Technology (IRTA), 17121 Monells, Spain; fatma.boukid@irta.cat
[2] Food Quality and Sensory Science Department, Teagasc Food Research Centre, Ashtown, D15 KN3K Dublin, Ireland
* Correspondence: mohammed.gagaoua@teagasc.ie or gmber2001@yahoo.fr

Abstract: Vegan eggs are designed with the aim to provide a healthier and more sustainable alternative to regular eggs. The major drivers of this industry are the increasing prevalence of egg allergies, awareness towards environmental sustainability, and the shift to vegan diets. This study intends to discuss, for the first time, the vegan egg market, including their formulation, nutritional aspects, and some applications (i.e., mayonnaise and bakery products). Recreating the complete functionality of eggs using plant-based ingredients is very challenging due to the complexity of eggs. Current, but scarce, research in this field is focused on making mixtures of plant-based ingredients to fit specific food formulations. Nutritionally, providing vegan eggs with similar or higher nutritional value to that of eggs can be of relevance to attract health-conscious consumers. Claims such as clean labels, natural, vegan, animal-free, gluten-free, and/or cholesterol-free can further boost the position of vegan eggs in the market in the coming year. At present, this market is still in its infancy stages, and clear regulations of labeling, safety, and risk assessment are deemed mandatory to organize the sector, and protect consumers.

Keywords: future foods; sustainability; egg; algae; starches; plant proteins; bakery products; mayonnaise; food formulation

Citation: Boukid, F.; Gagaoua, M. Vegan Egg: A Future-Proof Food Ingredient? *Foods* **2022**, *11*, 161. https://doi.org/10.3390/foods11020161

Academic Editors: Luca Serventi, Charles Brennan and Rana Mustafa

Received: 9 December 2021
Accepted: 6 January 2022
Published: 8 January 2022

Publisher's Note: MDPI stays neutral with regard to jurisdictional claims in published maps and institutional affiliations.

1. Introduction

Eggs play significant roles in foods, owing to its gelling, foaming, and emulsification features [1,2]. Eggs are versatile products available in the market in dry and liquid forms as whole eggs, egg whites, and egg yolks. Besides their functionality, eggs are of particular interest from a nutritional point of view since they contain proteins, vitamins, minerals, essential fatty acids, and other micronutrients [3]. These components are distributed between the egg yolk and egg white. Egg yolk is rich in lipids (65–70% on dry basis) and proteins (30% on dry basis), and it is a good source of lutein, zeaxanthin, and vitamins [4]. Egg white is rich in proteins, namely fibrous structural proteins (ovomucins), glycoproteins (ovalbumin, protease inhibitors), antibacterial proteins (lysozyme), and peptides [3].

Despite the nutritional value and functionality of eggs as a food ingredient, a high consumption was reported to be related to chronic degenerative diseases that can cause cardiovascular disease and mortality due to its high cholesterol content [5,6]. Moderate egg consumption (up to one egg per day) was found to be not associated with a cardiovascular disease risk [5]. Overall, it was recommended to limit cholesterol intake, and replace whole eggs with egg whites/substitutes for facilitating cardiovascular health and long-term survival [7]. In low- and middle-income countries, due to the price increase in meat, there has also been a shift towards increased egg consumption as a response to maintaining animal-based protein consumption in the diet [8]. On the other hand, consumers are turning towards plant-based food substitutes as a response to rising food safety concerns related to eggs, among other ethical concerns in European countries such as Germany,

France, and Italy [9–11]. Zoonotic diseases related to poultry and eggs, such as avian tuberculosis, erysipelas, and salmonellosis, and flus, is one of the main factors fueling the market growth of vegan eggs [12]. Over the last years, increased public awareness over numerous foodborne diseases transmittable through egg consumption are increasingly alarming for consumers, and negatively impacting the consumption of poultry eggs [13]. Several safety scandals have heavily affected the egg industry, such as fipronil in eggs in Europe (2017), and a salmonella outbreak in poultry in the USA (2015) and UK (2018) [13,14]. The use of antibiotics and hormones in poultry products to ensure rapid growth and health is another contentious issue in the poultry industry [15]. Nevertheless, the accumulation of these drug residues in eggs can cause significant health concerns by triggering allergic reactions and transmitting antibiotic-resistant microbial infections [16]. Egg allergies are one of the most common food allergies in young children, and tends to persist into adulthood [17,18]. This allergy is triggered by ovomucoid (Gal d 1), ovalbumin (Gal d 2), ovotransferrin/conalbumin (Gal d 3), and lysozyme (Gal d 4), which are mainly located in egg whites and, to a lesser extent, in yolk [19,20]. This allergy can cause serious skin reactions, nasal congestion, and gastrointestinal and respiratory symptoms [21]. Egg allergies may also coexist with other food allergies, such as nuts and fish [22,23]. A diet exempt of egg is the only solution to manage an egg allergy [24]. Different approaches were applied to reduce cholesterol and to mitigate epitopes present in eggs by chemical (solvents and biopolymers), physical (fractionation and separation), and biological (enzymes) processing [25,26]. These methods showed variable degrees of success, but they have not been scaled up due to safety, consumer acceptance, repeatability, and economic reasons [27]. These issues have increased consumers' concern toward the consumption of eggs for health, safety, or environmental reasons, and gave room to a new variety of alternative products, such as egg substitutes.

Creating egg substitutes to replace egg functionality and nutritional properties is a challenging task due to the multi-functionality of eggs that impact the taste, texture, and the aspect of food products. The first studies on egg substitutes go back to the seventies [28,29], where particular focus was attributed to replace eggs in bakery products. At first, animal proteins were mostly used, such as milk powder, casein, whey, and bovine plasma protein [30,31]. This is due to their excellent functional properties, such as solubility, emulsification, foaming, and heat-induced gelation properties [32]. Despite the occurrence of animal ingredients in the human diet, plant-based foods are gaining popularity around the world due to their health benefits, environmental sustainability, and ethical merit [33,34]. The coronavirus (COVID-19) outbreak consolidated this transition due to changes in consumers' dietary habits, associating plant-based diets to be healthier alternatives to animal products [35]. COVID-19 caused also a decrease in the demand for chickens and eggs, resulting in prices fluctuations due to lockdown restrictions limiting business opportunities and customer incomes [36]. Furthermore, there are false rumors suggesting zoonotic origins of COVID-19 or poultry products contributing to the spread of COVID-19 [12,37].

To cater for the growing vegan and health-conscious market, manufacturers have created vegan eggs using different types of plant-based ingredients (e.g., proteins, polysaccharide hydrocolloids, or emulsifiers), alone or combined, to replace regular eggs in food products. These ingredients might present nutritional benefits such as low allergenicity, reduced price, and high production volumes. Nevertheless, the functional properties are highly variable among the different ingredients in terms of composition, purity, and source [38]. The vegan egg market keeps growing to deliver different products with different properties to fit a wide range of applications. In this context, this study aimed, for the first time, to: (i) enable an overview about the current market landscape of vegan eggs, with a focus on drivers and barriers; (ii) address the main ingredients used in formulating vegan eggs; (iii) discuss their nutritional properties in comparison to conventional egg products; and (iv) confer their impact on food products, with a focus on mayonnaise and bakery products as examples among other potential products.

2. Global Market Landscape of Vegan Eggs

The global vegan egg market is moving upward, and accounted for US$1.5 billion in 2021, and is expected to witness a high compound annual growth rate of 8.3% through 2031 [39]. Due to the absence of exhaustive market reports about vegan eggs launched in the global market, the authors made their search using Mintel's GNPD database [40], with a focus on the period 2016–2021 to capture the current market landscape of vegan eggs. From 2016 to 2021, 102 vegan egg products were launched in the global market. The authors gathered all information on the front-of-pack labelling. Table 1 summarizes the main market trends in the vegan egg industry, relying on the main claims used on the retrieved products. Health and well-being, naturalness, sustainability, and convenience are the main trends, with "vegan/no animal" and "vegetarian" ingredients (related to health and well-being and sustainability), and "low/no/reduced allergen" and "gluten-free" (related to health and well-being) being the top four sub-trends [40]. In the last decade, consumers have become more concerned on health and well-being, and are paying more attention to what they eat. As a result, manufactures of vegan eggs consider the use of a large spectrum of ingredients to offer a portfolio of products to accommodate all consumers, including those with special needs. Indeed, 80.4% of marketed vegan eggs claim to have low/no/reduced allergens, including 65.7% and 18.6% claiming to be gluten-free and low/no/reduced lactose, respectively. Increasing niches with particular lifestyles, such as vegan, vegetarian, and flexitarian, contributed to the reduction of animal-derived products, such as alternative meat, vegan dairy, and vegan eggs [41–43]. In addition, 100% claim to be suitable for vegans and vegetarians due the absence of animal-based ingredients in egg formulations, such as whey protein, milk, or casein. Emphasizing that these products are made with plant-based ingredients was reflected by the use of term "plant-based" on 31.4% of products. This shift to non-animal ingredients seemed to continue to benefit the industry of vegan eggs [33,44].

Vegan eggs are also rising as a healthier alternative to eggs, since they contain no cholesterol. This aligns with market trends reporting around 41.2% of launches claimed to have low/no/reduced cholesterol. This industry is further focused on designing products with reduced sugar, fat, saturated fat, and sodium. Since consumers have a strong preference for food products free from additives and preservatives, there is a growing trend boosting the use of natural and clean label ingredients [45]. This was reflected by declarations, such as genetically modified organisms (GMO)-free (35.3%), organic (27.4%), no additives/preservatives (22.5%), free from added/artificial preservatives (7.8%), and free from added/artificial colorings (5.9%). Consumers' awareness towards contaminants is also considered where terms like "toxin-free" were used to describe 1.96% of the products. Convenience is an important driver of this market, in which 34.3% of products were declared as easily used. This aligns with a general trend in the food sector seeking quick and convenient meal solutions [46]. Finally, sustainability is becoming an essential criterion in the food sector [47], and the sustainability of vegan eggs is reflected by the fact that 83.3% of the products seemed to have environmental or ethical claims, including recycling food waste, and the use of sustainable packaging.

Like other emerging alternative products, the current market barriers of vegan eggs are the lack of high production volumes, targeted marketing, and clear regulations. A recent study based on in-depth interviews with egg industries and retailers and plant-based egg manufacturers revealed that replicating all eggs' nutrients and functionalities is not realistic, and considering plant-based eggs as potential competitors to conventional products is impossible. Also, there is uncertainty on how to present the labeling of plant-based eggs [48]. Consumer perception and acceptance is also an important factor for the growth of such a novel food sector. Consumer expectations from vegan egg products were found to be depending on product-related (color, shape, taste, ingredients, nutrients, method of production, and shelf life) and non-product-related attributes (price, packaging, country of origin, and product naming) [49]. More in-depth quantitative and quantitative studies are required for a deeper understanding of this first screening based on country surveys. From a manufacturer perspective, the main challenge of vegan eggs can be related to the difficulty in delivering similar nutrition, taste, and functionality to eggs [48].

Table 1. Current trends in vegan eggs launched in the global market (2016–2021) [1].

Trends	Sub-Trends	Number of Products	Percentage Products Out of Total Launches (%)
	Minus		
	Low/no/reduced fat	4	3.92%
	Low/no/reduced trans-fat	1	0.98%
	Low/no/reduced sodium	4	3.92%
	Low/no/reduced calorie	2	1.96%
	Low/no/reduced cholesterol	42	41.2%
	Sugar free	3	2.94%
	No added sugar	3	2.94%
	Low/no/reduced saturated fat	1	0.98%
	Plus		
	High/added protein	5	4.90%
	Vitamin/mineral fortified	1	0.98%
Health and well-being	High/added fiber	8	7.84%
	Free from		
	Hormone free	2	1.96%
	Dairy free	39	38.20%
	Functional		
	Functional—other	1	0.98%
	Functional—digestive	1	0.98%
	Suitability		
	Low/no/reduced allergen	82	80.44%
	Gluten free	67	65.69%
	Kosher	43	42.16%
	Low/no/reduced lactose	19	18.63%
	Suitable for vegan and vegetarian	102	100%
	Plant based	32	31.37%
	Microwaveable	4	3.92%
Convenience	Ease of use	35	34.31%
	Convenient packaging	3	2.94%
	Time/speed	1	0.98%
	No additives/preservatives	23	22.55%
	Free from added/artificial preservatives	8	7.84%
	Organic	28	27.45%
	Free from added/artificial colorings	6	5.88%
Naturalness	GMO-free	36	35.29%
	Free from added/artificial flavorings	6	5.88%
	Natural product	3	2.94%
	Wholegrain	1	0.98%
	Free from added/artificial additives	3	2.94%
	Environmentally friendly package	33	32.35%
	Recycling	26	25.49%
	Sustainable (habitat/resources)	6	5.88%
Ethical & environmental	Environmentally friendly	8	7.84%
	Animal welfare	6	5.88%
	Toxins free	2	1.96%
	Biodegradable packaging	4	3.92%

[1] Data based on Mintel's GNPD database [40]. The query was conducted on 11 November 2021, and retrieved 102 vegan egg products in the global market from January 2016 to October 2021.

Considering the current global market landscape (Table 2), vegan eggs are marketed in different forms (powder, liquid, and egg-shaped) [39]. The powder segment is the most dominant, and was estimated to be US$815.4 million in 2019 [50]. The demand of powder vegan eggs is expected to keep increasing at a high rate, with the on-the-go nature or ease of use claims establishing these products as convenient healthy snacks [51]. The most sold vegan eggs are made from starches, plant-based proteins, soy products (lecithin, tofu, and tahini), algae flours, and other ingredients (e.g., fruit purees and vinegar) [39,50]. By region, North America is estimated to account for 47.8% of the global market, and is expected to remain the dominant one until 2026 [50]. By the end of 2021, sales of vegan eggs in North America are expected to reach US$476.6 million, corresponding to 32% of global sales [39]. The vegan eggs market is trending in Europe due to their applications in reformulating snacks and meat alternatives [39]. The most important producers of vegan eggs are Corbion NV (Amsterdam, The Netherlands), Glanbia Plc (Kilkenny, Ireland), Tate & Lyle Plc (London, UK), Ingredion Incorporated (Westchester, IL, USA), Ener-G Foods, Inc. (Seattle, WA, USA), Natural Products, Inc. (Grinnell, IA, USA), Orchard Valley Foods Limited (Tenbury Wells, UK), Puratos Group (Dilbeck, Belgium), TerraVia Holdings, Inc. (San Francisco, CA, USA), and Archer Daniels Midland Company (Chicago, IL, USA) [50]. Vegan eggs are mostly used as substitutes of eggs in bakery products, desserts, and confectionary [39]. The mayonnaise segment is estimated to account for a value share of 38.2% in the global market, whereas bakery products are estimated to account for over 26% [50]. Claims such as natural, organic, clean label, Halal or Kosher certified, dairy-free, GMO-free, and gluten-free are also boosting the market of vegan eggs [50].

Table 2. Segmentation of the global market of vegan eggs adapted from [39,50].

Segment	Segmentation
Form	Powder Liquid Egg shape
Type	Starch Soy products (lecithin, tofu, and tahini) Plant proteins, such as pea and chickpea Algal flour Others (fruit purees and vinegar)
Application	Mayonnaise Biscuits and Cookies Cakes/Pastries/Muffins/Breads Chocolates Noodles and Pasta
Main players	Glanbia plc Ingredion Incorporated Cargill Bob's Red Mill Natural Foods, Inc. House Foods America Corporation EVO Foods Mantiqueira (N.Ovo) JUST Inc. Orgran Foods Terra Vegane Free and Easy Follow Your Heart The Vegg Vezlay Foods Private Limited Now Foods The Neat Egg Conagra Brands, Inc. Ener-G

Table 2. *Cont.*

Segment	Segmentation
Region	North America Latin America Europe, Middle East, and Africa Asia Pacific

3. Major Components of Vegan Eggs

Vegan eggs can be formulated by one plant-based ingredient or a combination of ingredients to recreate the functionality of eggs Pulses are ingredients rich in proteins, starches, and fibers, as well several health beneficial ingredients [52]. Proteins deriving from pea, lentil, lupine, and chickpea can confer in their native and modified forms interesting functionalities, such as gelling, emulsification, and foaming for formulating vegan eggs [53–56]. The proteins can be used in different forms, namely flours, protein concentrate, or isolates. Besides their high nutritional value, pulses are known for their affordability and sustainability [52]. Furthermore, pulses are recognizable products by the consumers, and their inclusion in vegan egg formulations might contribute to their acceptability. The proteins of pulses have plenty of pros, but they have some nutritional limitations, such as their low content in sulfur amino acids, which can be overcome by blending them with cereals. Also, plant proteins have a globular structure that impacts the functionality and, more specifically, the solubility. To overcome such concerns, the addition of hydrocolloids was suggested to improve the functionality of proteins [57,58]. As an alternative, these proteins can be improved by postprocessing using thermal treatments, fermentation, and crosslinking by means of enzymes to improve the emulsification, gelling, and foaming abilities [59–61]. Pulses also present flavors described as "beany" or "green", attributed to their content in saponins, ketones, and aldehyde compounds [62]. Several solutions are being applied to attenuate these flavors, such as using masking agents and mitigation processing [63,64]. Starches from pulses are also increasingly used in formulating vegan eggs to play the role of binding and thickening [65]. Native starches from pulses have some functional limitations compared to those usually used, such as tapioca and corn starches [66,67]. Nevertheless, several postprocessing methods are being developed to produce modified starch with high quality, likely-modified pea starch [68,69]. Another ingredient, aquafaba, derived from cooked chickpea, is gaining interest as an egg substitute due to its foaming, emulsifying, thickening, and gelling properties [70–72]. This is attributed to its composition, namely protein, water-soluble/insoluble carbohydrates, coacervates, saponins, and phenolic compounds [62,73]. The main limitation for the commercialization of aquafaba is the lack of product standardization due to the high variability in chickpea properties (differences in the composition and genotypes) and processing conditions (temperature, pressure, and cooking time) [38,74–76].

Different types of hydrocolloids, such as carrageenan, pectin, and guar gum, have been used as natural foaming, thickening, and emulsifier agents to further reinforce the structure made by plant-based proteins and starches, and for an improved mouthfeel [77–79]. Fibers from pulses are also of relevance in vegan egg formulations due to their gelling, binding, and thickening properties. Nevertheless, the most used fibers derive from apple, citrus, and oat fibers. Cellulose derivatives, such as carboxymethyl cellulose or hydroxypropyl methylcellulose, can be used as thickeners or emulsifiers.

Oilseeds (mainly soybeans) are also used in different forms, such as proteins, flour, or milk, owing to their high protein content, complete essential amino acids, and protein digestibility that can be comparable to that of animal proteins [33]. In recent years, consumers have been concerned about soy ingredients for their genetically modified reputation and allergenicity [80,81]. This has given room for more emerging sources, such as oat, mung bean, lentil, and faba bean [82–84].

Emerging ingredients, such as algal flours, are also of interest as food ingredients due to their high nutritional quality and sustainability [85]. They are a rich source of proteins, lipids, fibers, and vitamins [86]. Compared to plant ingredients, algae are also a good source of vitamin B12 for vegetarians and vegans [87]. They also contain functional ingredients, such as monoglycerides, diglycerides, and phospholipids, mainly acting as emulsifiers [88,89]. Indeed, the first vegan egg (VeganEgg) using algal flours was launched in 2017.

Vegetable oils, such as canola and sunflower oils, are also important as structuring agents in vegan egg formulations, hence contributing to the creation of the textural attributes, flavor profile, and mouthfeel of the final products [70]. Flavoring agents such as Himalayan black salt or "Kala namak" are commercially available to mimic the sulfur flavor of egg [90]. Other ingredients can be added, such as spices (e.g., garlic powder, sugar, and salt), buffers (e.g., bicarbonates or phosphates), and preservatives (e.g., nisin) [47].

4. Nutritional Value of Vegan Eggs

This section provides an overview of the nutritional composition of vegan eggs, yolks, whites, and whole eggs launched in the global market from January 2016 to October 2021 (Table 3). Based on Mintel's GNPD database, 102 new vegan egg products were launched to the global market [40]. The major ingredients in egg products are proteins and fat, whereas vegan eggs have a profile rich in carbohydrates, proteins, and fibers. Vegan eggs provide the highest calories, followed by whole eggs, yolks, and egg whites. This is due to their high content in carbohydrates (41.89 g/100 g), in which starch content (66.73 g/100 g) is the main contributor due to starchy ingredients (in the form of starchers and flours) used in vegan products. Egg carbohydrates were found mostly in egg yolks, whereas a lower amount was found in egg whites and whole eggs. It was reported that glucose is the dominant free sugar in the eggs, with traces of fructose, lactose, maltose, and galactose [91]. Total fat and saturated fat contents were found lower in vegan eggs compared to the whole egg and yolk, but higher than the egg white. This can be attributed to the use of vegetable oils rich in saturated fats, such as palm oil. Noteworthy, vegan eggs are cholesterol-free, whereas whole eggs have the highest value, followed by egg yolks, and egg whites. The whole egg and yolk have the high cholesterol content, exceeding the limits set by the American Heart Association of <300 mg/day [92].

Table 3. Nutritional composition of eggs (per 100 g) and their alternatives in the global market [1].

	Vegan Egg	Egg Yolk	Egg White	Whole Egg
Number of retrieved products	102	37	54	6517
Average values of nutrients				
Energy (kcal/100 g)	298.55	153.66	98.36	152.18
Fat (g/100 g)	6.10	10.40	2.35	9.97
Of which saturated (g/100 g)	2.10	2.72	1.19	3.26
Carbohydrates (g/100 g)	41.89	3.77	2.59	2.32
Of which sugars (g/100 g)	1.77	3.77	0.53	0.45
Fiber (g/100 g)	8.56	0.00	0.00	0.00
Protein (g/100 g)	11.60	13.69	16.53	12.39
Sodium (mg/100 g)	912.59	682.67	353.01	385.74
Vitamin B12 (µg per 100 g/mL)	0.75	nr	nr	21,844.4
Cholesterol (mg per 100 g/mL)	0.00	339.26	11.64	1509.53
Calcium (mg per 100 g/mL)	286.59	39.88	122.23	159.80

[1] Data based on Mintel's GNPD database [40]. The query was conducted on 11 November 2021, and retrieved egg products launched in the global market from January 2016 to October 2021. nr: not reported.

Eggs do not contain any fibers. However, vegan eggs provide high amounts of fibers that are added to mimic the emulsification properties of eggs. Egg whites and egg yolks are almost equality concentrated in proteins, but slightly higher than vegan eggs and whole eggs. Vegan eggs are made with different proteins to reach similar content to that of the conventional product, but little is known about their amino acid profiles. This underlines the great efforts being made to have a similar protein content to animal counterparts, which is usually known as a limitation of vegan products, including meat and dairy alternatives [41,43]. Vegetable proteins are the most used sources for compensating the protein content reduced by egg removal. It is well-known that animal proteins have a complete composition of essential amino acids and high digestibility compared to plant-based products [33]. It will be of interest to investigate such parameters in vegan products to address it in future product development projects.

Sodium was found to be higher in vegan eggs compared to regular eggs, egg yolks, and egg whites. A lower amount of sodium was previously reported in whole eggs (142 mg per 100 g of whole egg) [91]. This can be attributed to the increase of yolk-to-egg-white ratio [3,91]. Vitamin B12 is a big limitation in vegan eggs compared to whole eggs. For these reasons, fortifying vegan egg products with bioavailable forms of these micronutrients is required [93]. However, the nutritional facts of commercial yolks and whites did not present the amounts of B12, since it is not mandatory information. Surprisingly, calcium was found the highest in vegan eggs, showing the direction in new product development focusing on upgrading the nutritional value of vegan products.

5. Main Food Applications of Vegan Eggs

5.1. Egg-Free and Egg-Reduced Mayonnaise

Mayonnaise is one of the most popular condiments worldwide, providing a creamy texture and special flavor [94]. Mayonnaise is a colloidal system (oil-in-water emulsion) made from vegetable oil (70–80%), egg yolk, vinegar, salt, and spices [95]. Egg yolk is a key ingredient for emulsion stability due to its high emulsifying capacity attributed to the phospholipids and lipoproteins (high-density lipoprotein and low-density lipoprotein), and non-bonded proteins (phosvitin and livetin) [96]. Egg yolk also provides forming properties and prevents flocculation to ensure an appropriate texture of mayonnaise [1,97]. Nevertheless, the use of raw eggs in mayonnaise might present some inconveniences, such as possible contamination with Salmonella sp., and high cholesterol content [98]. As an alternative, egg-free mayonnaise is gaining traction as a healthier option for consumers, and is suitable for vegan customers, as well as being more cost-effective (no pasteurization is required). Several vegan eggs were used in single and combined forms to mimic the quality, taste, and color of conventional mayonnaise [98].

Vegetable protein isolates deriving from soy, pea, lentil, and rapeseed have been considered as suitable egg alternatives [33,99,100]. Egg-free mayonnaise designed using 6% soy protein concentrate (as an emulsifier to replace egg yolk) was accepted by consumers [101]. A mayonnaise was made with a 10% substitution level of eggs, using a vegan egg made by a combination of soy milk and a blend of 6.7% mono- and di-glycerides, 36.7% guar gum, and 56.7% xanthan gum. This low substitution level produced a low cholesterol-low fat mayonnaise with improved properties (i.e., the stability, heat stability, consistency coefficient, viscosity, firmness, adhesiveness, adhesive force, and overall acceptance) [102]. Eggs were replaced with soy milk at levels of 25, 50, 75, and 100%. Results showed that up to a 75% egg substitution level, viscosity was not affected, whereas stability was decreased. The sensory acceptability of the products was not impacted until 50% substitution level. This suggests that soy milk can be a good candidate to partially substitute egg (up to 50%) without hampering product viscosity and taste [103]. Nevertheless, combining soy milk with different hydrocolloids (i.e., xanthan gum and zodo gum) increased in the apparent viscosity, the consistency coefficient, and the firmness/emulsion stability of the mayonnaise, whereas the mayonnaise flow index was reduced. The optimal formulation of vegan eggs was 0.25% xanthan gum, 3.84% zodo gum, 37.50% oil, and 63.61% soy milk [104]. Egg yolk replaced with sesame-peanut meal milk decreased product quality, including pH, color, thermal stability, and acidity, with increasing substitution levels (0, 25, 50, 75, and 100%). Mayonnaise made with vegan eggs at 50% had desirable physical and thermal stability, and reduced cholesterol content [105].

Raikos et al. [106] reported that the use of liquid aquafaba (up to 70%) was capable of forming a stable emulsion resulting in mayonnaise with a desirable consistency and unaffected oxidative stability during storage. Using dry aquafaba resulted in a high stable mayonnaise for 28 days of storage at 4 °C. These results suggest that the use of dry aquafaba can be the solution to overcome standardization issues of aquafaba, and can be effectively used in mayonnaise manufacturing [75].

Starches were also used as ingredients to replace egg yolk in mayonnaise formulation, owing to their thickening properties. Native starches were found undesirable due to their unfavorable effect on the texture and flavor. Modified starches, such as octenyl succinic anhydride-modified potato starch, showed better emulsification properties when partially replacing eggs (0, 25, 50, 75, and 100%). Products made with 75% octenyl succinic anhydride-modified potato starch resulted in high emulsion stability even after two months of storage, and it also reduced cholesterol content, improved oil droplets particle size (maximum at 70 μm), and resulted in a consistent texture with no agglomerates. This is due to the formation of a stable cohesive layer of starch surrounding the oil droplets [97]. Mayonnaises were prepared with 35% freeze-dried chia mucilage instead of egg yolk due to their emulsification properties [107,108]. The resulting mayonnaise had similar stability and texture parameters, as well as sensory acceptance, to the control mayonnaise [109].

Several thickeners, such as gums (xanthan and guar gums, and Arabic gum), were also used for egg reduction, or complete removal, due to their emulsifying ability and stability [110]. Durian seed gum used at a level of 4% resulted in vegan mayonnaise with textural and sensory properties comparable to egg-based products [111]. As such, this substitution was able to generate a stable emulsion, and to prevent coalescence and flocculation for prolonged periods of storage (up to 5 months). Arabic gum may have inhibitory effects towards lipid oxidation and microbial contamination, owing to its high antioxidant activity [98]. Overall, these hydrocolloids improve emulsification, antibacterial activity, and sensory quality of the final product [110], whereas guar gum and/or xanthan are considered additives that are not fully appreciated by consumers seeking "clean" labels [94].

Algal ingredients, such as *Chlorella vulgaris*, were also used in partially substituted yolk in combination with acid casein curd. A mix of *Chlorella vulgaris* (10 and 15%) and casein curd (90 and 95%) improved the nutritional value, rheological properties, and sensory scores of mayonnaise at 25 and 50% of egg replacement [112].

5.2. Egg-Free and Egg-Reduced Bakery Products

Egg exclusion or reduction comes in the optic to promote the healthiness of egg-free and/or cholesterol-free bakery products. The total substitution of eggs by lupine protein isolates resulted in the collapse of the cakes. This can be explained by the lower functional properties compared to egg [54]. Thus, besides lupine isolates, soy lecithin, mono- and diglycerides, and xanthan gum were used as vegan egg substitutes. The resulting cakes had an improved structure of crumb, reduced shrinkage, and led to high height [113]. Likewise, a blend of soy protein isolate and 1% mono- and di-glycerides produced an egg-free cake with similar specific volume and gravity, firmness, and moisture content compared with egg-containing cakes [114]. Similarly, the use of only soymilk to replace eggs resulted in a batter with low density and viscosity, resulting in a firm, dark, and compact cake, whereas combining soymilk and soy lecithin improved the quality of egg-free cakes [115]. In another study, egg-free and egg-less cakes were successfully produced by replacing eggs with a mix of lupine protein, whey proteins, and soy lecithin [116]. The complete substitution of egg whites by *Chlorella vulgaris* decreased the consistency of the batter, which imparted the cake with low specific volume, and a high weight loss and hardness. Nevertheless, a partial substitution level (25%) did not affect the taste, color, odor, texture, and overall acceptability compared to the conventional preparation [117]. Aquafaba-based cakes resulted in a similar color and texture, as well sensory acceptability compared to egg-white-based cakes. This is due to the good foaming and emulsifying properties of aquafaba [38,70,71]. The main defect of these eggless cakes is their low springiness and cohesiveness [74]. A potential approach might be adding other ingredients together with aquafaba to overcome this defect.

Egg-free and egg-reduced cakes were also produced through the complete use of hydrocolloids, such as hydroxypropyl methylcellulose in combination with sodium stearoyl lactylate [118,119]. Depending on the level of substitution and the type of additives, cake attributes, including color, texture, and volume, significantly changed, but in some cases, such changes were not perceived at the sensory level [115]. Nevertheless, this type of substitution is decreasingly desired due to the market shift towards natural ingredients.

6. Conclusions

The market of vegan eggs is steadily growing as healthier, more sustainable, and ethical alternatives to regular eggs. Affordable and available ingredients are required to develop cost-effective vegan eggs. Although there is plenty of ingredients that can mimic the functionality of eggs, the nutritional value of vegan eggs must be carefully considered. Protein rich ingredients and vitamin/minerals fortification(s) are required to avoid nutritional deficiencies, especially in the case of vegan consumers. Natural and clean label ingredients are becoming a must by health-conscious consumers. There is no vegan egg fitting all food formulations. Therefore, the selection of egg replacements needs to be made based on the functionality required for each food product. At present, the market for vegan eggs is still a new commodity where clear regulation is required to organize the sector. Furthermore, in-depth market studies are required to capture this emerging market's challenges and opportunities. Qualitative and quantitative surveys considering different countries, continents, gender, age, education level, and income are for interest to understand consumers' behaviors toward such a new market. Consumer studies are needed to evaluate the sensorial properties of vegan eggs (different formulations) in comparison to regular eggs to provide a further understanding of the preferences and acceptability of consumers.

Author Contributions: Conceptualization, F.B. and M.G.; investigation, F.B. and M.G.; resources, F.B. and M.G.; writing—review and editing, F.B. and M.G. All authors have read and agreed to the published version of the manuscript.

Funding: This work was supported by CERCA Programme (Generalitat de Catalunya).

Institutional Review Board Statement: Not applicable.

Informed Consent Statement: Not applicable.

Data Availability Statement: Not applicable.

Conflicts of Interest: The authors declare no conflict of interest.

References

1. Mirzanajafi-Zanjani, M.; Yousefi, M.; Ehsani, A. Challenges and approaches for production of a healthy and functional mayonnaise sauce. *Food Sci. Nutr.* **2019**, *7*, 2471–2484. [CrossRef] [PubMed]
2. Bhat, Z.F.; Morton, J.D.; Bekhit, A.E.D.A.; Kumar, S.; Bhat, H.F. Effect of processing technologies on the digestibility of egg proteins. *Compr. Rev. Food Sci. Food Saf.* **2021**, *20*, 4703–4738. [CrossRef]
3. Réhault-Godbert, S.; Guyot, N.; Nys, Y. The golden egg: Nutritional value, bioactivities, and emerging benefits for human health. *Nutrients* **2019**, *11*, 684. [CrossRef]
4. Maeta, A.; Matsushima, M.; Katahira, R.; Sakamoto, N.; Takahashi, K. Diets Supplemented with 1% Egg White Induce Oral Desensitization and Immune Tolerance in an Egg White-Specific Allergic Mouse Model. *Int. Arch. Allergy Immunol.* **2018**, *176*, 205–214. [CrossRef] [PubMed]
5. Drouin-Chartier, J.P.; Chen, S.; Li, Y.; Schwab, A.L.; Stampfer, M.J.; Sacks, F.M.; Rosner, B.; Willett, W.C.; Hu, F.B.; Bhupathiraju, S.N. Egg consumption and risk of cardiovascular disease: Three large prospective US cohort studies, systematic review, and updated meta-analysis. *BMJ* **2020**, *368*, 1986–2012. [CrossRef]
6. Xia, P.F.; Pan, X.F.; Chen, C.; Wang, Y.; Ye, Y.; Pan, A. Dietary intakes of eggs and cholesterol in relation to all-cause and heart disease mortality: A prospective cohort study. *J. Am. Heart Assoc.* **2020**, *9*, 15743. [CrossRef]

7. Zhuang, P.; Wu, F.; Mao, L.; Zhu, F.; Zhang, Y.; Chen, X.; Jiao, J.; Zhang, Y. Egg and cholesterol consumption and mortality from cardiovascular and different causes in the United States: A population-based cohort study. *PLoS Med.* **2021**, *18*, e1003508. [CrossRef] [PubMed]

8. FAO. *The Future of Food and Agriculture*; FAO: Rome, Italy, 2017.

9. Hafez, H.M.; Attia, Y.A. Challenges to the Poultry Industry: Current Perspectives and Strategic Future After the COVID-19 Outbreak. *Front. Vet. Sci.* **2020**, *7*, 516. [CrossRef]

10. Sanchez-Sabate, R.; Sabaté, J. Consumer attitudes towards environmental concerns of meat consumption: A systematic review. *Int. J. Environ. Res. Public Health* **2019**, *16*, 1220. [CrossRef]

11. Kopplin, C.S.; Rausch, T.M. Above and beyond meat: The role of consumers' dietary behavior for the purchase of plant-based food substitutes. *Rev. Manag. Sci.* **2021**, 1–30. [CrossRef]

12. Al Sattar, A.; Mahmud, R.; Mohsin, M.A.S.; Chisty, N.N.; Uddin, M.H.; Irin, N.; Barnett, T.; Fournie, G.; Houghton, E.; Hoque, M.A. COVID-19 Impact on Poultry Production and Distribution Networks in Bangladesh. *Front. Sustain. Food Syst.* **2021**, *5*, 306. [CrossRef]

13. Whiley, H.; Ross, K. Salmonella and Eggs: From Production to Plate. *Int. J. Environ. Res. Public Health* **2015**, *12*, 2543. [CrossRef] [PubMed]

14. Reich, H.; Triacchini, G.A. Occurrence of residues of fipronil and other acaricides in chicken eggs and poultry muscle/fat. *EFSA J.* **2018**, *16*, e05164. [CrossRef]

15. Mund, M.D.; Khan, U.H.; Tahir, U.; Mustafa, B.E.; Fayyaz, A. Antimicrobial drug residues in poultry products and implications on public health: A review. *Int. J. Food Prop.* **2016**, *20*, 1433–1446. [CrossRef]

16. Kang, J.; Hossain, M.A.; Park, H.C.; Jeong, O.M.; Park, S.W.; Her, M. Cross-Contamination of Enrofloxacin in Veterinary Medicinal and Nutritional Products in Korea. *Antibiotics* **2021**, *10*, 128. [CrossRef] [PubMed]

17. Dang, T.D.; Peters, R.L.; Koplin, J.J.; Dharmage, S.C.; Gurrin, L.C.; Ponsonby, A.-L.; Martino, D.J.; Neeland, M.; Tang, M.L.K.; Allen, K.J.; et al. Egg allergen specific IgE diversity predicts resolution of egg allergy in the population cohort HealthNuts. *Allergy* **2019**, *74*, 318–326. [CrossRef]

18. Sabouraud, M.; Biermé, P.; Andre-Gomez, S.A.; Villard-Truc, F.; Corréard, A.K.; Garnier, L.; Payot, F.; Braun, C. Oral immunotherapy in food allergies: A practical update for pediatricians. *Arch. Pediatr.* **2021**, *28*, 319–324. [CrossRef]

19. Mills, E.N.C. *Allergies: Public Health: From Encyclopedia Food and Health*; Elsevier Ltd.: Amsterdam, The Netherlands, 2016; pp. 115–121. [CrossRef]

20. Sakai, S.; Adachi, R.; Teshima, R. Detection and control of eggs as a food allergen. *Handb. Food Allerg. Detect. Control* **2015**, 313–340. [CrossRef]

21. Uneoka, K.; Horino, S.; Ozaki, A.; Aki, H.; Toda, M.; Miura, K. Differences in allergic symptoms after the consumption of egg yolk and egg white. *Allergy Asthma. Clin. Immunol.* **2021**, *17*, 97. [CrossRef]

22. Ngai, N.A.; Leung, A.S.Y.; Leung, J.C.H.; Chan, O.M.; Leung, T.F. Identification of predictors for persistence of immediate-type egg allergy in Chinese children. *Asia Pac. Allergy* **2021**, *11*, e41. [CrossRef]

23. Murai, H.; Irahara, M.; Sugimoto, M.; Takaoka, Y.; Takahashi, K.; Wada, T.; Yamamoto-Hanada, K.; Okafuji, I.; Yamada, Y.; Futamura, M.; et al. Is oral food challenge useful to avoid complete elimination in Japanese patients diagnosed with or suspected of having IgE-dependent hen's egg allergy? A systematic review. *Allergol. Int.* **2021**, in press. [CrossRef]

24. Han, P.; Gu, J.Q.; Li, L.S.; Wang, X.Y.; Wang, H.T.; Wang, Y.; Chang, C.; Sun, J.L. The Association Between Intestinal Bacteria and Allergic Diseases—Cause or Consequence? *Front. Cell. Infect. Microbiol.* **2021**, *11*, 284. [CrossRef]

25. Ma, X.; Liang, R.; Xing, Q.; Lozano-Ojalvo, D. Can food processing produce hypoallergenic egg? *J. Food Sci.* **2020**, *85*, 2635–2644. [CrossRef]

26. Tong, P.; Chen, S.; Gao, J.; Li, X.; Wu, Z.; Yang, A.; Yuan, J.; Chen, H. Caffeic acid-assisted cross-linking catalyzed by polyphenol oxidase decreases the allergenicity of ovalbumin in a Balb/c mouse model. *Food Chem. Toxicol.* **2018**, *111*, 275–283. [CrossRef] [PubMed]

27. Puertas, G.; Vázquez, M. Advances in techniques for reducing cholesterol in egg yolk: A review. *Crit. Rev. Food Sci. Nutr.* **2019**, *59*, 2276–2286. [CrossRef] [PubMed]

28. Gardner, F.A.; Beck, M.L.; Denton, J.H. Functional Quality Comparison of Whole Egg and Selected Egg Substitute Products. *Poult. Sci.* **1982**, *61*, 75–78. [CrossRef]

29. Childs, M.T.; Ostrander, J. Egg substitutes: Chemical and biologic evaluations. *J. Am. Diet. Assoc.* **1976**, *68*, 229–234. [CrossRef]

30. Mohamed, S.; Lajis, S.M.M.; Hamid, N.A. Effects of protein from different sources on the characteristics of sponge cakes, rice cakes (apam), doughnuts and frying batters. *J. Sci. Food Agric.* **1995**, *68*, 271–277. [CrossRef]

31. Myhara, R.M.; Kruger, G. The performance of decolorized bovine plasma protein as a replacement for egg white in high ratio white cakes. *Food Qual. Prefer.* **1998**, *9*, 135–138. [CrossRef]

32. Paraskevopoulou, A.; Donsouzi, S.; Nikiforidis, C.V.; Kiosseoglou, V. Quality characteristics of egg-reduced pound cakes following WPI and emulsifier incorporation. *Food Res. Int.* **2015**, *69*, 72–79. [CrossRef]

33. Boukid, F.; Rosell, C.M.; Rosene, S.; Bover-Cid, S.; Castellari, M. Non-animal proteins as cutting-edge ingredients to reformulate animal-free foodstuffs: Present status and future perspectives. *Crit. Rev. Food Sci. Nutr.* **2021**, *137*, 1–31. [CrossRef] [PubMed]

34. Fehér, A.; Gazdecki, M.; Véha, M.; Szakály, M.; Szakály, Z. A comprehensive review of the benefits of and the barriers to the switch to a plant-based diet. *Sustainability* **2020**, *12*, 4136. [CrossRef]

35. Di Renzo, L.; Gualtieri, P.; Pivari, F.; Soldati, L.; Attinà, A.; Cinelli, G.; Leggeri, C.; Caparello, G.; Barrea, L.; Scerbo, F.; et al. Eating habits and lifestyle changes during COVID-19 lockdown: An Italian survey. *J. Transl. Med.* **2020**, *18*, 229. [CrossRef]

36. GlobeNewsWire Global Egg Market Forecasts 2020–2030: COVID-19 Impact and Recovery. Available online: https://www.globenewswire.com/news-release/2020/05/08/2030249/0/en/Global-Egg-Market-Forecasts-2020-2030-COVID-19-Impact-and-Recovery.html (accessed on 21 November 2021).

37. Marchant-Forde, J.N.; Boyle, L.A. COVID-19 Effects on Livestock Production: A One Welfare Issue. *Front. Vet. Sci.* **2020**, *7*, 625372. [CrossRef]

38. Buhl, T.F.; Christensen, C.H.; Hammershøj, M. Aquafaba as an egg white substitute in food foams and emulsions: Protein composition and functional behavior. *Food Hydrocoll.* **2019**, *96*, 354–364. [CrossRef]

39. FACTMR Vegan Egg Market Size, Share, Trends & Forecast, 2021–2031. Available online: https://www.factmr.com/report/vegan-eggs-market (accessed on 21 November 2021).

40. Mintel Mintel Global New Product Database. Available online: https://www.gnpd.com/sinatra/search_results/?search_id=3tMdwh5e47&page=0 (accessed on 1 October 2021).

41. Boukid, F.; Castellari, M. Veggie burgers in the EU market: A nutritional challenge? *Eur. Food Res. Technol.* **2021**, *247*, 2445–2453. [CrossRef]

42. Boukid, F. Plant-based meat analogues: From niche to mainstream. *Eur. Food Res. Technol.* **2020**, *247*, 297–308. [CrossRef]

43. Boukid, F.; Lamri, M.; Dar, B.N.; Garron, M.; Castellari, M. Vegan Alternatives to Processed Cheese and Yogurt Launched in the European Market during 2020: A Nutritional Challenge? *Foods* **2021**, *10*, 2782. [CrossRef]

44. Lentz, G.; Connelly, S.; Mirosa, M.; Jowett, T. Gauging attitudes and behaviours: Meat consumption and potential reduction. *Appetite* **2018**, *127*, 230–241. [CrossRef]

45. Román, S.; Sánchez-Siles, L.M.; Siegrist, M. The importance of food naturalness for consumers: Results of a systematic review. *Trends Food Sci. Technol.* **2017**, *67*, 44–57. [CrossRef]

46. Janssen, M. Determinants of organic food purchases: Evidence from household panel data. *Food Qual. Prefer.* **2018**, *68*, 19–28. [CrossRef]

47. Alcorta, A.; Porta, A.; Tárrega, A.; Alvarez, M.D.; Vaquero, M.P. Foods for Plant-Based Diets: Challenges and Innovations. *Foods* **2021**, *10*, 293. [CrossRef] [PubMed]

48. Rondoni, A.; Millan, E.; Asioli, D. Plant-based Eggs: Views of Industry Practitioners and Experts. *J. Int. Food Agribus. Mark.* **2021**, 1–24. [CrossRef]

49. Rondoni, A.; Millan, E.; Asioli, D. Consumers' preferences for intrinsic and extrinsic product attributes of plant-based eggs: An exploratory study in the United Kingdom and Italy. *Br. Food J.* **2021**, *123*, 3704–3725. [CrossRef]

50. FutureMarketInsights Egg Replacement Ingredient Market Analysis and Review 2019–2026 | Future Market Insights (FMI). Available online: https://www.futuremarketinsights.com/reports/egg-replacement-ingredient-market (accessed on 26 October 2021).

51. Mintel How Plant-Based Eggs will Crack into Mainstream Food—Mintel. Available online: https://clients.mintel.com/insight/how-plant-based-eggs-will-crack-into-mainstream-food?fromSearch=%3Ffreetext%3Dvegan%2520egg (accessed on 28 October 2021).

52. Boukid, F.; Zannini, E.; Carini, E.; Vittadini, E. Pulses for bread fortification: A necessity or a choice? *Trends Food Sci. Technol.* **2019**, *88*, 416–428. [CrossRef]

53. Boukid, F.; Rosell, C.M.; Castellari, M. Pea protein ingredients: A mainstream ingredient to (re)formulate innovative foods and beverages. *Trends Food Sci. Technol.* **2021**, *110*, 729–742. [CrossRef]

54. Boukid, F.; Pasqualone, A. Lupine (*Lupinus* spp.) proteins: Characteristics, safety and food applications. *Eur. Food Res. Technol.* **2021**, *1*, 3. [CrossRef]

55. Boukid, F. Chickpea (*Cicer arietinum* L.) protein as a prospective plant-based ingredient: A review. *Int. J. Food Sci. Technol.* **2021**, *56*, 5435–5444. [CrossRef]

56. Ladjal-Ettoumi, Y.; Boudries, H.; Chibane, M.; Romero, A. Pea, Chickpea and Lentil Protein Isolates: Physicochemical Characterization and Emulsifying Properties. *Food Biophys.* **2016**, *11*, 43–51. [CrossRef]

57. Rachwa-Rosiak, D.; Nebesny, E.; Budryn, G. Chickpeas—Composition, Nutritional Value, Health Benefits, Application to Bread and Snacks: A Review. *Crit. Rev. Food Sci. Nutr.* **2015**, *55*, 1137–1145. [CrossRef] [PubMed]

58. Moreno, H.M.; Herranz, B.; Borderías, A.J.; Tovar, C.A. Effect of high pressure treatment on the structural, mechanical and rheological properties of glucomannan gels. *Food Hydrocoll.* **2016**, *60*, 437–444. [CrossRef]

59. Gharibzahedi, S.M.T.; Roohinejad, S.; George, S.; Barba, F.J.; Greiner, R.; Barbosa-Cánovas, G.V.; Mallikarjunan, K. Innovative food processing technologies on the transglutaminase functionality in protein-based food products: Trends, opportunities and drawbacks. *Trends Food Sci. Technol.* **2018**, *75*, 194–205. [CrossRef]

60. Cai, Q.; Zhang, W.J.; Zhu, Q.Q.; Chen, Q. Influence of heat treatment on the structure and core IgE-binding epitopes of rAra h 2.02. *Food Chem.* **2016**, *202*, 404–408. [CrossRef]

61. Ma, Z.; Boye, J.I.; Hu, X. Nutritional quality and techno-functional changes in raw, germinated and fermented yellow field pea (*Pisum sativum* L.) upon pasteurization. *LWT-Food Sci. Technol.* **2018**, *92*, 147–154. [CrossRef]

62. Damian, J.J.; Huo, S.; Serventi, L. Phytochemical content and emulsifying ability of pulses cooking water. *Eur. Food Res. Technol.* **2018**, *244*, 1647–1655. [CrossRef]

63. Fischer, E.; Cachon, R.; Cayot, N. Pisum sativum vs Glycine max, a comparative review of nutritional, physicochemical, and sensory properties for food uses. *Trends Food Sci. Technol.* **2020**, *95*, 196–204. [CrossRef]

64. Trikusuma, M.; Paravisini, L.; Peterson, D.G. Identification of aroma compounds in pea protein UHT beverages. *Food Chem.* **2020**, *312*, 126082. [CrossRef] [PubMed]

65. Ali, R.; Saeed, S.M.G.; Ali, S.A.; Sayed, S.A.; Ahmed, R.; Mobin, L. Effect of black gram flour as egg replacer on microstructure of biscuit dough and its impact on edible qualities. *J. Food Meas. Charact.* **2018**, *12*, 1641–1647. [CrossRef]

66. Martinez, M.M.; Boukid, F. Future-Proofing Dietary Pea Starch. *ACS Food Sci. Technol.* **2021**, *1*, 1371–1372. [CrossRef]

67. Ashogbon, A.O.; Akintayo, E.T.; Oladebeye, A.O.; Oluwafemi, A.D.; Akinsola, A.F.; Imanah, O.E. Developments in the isolation, composition, and physicochemical properties of legume starches. *Crit. Rev. Food Sci. Nutr.* **2021**, *61*, 2938–2959. [CrossRef]

68. Dong, H.; Vasanthan, T. Amylase resistance of corn, faba bean, and field pea starches as influenced by three different phosphorylation (cross-linking) techniques. *Food Hydrocoll.* **2020**, *101*, 105506. [CrossRef]

69. Yildiz, G.; Ding, J.; Andrade, J.; Engeseth, N.J.; Feng, H. Effect of plant protein-polysaccharide complexes produced by mano-thermo-sonication and pH-shifting on the structure and stability of oil-in-water emulsions. *Innov. Food Sci. Emerg. Technol.* **2018**, *47*, 317–325. [CrossRef]

70. Saget, S.; Costa, M.; Styles, D.; Williams, M. Does Circular Reuse of Chickpea Cooking Water to Produce Vegan Mayonnaise Reduce Environmental Impact Compared with Egg Mayonnaise? *Sustainability* **2021**, *13*, 4726. [CrossRef]

71. Yazici, G.N.; Ozer, M.S. A review of egg replacement in cake production: Effects on batter and cake properties. *Trends Food Sci. Technol.* **2021**, *111*, 346–359. [CrossRef]

72. Serventi, L. *Upcycling Legume Water: From Wastewater to Food Ingredients*; Springer Nature: Berlin/Heidelberg, Germany, 2020; pp. 1–174. [CrossRef]

73. Stantiall, S.E.; Dale, K.J.; Calizo, F.S.; Serventi, L. Application of pulses cooking water as functional ingredients: The foaming and gelling abilities. *Eur. Food Res. Technol.* **2018**, *244*, 97–104. [CrossRef]

74. Mustafa, R.; He, Y.; Shim, Y.Y.; Reaney, M.J.T. Aquafaba, wastewater from chickpea canning, functions as an egg replacer in sponge cake. *Int. J. Food Sci. Technol.* **2018**, *53*, 2247–2255. [CrossRef]

75. He, Y.; Purdy, S.K.; Tse, T.J.; Tar'an, B.; Meda, V.; Reaney, M.J.T.; Mustafa, R. Standardization of Aquafaba Production and Application in Vegan Mayonnaise Analogs. *Foods* **2021**, *10*, 1978. [CrossRef]

76. Shim, Y.Y.; Mustafa, R.; Shen, J.; Ratanapariyanuch, K.; Reaney, M.J.T. Composition and properties of aquafaba: Water recovered from commercially canned chickpeas. *J. Vis. Exp.* **2018**, *132*, e56305. [CrossRef]

77. Hedayatnia, S.; Tan, C.P.; Joanne Kam, W.L.; Tan, T.B.; Mirhosseini, H. Modification of physicochemical and mechanical properties of a new bio-based gelatin composite films through composition adjustment and instantizing process. *LWT* **2019**, *116*, 108575. [CrossRef]

78. Larrosa, V.; Lorenzo, G.; Zaritzky, N.; Califano, A. Dynamic rheological analysis of gluten-free pasta as affected by composition and cooking time. *J. Food Eng.* **2015**, *160*, 11–18. [CrossRef]

79. Varela, M.S.; Navarro, A.S.; Yamul, D.K. Effect of hydrocolloids on the properties of wheat/potato starch mixtures. *Starch/Staerke* **2016**, *68*, 753–761. [CrossRef]

80. Li, H.; Zhu, K.; Zhou, H.; Peng, W.; Guo, X. Comparative study of four physical approaches about allergenicity of soybean protein isolate for infant formula. *Food Agric. Immunol.* **2016**, *27*, 604–623. [CrossRef]

81. Lam, A.C.Y.; Can Karaca, A.; Tyler, R.T.; Nickerson, M.T. Pea protein isolates: Structure, extraction, and functionality. *Food Rev. Int.* **2018**, *34*, 126–147. [CrossRef]

82. Boukid, F. Oat proteins as emerging ingredients for food formulation: Where we stand? *Eur. Food Res. Technol.* **2021**, *247*, 535–544. [CrossRef]

83. Yi-Shen, Z.; Shuai, S.; Fitzgerald, R. Mung bean proteins and peptides: Nutritional, functional and bioactive properties. *Food Nutr. Res.* **2018**, *62*, 1–11. [CrossRef]

84. Jeske, S.; Bez, J.; Arendt, E.K.; Zannini, E. Formation, stability, and sensory characteristics of a lentil-based milk substitute as affected by homogenisation and pasteurisation. *Eur. Food Res. Technol.* **2019**, *245*, 1519–1531. [CrossRef]

85. Boukid, F.; Castellari, M. Food and Beverages Containing Algae and Derived Ingredients Launched in the Market from 2015 to 2019: A Front-of-Pack Labeling Perspective with a Special Focus on Spain. *Foods* **2021**, *10*, 173. [CrossRef]

86. Lafarga, T.; Acién-Fernández, F.G.; Garcia-Vaquero, M. Bioactive peptides and carbohydrates from seaweed for food applications: Natural occurrence, isolation, purification, and identification. *Algal Res.* **2020**, *48*, 101909. [CrossRef]

87. Edelmann, M.; Aalto, S.; Chamlagain, B.; Kariluoto, S.; Piironen, V. Riboflavin, niacin, folate and vitamin B12 in commercial microalgae powders. *J. Food Compos. Anal.* **2019**, *82*, 103226. [CrossRef]

88. Ahmmed, M.K.; Ahmmed, F.; Tian, H.; Carne, A.; Bekhit, A.E.D. Marine omega-3 (n-3) phospholipids: A comprehensive review of their properties, sources, bioavailability, and relation to brain health. *Compr. Rev. Food Sci. Food Saf.* **2020**, *19*, 64–123. [CrossRef]

89. Ferreira, G.F.; Pessoa, J.G.B.; Ríos Pinto, L.F.; Maciel Filho, R.; Fregolente, L.V. Mono- and diglyceride production from microalgae: Challenges and prospects of high-value emulsifiers. *Trends Food Sci. Technol.* **2021**, *118*, 589–600. [CrossRef]

90. Ismail, B.P.; Senaratne-Lenagala, L.; Stube, A.; Brackenridge, A. Protein demand: Review of plant and animal proteins used in alternative protein product development and production. *Anim. Front. Rev. Mag. Anim. Agric.* **2020**, *10*, 53. [CrossRef] [PubMed]

91. Kuang, H.; Yang, F.; Zhang, Y.; Wang, T.; Chen, G. The Impact of Egg Nutrient Composition and Its Consumption on Cholesterol Homeostasis. *Cholesterol* **2018**, *2018*, 6303810. [CrossRef]

92. Miranda, J.M.; Anton, X.; Redondo-Valbuena, C.; Roca-Saavedra, P.; Rodriguez, J.A.; Lamas, A.; Franco, C.M.; Cepeda, A. Egg and egg-derived foods: Effects on human health and use as functional foods. *Nutrients* **2015**, *7*, 706–729. [CrossRef] [PubMed]

93. McClements, D.J.; Grossmann, L. A brief review of the science behind the design of healthy and sustainable plant-based foods. *npj Sci. Food* **2021**, *5*, 17. [CrossRef]

94. Carcelli, A.; Crisafulli, G.; Carini, E.; Vittadini, E. Can a physically modified corn flour be used as fat replacer in a mayonnaise? *Eur. Food Res. Technol.* **2020**, *246*, 2493–2503. [CrossRef]

95. Sun, C.; Liu, R.; Liang, B.; Wu, T.; Sui, W.; Zhang, M. Microparticulated whey protein-pectin complex: A texture-controllable gel for low-fat mayonnaise. *Food Res. Int.* **2018**, *108*, 151–160. [CrossRef] [PubMed]

96. Villarreal, M.B.; Gallardo Rivera, C.T.; Márquez, E.G.; Rodríguez, J.R.; González, M.A.N.; Montes, A.C.; Báez González, J.G. Comparative Reduction of Egg Yolk Cholesterol Using Anionic Chelating Agents. *Molecules* **2018**, *23*, 3204. [CrossRef]

97. Ghazaei, S.; Mizani, M.; Piravi-Vanak, Z.; Alimi, M. Particle size and cholesterol content of a mayonnaise formulated by OSA-modified potato starch. *Food Sci. Technol.* **2015**, *35*, 150–156. [CrossRef]

98. Ali, M.R.; EL Said, R.M. Assessment of the potential of Arabic gum as an antimicrobial and antioxidant agent in developing vegan "egg-free" mayonnaise. *J. Food Saf.* **2020**, *40*, 40. [CrossRef]

99. Sridharan, S.; Meinders, M.B.J.; Bitter, J.H.; Nikiforidis, C.V. Pea flour as stabilizer of oil-in-water emulsions: Protein purification unnecessary. *Food Hydrocoll.* **2020**, *101*, 105533. [CrossRef]

100. Yerramilli, M.; Longmore, N.; Ghosh, S. Improved stabilization of nanoemulsions by partial replacement of sodium caseinate with pea protein isolate. *Food Hydrocoll.* **2017**, *64*, 99–111. [CrossRef]

101. Garcia, K.; Sriwattana, S.; No, H.K.; Corredor, J.A.H.; Prinyawiwatkul, W. Sensory optimization of a mayonnaise-type spread made with rice bran oil and soy protein. *J. Food Sci.* **2009**, *74*, S248–S254. [CrossRef] [PubMed]

102. Nikzade, V.; Tehrani, M.M.; Saadatmand-Tarzjan, M. Optimization of low-cholesterol-low-fat mayonnaise formulation: Effect of using soy milk and some stabilizer by a mixture design approach. *Food Hydrocoll.* **2012**, *28*, 344–352. [CrossRef]

103. Rahmati, K.; Mazaheri Tehrani, M.; Daneshvar, K. Soy milk as an emulsifier in mayonnaise: Physico-chemical, stability and sensory evaluation. *J. Food Sci. Technol.* **2014**, *51*, 3341–3347. [CrossRef]

104. Mozafari, H.R.; Hosseini, E.; Hojjatoleslamy, M.; Mohebbi, G.H.; Jannati, N. Optimization low-fat and low cholesterol mayonnaise production by central composite design. *J. Food Sci. Technol.* **2017**, *54*, 591. [CrossRef]

105. Karshenas, M.; Goli, M.; Zamindar, N. The effect of replacing egg yolk with sesame–peanut defatted meal milk on the physico-chemical, colorimetry, and rheological properties of low-cholesterol mayonnaise. *Food Sci. Nutr.* **2018**, *6*, 824. [CrossRef]

106. Raikos, V.; Hayes, H.; Ni, H. Aquafaba from commercially canned chickpeas as potential egg replacer for the development of vegan mayonnaise: Recipe optimisation and storage stability. *Int. J. Food Sci. Technol.* **2020**, *55*, 1935–1942. [CrossRef]

107. Timilsena, Y.P.; Adhikari, R.; Barrow, C.J.; Adhikari, B. Physicochemical and functional properties of protein isolate produced from Australian chia seeds. *Food Chem.* **2016**, *212*, 648–656. [CrossRef] [PubMed]

108. Timilsena, Y.P.; Wang, B.; Adhikari, R.; Adhikari, B. Preparation and characterization of chia seed protein isolate-chia seed gum complex coacervates. *Food Hydrocoll.* **2015**, *52*, 554–563. [CrossRef]

109. Fernandes, S.S.; Mellado, M.D.L.M.S. Development of Mayonnaise with Substitution of Oil or Egg Yolk by the Addition of Chia (*Salvia Hispânica* L.) Mucilage. *J. Food Sci.* **2018**, *83*, 74–83. [CrossRef]

110. Hashemi, M.M.; Aminlari, M.; Forouzan, M.M.; Moghimi, E.; Tavana, M.; Shekarforoush, S.; Mohammadifar, M.A. Production and Application of lysozyme-Gum Arabic Conjugate in Mayonnaise as a natural Preservative and Emulsifier. *Pol. J. Food Nutr. Sci.* **2018**, *68*, 33–43. [CrossRef]

111. Cornelia, M.; Siratantri, T.; Prawita, R. The Utilization of Extract Durian (*Durio zibethinus* L.) Seed Gum as an Emulsifier in Vegan Mayonnaise. *Procedia Food Sci.* **2015**, *3*, 1–18. [CrossRef]

112. El-Razik, M.A.; Mohamed, A. Utilization of Acid Casein Curd Enriched with Chlorella vulgaris Biomass as Substitute of Egg in Mayonnaise Production. *World Appl. Sci. J.* **2013**, *26*, 917–925. [CrossRef]

113. Arozarena, I.; Bertholo, H.; Empis, J.; Bunger, A.; de Sousa, I. Study of the total replacement of egg by white lupine protein, emulsifiers and xanthan gum in yellow cakes. *Eur. Food Res. Technol.* **2014**, *213*, 312–316. [CrossRef]

114. Lin, M.; Tay, S.H.; Yang, H.; Yang, B.; Li, H. Replacement of eggs with soybean protein isolates and polysaccharides to prepare yellow cakes suitable for vegetarians. *Food Chem.* **2017**, *229*, 663–673. [CrossRef] [PubMed]

115. Hedayati, S.; Mazaheri Tehrani, M. Effect of total replacement of egg by soymilk and lecithin on physical properties of batter and cake. *Food Sci. Nutr.* **2018**, *6*, 1154. [CrossRef] [PubMed]

116. Abdul Hussain, S.S. Studying the Possibility of Preparing An Egg-Free Or Egg-Less Cake. *Int. J. Eng. Technol.* **2009**, *1*, 324–329. [CrossRef]

117. Hesarinejad, M.A.; Rezaiyan Attar, F.; Mossaffa, O.; Shokrolahi, B. The effect of incorporation of chlorella vulgaris into cake as an egg white substitute on physical and sensory properties. *Iran. J. Food Sci. Technol.* **2017**, *14*, 61–72.

118. Ashwini, A.; Jyotsna, R.; Indrani, D. Effect of hydrocolloids and emulsifiers on the rheological, microstructural and quality characteristics of eggless cake. *Food Hydrocoll.* **2009**, *23*, 700–707. [CrossRef]

119. Ratnayake, W.S.; Geera, B.; Rybak, D.A. Effects of egg and egg replacers on yellow cake product quality. *J. Food Process. Preserv.* **2012**, *36*, 21–29. [CrossRef]

foods

Article

Standardization of Aquafaba Production and Application in Vegan Mayonnaise Analogs

Yue He [1], Sarah K. Purdy [2], Timothy J. Tse [2], Bunyamin Tar'an [2], Venkatesh Meda [1], Martin J. T. Reaney [2,3,4] and Rana Mustafa [2,4,*]

1 Department of Chemical and Biological Engineering, University of Saskatchewan, Saskatoon, SK S7N 5A9, Canada; yuh885@mail.usask.ca (Y.H.); venkatesh.meda@usask.ca (V.M.)
2 Department of Plant Sciences, University of Saskatchewan, Saskatoon, SK S7N 5A8, Canada; sarah.purdy@usask.ca (S.K.P.); timothy.tse@usask.ca (T.J.T.); bunyamin.taran@usask.ca (B.T.); martin.reaney@usask.ca (M.J.T.R.)
3 Prairie Tide Diversified Inc., 102 Melville Street, Saskatoon, SK S7J 0R1, Canada
4 Guangdong Saskatchewan Oilseed Joint Laboratory, Department of Food Science and Engineering, Jinan University, Guangzhou 510632, China
* Correspondence: rana.mustafa@usask.ca; Tel.: +1-(306)-250-1301

Abstract: Canning or boiling pulse seeds in water produces a by-product solution, called "aquafaba", that can be used as a plant-based emulsifier. One of the major problems facing the commercialization of aquafaba is inconsistency in quality and functionality. In this study, chickpea aquafaba production and drying methods were optimized to produce standardized aquafaba powder. Aquafaba samples, both freeze-dried and spray-dried, were used to make egg-free, vegan mayonnaise. Mayonnaise and analog physicochemical characteristics, microstructure, and stability were tested and compared to mayonnaise prepared using egg yolk. Chickpeas steeped in water at 4 °C for 16 h, followed by cooking at 75 kPa for 30 min at 116 °C, yielded aquafaba that produced the best emulsion qualities. Both lyophilization and spray drying to dehydrate aquafaba resulted in powders that retained their functionality following rehydration. Mayonnaise analogs made with aquafaba powder remained stable for 28 days of storage at 4 °C, although their droplet size was significantly higher than the reference sample made with egg yolk. These results show that aquafaba production can be standardized for optimal emulsion qualities, and dried aquafaba can mimic egg functions in food emulsions and has the potential to produce a wide range of eggless food products.

Keywords: aquafaba; chickpea; emulsifiers; egg replacement; mayonnaise

Citation: He, Y.; Purdy, S.K.; Tse, T.J.; Tar'an, B.; Meda, V.; Reaney, M.J.T.; Mustafa, R. Standardization of Aquafaba Production and Application in Vegan Mayonnaise Analogs. *Foods* 2021, 10, 1978. https://doi.org/10.3390/foods10091978

Academic Editor: Alex Martynenko

Received: 28 June 2021
Accepted: 20 August 2021
Published: 24 August 2021

Publisher's Note: MDPI stays neutral with regard to jurisdictional claims in published maps and institutional affiliations.

1. Introduction

Mayonnaise is a popular semisolid condiment that can improve the texture and flavour of foods such as salads, dips, and sandwiches. In recent years, because of health and environmental concerns, there has been an upward trend towards replacing egg with plant-based ingredients, especially in the formulation of mayonnaise analogs. Plant-based proteins [1,2], soymilk [3–5], starch, and modified starch [6] are reported to function as egg replacers that act as emulsifiers in mayonnaise analogs. To develop a vegan mayonnaise analog, one of the most difficult problems to solve is to create a stable emulsion structure that can withstand prolonged storage without coalescence or flocculation [7–9]. Emulsions are thermodynamically unstable systems, necessitating the use of emulsifiers to improve their storage stability [10].

Chickpea cooking water, commonly known as aquafaba, has recently been utilized as a vegan emulsifier in culinary formulations and as an egg replacement in vegan mayonnaise analogs. Aquafaba's functional properties (emulsibility, foamability, gelation, and thickening properties) are attributed to its composition of protein, water-soluble/insoluble carbohydrates (oligosaccharide, starch, cellulose, hemicellulose, or lignin), polysaccharide-protein complexes, coacervates, saponins, and phenolic compounds [7,8,11–15]. Aquafaba is a by-product of pulse canning/boiling and freezing processes and hummus production. Using aquafaba in food products expands the market for plant-based foods, increases the demand for pulses, and reduces wastewater generated from some bean processes [7]. However, because most boiling and canning processes are designed to produce cooked pulse seed, the quality of aquafaba recovered varies significantly between manufacturers and within batches [16,17]. To assure aquafaba consistency and the quality of products made from it, standardization of aquafaba production is required. Different parameters, such as chickpea cultivars selected for aquafaba production and production conditions (i.e., water to seed ratio, temperature, pressure, time, and additives) should be addressed when standardizing aquafaba composition and functionality [16,18,19]. As aquafaba has a moisture content of more than 90% [16], it is also preferred to concentrate or dry aquafaba to improve transport efficiency by minimizing shipping costs, decreasing space required for storage, and preventing undesirable microbial growth [20]. To our knowledge, the impact of the drying process on the functional qualities of aquafaba has not yet been explored.

The purpose of this study is to standardize aquafaba production and drying processes, as well as to determine conditions that improve aquafaba powder's function as an emulsifier in mayonnaise analog production. The emulsion properties of aquafaba produced from different cooking and drying methods were measured, and the physicochemical characteristics and stability of aquafaba-based mayonnaise analog and egg-based mayonnaise were examined and compared.

2. Materials and Methods

2.1. Materials

In our previous research, the Kabuli chickpea cultivar 'CDC Leader' was identified to produce a more favourable aquafaba with superior emulsion properties when compared with aquafaba produced from other cultivars [16]. Chickpea seeds (CDC Leader) were generously provided by Dr. Bunyamin Tar'an from the University of Saskatchewan, Crop Development Centre (CDC) (Saskatoon, SK, Canada). Chickpea seed was manually cleaned to remove broken seed, dust, and other foreign materials. Canola oil (purity 100%; ACH Food Companies, Inc., Terrace, IL, USA), eggs (Great Value large sized, Canada), baking soda (NaHCO$_3$; Arm & Hammer by Church & Dwight Co., Inc, Mississauga, ON, Canada), and table salt (Windsor Salt, Pointe-Claire, QC, Canada) were purchased from a local supermarket (Walmart, Saskatoon, SK, Canada). Whole eggs were kept refrigerated at 4 °C, and before preparing mayonnaise, the yolk was separated from the egg white using an egg separator. White vinegar (No Name, Loblaws Inc., Toronto, ON, Canada) and sugar (Rogers granulated white sugar, Lantic Inc., Vancouver, BC, Canada) were purchased from a local supermarket (Real Canadian Superstore, Saskatoon, SK, Canada). Sodium dodecyl sulphate (SDS) was purchased from GE Healthcare (Mississauga, ON, Canada), and Nile red pigment was supplied by Sigma-Aldrich (Oakville, ON, Canada).

2.2. Aquafaba Production and Drying Method Standardization

Our goal was to identify the best conditions to produce aquafaba with superior emulsion properties. Tests were conducted in sequence by selecting the top-performing parameters for the next test. First, five aquafaba production conditions were evaluated, and the best aquafaba production conditions were selected based on the emulsifying activity index (EAI) and stability (ES). Second, aquafaba made with the optimized production methods was dried using five different drying methods. To make aquafaba powder, the drying methods that maintained aquafaba functionality were chosen. Third, the resulting aquafaba powder was used as an emulsifier to make analogs for mayonnaise.

2.2.1. Optimization of Aquafaba Production

Dry chickpea seed (approx. 100 g) was washed and rehydrated by soaking in distilled water at a ratio of 1:4 (*w/w*) over time intervals of 1 and 16 h, at temperatures of 4 and 85 °C (Table 1). The soaking water was then discarded. The soaked chickpea seed (100 g) was rinsed with distilled water and combined with 100 mL distilled water with and without 0.2% NaHCO$_3$ in 250 mL sealed glass jars and cooked for different times (20, 30, or 60 min) in a pressure cooker (70–80 kPa, and 115–118 °C; Instant Pot® 7-in-1 multi-use programmable pressure cooker, IP-DUO60 V2, 6 quart/litres, Ottawa, ON, Canada). After cooking, the jars were cooled to room temperature (21 ± 1 °C) for 24 h. Cooled aquafaba was separated from cooked chickpea seed using a stainless-steel strainer and stored in a freezer (−18 °C) until use. Frozen aquafaba was thawed at 4 °C overnight and then warmed to 22 °C for 2 h before use. The aquafaba EAI and ES were measured according to He et al. [16].

Table 1. Aquafaba production conditions.

Condition	A	B	C	D	E
Soaking time (h)	16	16	16	1	1
Soaking temperature (°C)	4	4	4	85	85
Soaking water additives (w/w)	NA	NA	NA	NA	0.2% NaHCO$_3$
Cooking time (min)	20	30	60	30	20

NA—no additives.

2.2.2. Comparison of Drying Methods

Liquid aquafaba sample (750 g) prepared from chickpea seed soaked in 4 °C water for 16 h then cooked for 30 min (condition B, Table 1) was divided into five equal parts (150 g) and dried using five different drying methods: freeze drying, spray drying, oven drying, Rotovap drying, and vacuum pressure drying. Freeze drying was performed by freezing the samples at −20 °C followed by drying in a FreeZone 12 Liter Console Freeze Dryer with Stoppering Tray Dryer (Labconco Corporation, Kansas City, MO, USA) until the sample temperature rose to −5 °C, indicating the sample had been thoroughly dried. Spray drying was completed at 150 °C, using a Büchi Mini Spray Dryer B-290 (Labortechnik AG, Flawil, Switzerland). Oven-dried samples were treated at 80 °C, using a VWR® Signature™ Forced Air Safety Oven (Radnor, PA, USA) until they reached a consistent weight. Samples for Rotovap drying were placed in round-bottom flasks connected to a rotary evaporator (Büchi® Model R-210 BUCHI Labortechnik AG, Switzerland) at 50 °C under vacuum pressure (50–100 mbar). Vacuum-dried samples were heated in a Fisherbrand™ Isotemp™ Model 281A vacuum oven (Fisher Scientific International, Ottawa, ON, Canada) at 60 °C for 12 h, under vacuum (33 mbar).

The residual moisture content of dried aquafaba samples was determined following the American Association of Cereal Chemists (AACC) method 44-15.02 [21]. Aquafaba powder was mixed with water to obtain the same concentration of solid materials as fresh aquafaba [16], and rehydrated aquafaba samples were mixed with canola oil to evaluate aquafaba emulsion properties.

2.3. Aquafaba Water Holding Capacity and Oil Absorption Capacity

Water holding capacity (WHC) and oil absorption capacity (OAC) of aquafaba powder were determined using the method described by Damian et al. (2018) [12] with minor modification. Aquafaba powder (1 g) was mixed with 20 g of distilled water and stirred for 1 min. The solution was then centrifuged for 10 min at 1860× g. After centrifugation, the supernatant was discarded, and the pellet weight was recorded. The WHC values were calculated as a ratio of the pellet weight to sample weight and were expressed as g water/g pulse cooking water (PCW). For OAC, distilled water was replaced with canola oil, and values were expressed as g oil/g PCW.

2.4. Development of Aquafaba Mayonnaise Analogs

Two formulations of egg-free analogs for mayonnaise were prepared using freeze-dried aquafaba and spray-dried aquafaba (analog A and B, respectively). Traditional mayonnaise, made with egg yolk, was used as a reference (mayonnaise C). The mayonnaise and analog formulations were modified from Raikos et al. (2019) [8] and included 80 mL canola oil, 4 mL vinegar (4% acidity), 0.5 g salt, 0.5 g sugar, and 15 g emulsifying agent. Canola oil was slowly added to the aqueous mixture (aquafaba/egg yolk, vinegar, sugar, and salt) and mixed for 5 min using a Kitchen Aid Ultra Power Mixer with a 4.3 L stationary bowl (Kitchen Aid, St. Joseph's, MI, USA). Mayonnaise and mayonnaise analog samples were aliquoted and stored in a refrigerator (4 °C) for further analysis.

2.4.1. Colour and pH

Mayonnaise and analog pH values were measured using a portable food and dairy pH meter (Hanna Instruments Ltd., Leighton Buzzard, UK). The colour characteristics were assessed using a Hunter Lab ColorFlex spectrophotometer (Hunter Associates Laboratory, Inc., Reston, VA, USA). Mayonnaise and analog colour, represented by lightness (L^*), redness/greenness ($\pm a^*$), and yellowness/blueness ($\pm b^*$), was also determined initially after preparation and after 28 days of cold storage (4 °C). Chroma (Ch), colour difference from the control ($\Delta E_1{}^*$), and total colour change ($\Delta E_2{}^*$) of mayonnaise and analog samples during cold storage, were calculated using the following equations [8,18]:

$$Ch = \sqrt{a^{*2} + b^{*2}} \tag{1}$$

$$\Delta E^* = \sqrt{(\Delta L^*)^2 + (\Delta a^*)^2 + (\Delta b^*)^2} \tag{2}$$

2.4.2. Mayonnaise and Analog Stability Test

Mayonnaise and analog samples (F_0 = 10 g) were transferred to 15 mL centrifuge tubes and centrifuged for 30 min at $1860\times g$. The weight of the emulsified fractions (upper layer, F_1) was measured after centrifugation, and emulsion stability (ES, %) was determined by Equation (3):

$$ES = \left(\frac{F_1}{F_0}\right) \times 100\% \tag{3}$$

To measure heat stability, mayonnaise and analog samples were stored in an 80 °C water bath for 30 min before centrifugation. The heat stability was then characterized using Equation (3).

2.4.3. Confocal Laser Scanning Microscopy

Mayonnaise and analog microstructure were analyzed with a Nikon C2 confocal laser scanning microscope (CLSM) (Nikon, Mississauga, ON, Canada) using a 543 nm laser with a $60\times$ Plan-Apochromat VC (numerical aperture 1.4) oil immersion objective lens and five times digital zoom. The oil phase was stained using Nile red dye (0.01 wt.%). A drop of emulsion was placed on a microscope slide (Fisher Scientific, Nepean, ON, Canada) with a glass rod, covered with a coverslip (VWR International, Edmonton, AB, Canada) and observed under the CLSM.

2.4.4. Droplet Size Distribution

Droplet size distribution of mayonnaise and analog samples was measured as a function of time (0, 7, 14, 21, and 28 days) using a static laser diffraction particle analyzer (Mastersizer 2000, Malvern Instrument, Montreal, QC, Canada) with a Hydro 2000S sample dispersion unit (containing water). The dispersion refractive index was 1.33, and the refractive index used for canola oil droplets was 1.47. Drops of samples were added to the sample dispersion unit until the obscuration index reached approximately 15%, and the average droplet size was reported in terms of volume-weighted mean diameter, d_{43}, defined by Equation (4):

$$d_{43} = \frac{\sum n_i d_i^4}{\sum n_i d_i^3} \tag{4}$$

where d_i is the diameter of a droplet and n_i is the number of droplets with the size of d_i.

2.5. Statistical Analysis

Experiments were conducted in triplicate and the data were presented as means ± standard deviation (SD). Graphical illustrations were processed with Microsoft Excel® 2018. Statistical analyses were completed using the Statistical Package for the Social Science (SPSS) version 25.0 (IBM Corp., Armonk, NY, USA). Analysis of variance (ANOVA) and Tukey's post hoc statistical tests were used to evaluate significant differences in aquafaba physicochemical properties and mayonnaise and analog characteristics. Statistical significance was accepted at $p < 0.05$.

3. Results and Discussion

3.1. Optimization of Aquafaba Production

The effects of soaking conditions (1 h at 85 °C; 16 h at 4 °C), cooking time (20, 30, and 60 min), and additive ($NaHCO_3$) on aquafaba EAI and ES are shown in Figure 1. The highest EAI (1.30 ± 0.05 m^2g^{-1}) ($p > 0.05$) was obtained from aquafaba prepared by soaking chickpea seed in 4 °C water for 16 h and cooking for 30 min without additives (condition B, Table 1). The EAI dropped by 27% and 46% when the cooking time increased to 60 min or decreased to 20 min without additives (conditions C and A, 0.944 ± 0.072 m^2g^{-1} and 0.699 ± 0.087 m^2g^{-1}, respectively). Soaking chickpea seed in 85 °C water for 1 h and cooking for 30 min (condition D, Table 1) saved soaking time but reduced the EAI of aquafaba to 0.843 ± 0.099 m^2 g^{-1}. By contrast, adding 0.2% (w/w) $NaHCO_3$ to the soaking water (85 °C) (condition E, Table 1) slightly improved the EAI (1.17 ± 0.06 m^2g^{-1}); the EAI value under these conditions remained significantly lower than that of aquafaba prepared under condition B ($p > 0.05$). Aquafaba prepared under long cooking time conditions (conditions B and C) had comparable emulsion stability ($77.1 \pm 0.5\%$ and $77.5 \pm 1.4\%$, respectively), indicating that no significant difference was observed when prolonging the cooking time from 30 min to 60 min. However, decreasing the cooking time to 20 min (condition A) slightly decreased the aquafaba emulsion stability to $72.0 \pm 2.1\%$.

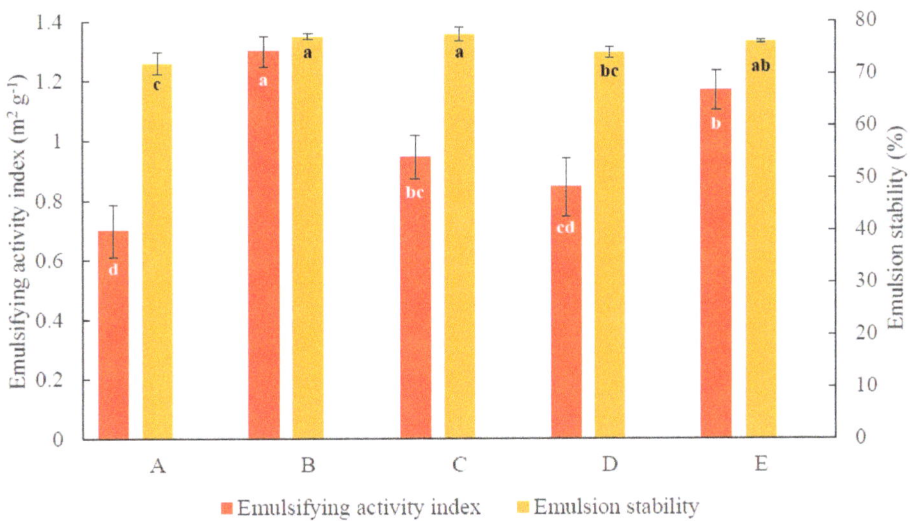

Figure 1. Emulsifying activity index and emulsion stability of aquafaba using different cooking conditions. A: Soaking seed in 4 °C water for 16 h then cooking for 20 min; B: soaking seed in 4 °C water for 16 h then cooking for 30 min; C: soaking seed in 4 °C water for 16 h and cooking for 60 min; D: soaking seed in 85 °C water for 1 h and cooking for 30 min; and E: soaking seed in 85 °C water with 0.2% (w/w) $NaHCO_3$ for 1 h and cooking for 20 min. Different letters (a–d) refer to significant differences among different aquafaba production conditions according to Tukey's test ($p < 0.05$).

The seed component hydration capacity was correlated with soaking time and temperature. During the soaking and cooking process, the outer cell layers of the seed coat transform into a selective membrane that regulates chemical diffusion to the cooking water (aquafaba). Sodium bicarbonate softens the seed coat and cotyledons and increases the concentration of compounds extracted from the seed into the cooking water [7]. Aquafaba emulsion properties are correlated to the concentration of protein, water-soluble/insoluble carbohydrates, polysaccharide–protein complexes, coacervates, saponins, and phenolic compounds [7,22,23]. Proteins in aquafaba, for example, are amphiphilic molecules with a low molecular weight (25 kDa) [17]. These molecules can aggregate at the water–oil interface, lowering the interfacial tension of the solution and forming an intermolecular cohesive film with enough elasticity to stabilize emulsions [7]. Polysaccharides enhance emulsion stability by gelling or changing the viscosity of the aqueous continuous phase, resulting in fewer droplet collisions. Previous research has also shown that the emulsion capacity of aquafaba from various pulses is proportional to their saponin and phenolic compounds concentration. These compounds bind with proteins and polysaccharides, changing their solubility and emulsifying properties [8,23]. During prolonged cooking times, heat and water damage the cell walls in the selective membrane layers, allowing larger molecular compounds to transfer into the aquafaba, lowering its emulsion properties [7]. On the other hand, short cooking time does not sufficiently soften chickpea seed to accelerate the leaching of compounds, limiting aquafaba solid material concentration and emulsion properties.

3.2. Comparison of Drying Methods

Given that aquafaba produced from chickpea 'CDC Leader' seed soaked in 4 °C water for 16 h then cooked for 30 min (condition B) had the highest EAI compared with other conditions, we chose aquafaba production condition B for further experiments. We investigated different drying methods in terms of drying time and aquafaba sensory and functional properties. Images of dried aquafaba powder prepared by different drying methods are provided in Figure 2. Freeze drying and spray drying methods resulted in a bright white and pale-yellow powder, respectively (Figure 2A,B). On the other hand, oven-dried aquafaba (Figure 2C) changed colour from pale yellow to dark brown, and its texture became brittle. Meanwhile, aquafaba dried via Rotovap drying (Figure 2D) resulted in a thick rubbery gel that adhered to the evaporator flask. Drying aquafaba using vacuum drying was slow and resulted in a rubbery sheet that had a higher moisture content compared with aquafaba dried using other methods (Figure 2E). Both freeze drying and spray drying methods produced aquafaba powders that would be preferable for home and industrial use because of their attractive appearance and good water solubility. However, freeze drying is not typically preferred on an industrial scale because of the high capital cost of equipment and the requirement for large amounts of energy.

The amount of water removed and the drying time for different drying methods were calculated and are reported in Table 2. Spray drying removed the largest amount of water (95.0 ± 0.03 g) in the shortest amount of time (0.287 ± 0.001 h) compared with other methods.

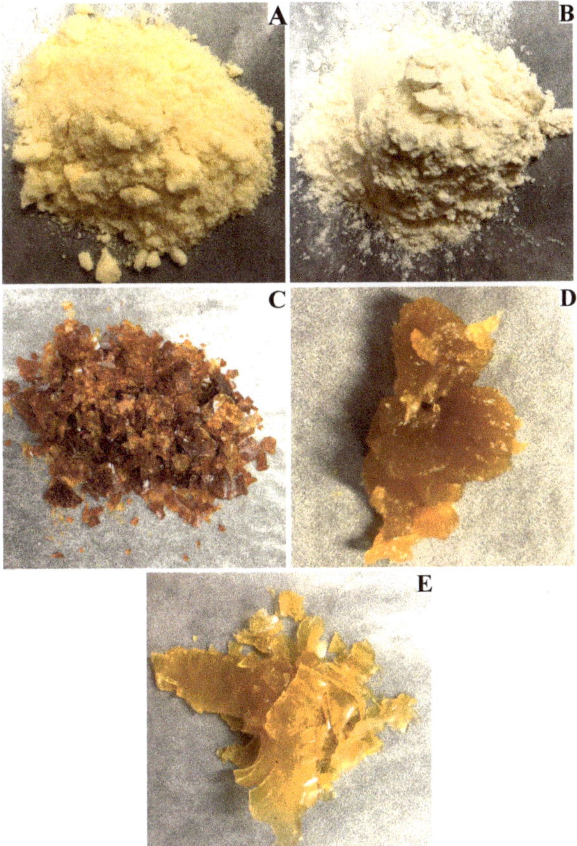

Figure 2. Aquafaba samples prepared by different drying methods: (**A**) freeze drying; (**B**) spray drying; (**C**) oven drying; (**D**) Rotovap drying; and (**E**) vacuum drying.

Table 2. Water removed, drying time, and dried aquafaba yield of different drying methods for 100 g fresh aquafaba and water added to rehydrate aquafaba.

Drying Methods	Water Removed (g)	Drying Time (h)	Dried Aquafaba Yield (g/100 g Fresh Aquafaba)	Water Added (g/10 g Dried Aquafaba)
Spray drying	95.0 ± 0.03 [a]	0.287 ± 0.001 [d]	5.01 ± 0.03 [e]	190
Freeze drying	92.9 ± 0.04 [b]	129 ± 5 [a]	7.06 ± 0.04 [d]	132
Oven drying	92.8 ± 0.06 [c]	29.0 ± 2.1 [c]	7.22 ± 0.06 [c]	129
Rotovap drying	91.2 ± 0.09 [e]	3.22 ± 0.08 [d]	8.78 ± 0.09 [a]	104
Vacuum drying	92.6 ± 0.01 [d]	45.6 ± 1.5 [b]	7.37 ± 0.01 [b]	126

Data are expressed as means $\pm$ standard deviation ($n = 3$). Different letters ([a–e]) refer to significant differences among different drying methods according to Tukey's test ($p < 0.05$).

The yields of dried aquafaba obtained by different drying methods are presented in Table 2. Rotovap drying provided the highest aquafaba yield by weight ($8.78 \pm 0.09\%$) ($p > 0.05$), but the final product had gel-like properties, which could be attributed to Maillard and caramelization reactions caused by the high temperature (50 °C) used in this drying method. The yield of aquafaba dried by vacuum drying was $7.37 \pm 0.01\%$, followed by oven drying ($7.22 \pm 0.06\%$) and freeze drying ($7.06 \pm 0.04\%$). Spray-dried aquafaba had the lowest yield ($5.01 \pm 0.03\%$) ($p > 0.05$) because of sample loss in the lab-scale spray dryer. Some parts of the resulting powder were not properly recovered from the spray dryer because it adhered to the spray dryer parts. These losses would become negligible in a commercial spray dryer.

The dried aquafaba was rehydrated using distilled water to obtain the same concentration of solid materials as fresh aquafaba [16] and mixed with canola oil to make emulsions. Figure 3 shows the emulsion properties of rehydrated aquafaba dried by different drying methods. Spray-dried aquafaba demonstrated comparable EAI (1.26 ± 0.07 m^2 g^{-1}) to freshly prepared aquafaba (1.30 ± 0.05 m^2 g^{-1}) ($p > 0.05$). There were no significant differences between the EAIs of spray-dried (1.30 ± 0.05 m^2 g^{-1}), freeze-dried (1.09 ± 0.06 m^2 g^{-1}), and vacuum-dried samples (1.10 ± 0.04 m^2 g^{-1}) ($p > 0.05$). Oven-dried (1.08 ± 0.03 m^2 g^{-1}) and Rotovap dried (0.942 ± 0.168 m^2 g^{-1}) aquafaba samples showed significantly lower EAIs compared with spray-dried and freshly prepared aquafaba.

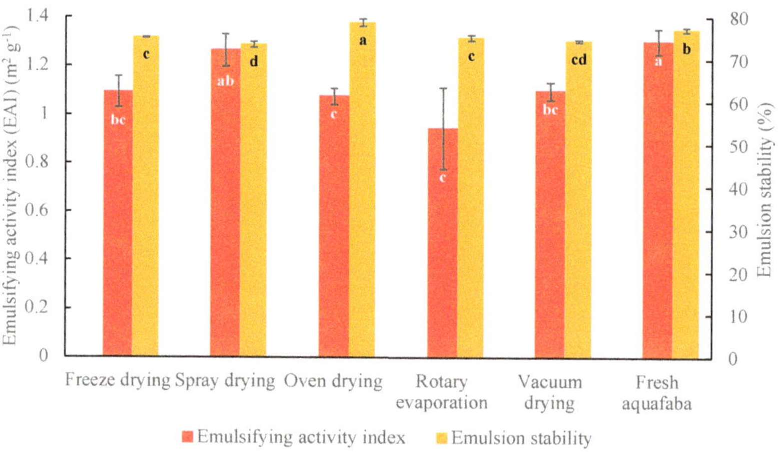

Figure 3. Emulsifying activity index and emulsion stability of rehydrated aquafaba dried by different drying methods. Different letters (a–d) refer to significant differences among different drying methods according to Tukey's test ($p < 0.05$).

Oven-dried aquafaba produced emulsions with the highest ES ($78.8 \pm 0.9\%$) and demonstrated slightly higher stability compared with fresh aquafaba emulsion ($77.1 \pm 0.5\%$) ($p > 0.05$) [16]. The formation of covalent conjugates between proteins and polysaccharides during drying for more than 12 h at 80 °C because of oxidation and thermal-induced reactions (Maillard and caramelization) of aquafaba components (e.g., polysaccharide and protein) may have contributed to higher emulsion stability [7]. However, there was evidence of unsatisfactory browning in oven-dried samples. The ES of freeze-dried ($75.2 \pm 0.1\%$), Rotovap dried ($75.1 \pm 0.7\%$) and vacuum-dried ($74.4 \pm 0.3\%$) samples did not differ significantly. Spray-dried and vacuum-dried aquafaba samples had similar ES, but the former showed a slightly lower ES ($73.6 \pm 0.7\%$) compared with other dried samples.

Freeze drying and spray drying methods were selected for further experiments on the basis of the yield of aquafaba powder, solubility in water, EAI, and ES.

3.3. Aquafaba Water Holding Capacity and Oil Absorption Capacity

Freeze-dried powder of aquafaba prepared under condition B demonstrated higher WHC (4.36 ± 0.20 g/g) and OAC (4.6 ± 0.26 g/g) than spray-dried aquafaba (WHC, 1.92 ± 0.09 g/g; OAC, 1.98 ± 0.12 g/g) ($p > 0.05$) (Table 3). Interestingly, Damian et al. [12] observed significantly lower WHC (1.5 g/g) and OAC (3.2 g/g) of freeze-dried chickpea aquafaba. Two main factors affecting these contradictory observations include various soaking and cooking conditions and differences in aquafaba composition and concentration. As the cooking time progresses, protein denaturation occurs, resulting in hydrophobic molecular regions becoming exposed and thus increasing the oil binding capacity, thereby changing its OAC and WAC properties. In addition, cooking under pressure has been shown to cause protein dissociation, exposing more water/oil-binding sites and increasing both WAC and OAC [24]. Comparatively, in Damian et al.'s [12] study, chickpea seed was boiled in water for 90 min, compared with 30 min in our study. Alsalman et al. [19] indicated that increasing cooking time from 15 to 60 min significantly reduced aquafaba WHC (from 2.4 g/g to 1.6 g/g), supporting the improved WHC observed with this shorter cooking time. In addition, they observed that aquafaba OAC increased with longer cooking time and higher chickpea/water ratio. Although the cooking time in our study was shorter, the higher chickpea/water ratio (1:1 vs. 1:1.75) might play a dominant role in increasing OAC.

Table 3. Physicochemical properties of freeze-dried and spray-dried aquafaba.

Dried Aquafaba	Freeze-Dried Aquafaba	Spray-Dried Aquafaba
Moisture content (%)	5.17 ± 0.21 [a]	2.50 ± 0.01 [b]
WHC (g/g)	4.36 ± 0.20 [a]	1.92 ± 0.09 [b]
OAC (g/g)	4.64 ± 0.26 [a]	1.98 ± 0.12 [b]

WHC, water holding capacity; OAC, oil absorption capacity. Data are expressed as means ± standard deviation ($n = 3$). Different letters ([a,b]) refer to significant differences according to Tukey's test ($p < 0.05$).

3.4. Mayonnaise and Analog Stability during Cold Storage

The stability of mayonnaise and analogs was evaluated by studying their microstructure and particle size distribution. The emulsion and heating stability are represented in Table 4. Freshly prepared analogs had 15% lower emulsion stability than egg yolk mayonnaise. The stability of egg yolk mayonnaise remained stable after 28 days of storage, while no significant differences in emulsion stability were observed for either analog A or B (73–85 %) stored up to 21 days, after which the emulsion stability of mayonnaise analog B decreased to 56%.

Table 4. Physicochemical properties of mayonnaise and analogs over 28 days of storage at 4 °C.

Mayonnaise and Analog	A	B	C
Emulsifier	Freeze-Dried Aquafaba (FA)	Spray-Dried Aquafaba (SA)	Egg Yolk (EY)
Day 0			
pH	3.99 ± 0.17 [Ab]	3.74 ± 0.10 [Bb]	4.66 ± 0.07 [Aa]
L^*	87.6 ± 0.03 [Ab]	85.6 ± 0.04 [Ac]	90.6 ± 0.1 [Aa]
a^*	−2.31 ± 0.02 [Bb]	−2.17 ± 0.01 [Bb]	−1.85 ± 0.02 [Ac]
b^*	12.8 ± 0.02 [Bc]	14.8 ± 0.05 [Bb]	20.8 ± 0.09 [Ba]
Ch	13.0 ± 0.03 [Bc]	14.9 ± 0.05 [Bb]	20.9 ± 0.09 [Ba]
ΔE_1^*	8.56 ± 0.13 [a]	7.82 ± 0.15 [b]	
Emulsion stability	85.0 ± 3.2 [Ab]	84.6 ± 2.0 [Ab]	100 ± 0 [Aa]
Heating stability	68.3 ± 5.0 [Aa]	62.8 ± 1.7 [Bab]	59.4 ± 1.0 [ABb]
Day 7			
pH	4.05 ± 0.03 [Ab]	4.00 ± 0.10 [Ab]	4.50 ± 0.06 [Ba]
Emulsion stability	83.4 ± 4.3 [ABb]	83.6 ± 1.6 [Ab]	100 ± 0 [Aa]
Heating stability	70.2 ± 2.4 [Ab]	76.2 ± 0.8 [Aa]	61.6 ± 1.3 [Ac]
Day 14			
pH	4.07 ± 0.02 [Ab]	4.04 ± 0.07 [Ab]	4.42 ± 0.01 [Ba]
Emulsion stability	82.9 ± 4.2 [ABb]	76.2 ± 3.7 [ABb]	100 ± 0 [Aa]
Heating stability	62.8 ± 8.0 [Aa]	58.7 ± 7.1 [Ba]	48.8 ± 1.4 [BCa]

Table 4. *Cont.*

Mayonnaise and Analog	A	B	C
Emulsifier	**Freeze-Dried Aquafaba (FA)**	**Spray-Dried Aquafaba (SA)**	**Egg Yolk (EY)**
Day 21			
pH	4.02 ± 0.03 [Ab]	3.98 ± 0.05 [Ab]	4.40 ± 0.01 [Ba]
Emulsion stability	75.4 ± 1.4 [Bb]	73.3 ± 2.3 [Bb]	100 ± 0 [Aa]
Heating stability	66.5 ± 5.2 [Aa]	57.6 ± 1.4 [Bab]	48.3 ± 8.1 [Cb]
Day 28			
pH	4.05 ± 0.09 [Ab]	3.94 ± 0.03 [Ab]	4.46 ± 0.02 [Ba]
L^*	82.4 ± 0.2 [Bb]	81.7 ± 0.01 [Bc]	88.2 ± 0.09 [Ba]
a^*	-2.67 ± 0.02 [Aa]	-2.33 ± 0.01 [Ab]	-0.587 ± 0.040 [Bc]
b^*	13.7 ± 0.08 [Ac]	17.0 ± 0.04 [Ab]	25.4 ± 0.1 [Aa]
Ch	14.0 ± 0.08 [Ac]	17.1 ± 0.04 [Ab]	25.4 ± 0.1 [Aa]
ΔE_1^*	13.2 ± 0.3 [a]	10.8 ± 0.2 [b]	
ΔE_2^*	5.30 ± 0.23 [a]	4.50 ± 0.04 [b]	5.35 ± 0.08 [a]
Emulsion stability	79.7 ± 2.7 [ABb]	55.9 ± 5.8 [Cc]	100 ± 0 [Aa]
Heating stability	66.4 ± 5.7 [Aa]	54.6 ± 3.9 [Bb]	48.5 ± 3.8 [BCb]

Ch, chroma; ΔE_1^*, colour difference from the control; ΔE_2^*, colour difference of mayonnaise and analog samples during cold storage. Data are expressed as means $\pm$ standard deviation (n = 3). Different letters ([a–c]) refer to significant differences according to Tukey's test ($p < 0.05$). Lower case letters show significant differences among different emulsifiers. Capital letters show significant differences among storage time.

Analog sample A, made from freeze-dried powder, exhibited similar heating stability compared with analog B (from spray-dried powder) on day 0, day 14, and day 21, and higher heating stability on day 28 ($p > 0.05$). The heating stability of mayonnaise analogs was higher compared with egg yolk mayonnaise. Aquafaba contains heat-stable proteins [17], which might contribute to heating stability. These results are comparable to the emulsion stability and heating stability of other egg-free mayonnaises made from both mono- and diglycerides emulsifier (MDG) and guar gum (GG)/xanthan gum (XG) or a mixture of MDG, GG, and XG [4].

To explain the previous results, the long-term stability of mayonnaise and our analogs were also evaluated in terms of pH and colour. The analog samples were acidic (pH ranging from 3.74 to 4.66) compared with egg yolk mayonnaise, and the pH of all samples remained stable up to 28 days (Table 4). The colour profiles of aquafaba mayonnaise analogs were similar, regardless of the drying method; however, a difference was observed between mayonnaise and the analogs ($\Delta E_1^* > 3$) [18]. Samples of freshly prepared mayonnaise analog A and B had lower L^* (87.6, 85.6 vs. 90.6) and *Ch* values (13.0, 14.9 vs. 20.9) ($p < 0.05$) when compared with mayonnaise C, suggesting a darker appearance and a lower colour intensity. Previous research revealed that emulsion colour can change from gray to an increasingly bright white colour with decreasing droplet size, likely due to an increase in light scattering [25–27]. This was further confirmed by the larger droplet sizes observed in analog samples A and B. Comparatively, mayonnaise (C) had the highest b^* value because of the higher content of pigments in the egg yolk. Nonetheless, ΔE_2^* between freshly prepared mayonnaise and analog, along with those that were stored, were not significantly different, indicating that samples A, B, and C all remained stable during storage. Altogether, the colour differences between mayonnaise and analog (ΔE_1^*) and colour change after storage (ΔE_2^*) were similar to those of aquafaba mayonnaise analogs with three different aquafaba-to-oil ratios in previous studies [8,18].

3.5. Mayonnaise and Analog Microstructure

Confocal laser scanning micrographs of freshly prepared mayonnaise, C, and analogs, A and B, are illustrated in Figure 4. The microstructures of analogs made with both freeze-dried and spray-dried aquafaba were similar, although the interspace voids of analog B (Figure 4B) were slightly larger than in the case of analog A (Figure 4A). The droplets of these aquafaba mayonnaise analog samples were densely packed and revealed polydisperse (oil droplets of different sizes) features. Aquafaba mayonnaise analogs consisted of a large fraction of irregular elliptic oil droplets with greater diameters and a small fraction of spherical oil droplets with smaller diameters. Mayonnaise C (Figure 4C) consisted of finely dispersed, evenly distributed, and significantly smaller spherical oil droplets. It is possible that the large oil droplets became distorted from spherical shape because of the effect of high oil content (80%) and close packing [8]. Moreover, the tightly packed droplets may have also contributed to the stability of the mayonnaise structure [28]. Additionally, coalescence was observed in analog sample A (Figure 4A). Because of the high fat content in mayonnaise and analogs, coalescence is a primary concern for emulsion stability, which is the result of oil droplet convergence [29]. The most effective way to limit coalescence is to generate strong repulsive forces between droplets [4].

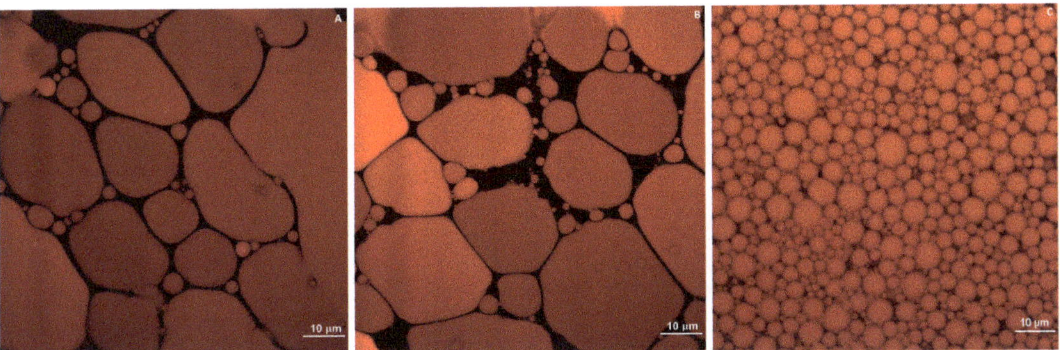

Figure 4. Confocal laser scanning micrographs of mayonnaise and analog prepared with (**A**) freeze-dried aquafaba, (**B**) spray-dried aquafaba, and (**C**) egg yolk. All images were captured at a working magnification of 600× with a 5 times digital zoom. Oil phase was stained with 0.01 wt.% Nile red. Scale bars represent 10 μm.

The results observed in this study partially contrast with previous research, in which finely diffused spherical oil drops in aquafaba mayonnaise analog were comparable to the microstructure of traditional mayonnaise [8]. Furthermore, Raikos et al. (2020) [8] obtained aquafaba from commercial cans of chickpeas, and mayonnaise analogs were prepared using a homogeniser rather than a hand mixer. The microstructure of mayonnaise and its analogs can be determined by different parameters, including the emulsifying and stabilizing agent types and their concentration, the size of the droplets, oil types and concentrations in mayonnaise and mayonnaise analog formulations, and production process methods [25,30]. Mustafa et al. (2018) [31] showed that aquafaba requires more mixing time to decrease the particle size and obtain functional properties comparable to egg-based foam and emulsion. Therefore, we predict that prolonging the mixing time and applying high-pressure homogenization could help to obtain an aquafaba-based mayonnaise analog with smaller droplet sizes and more stability during storage.

3.6. Mayonnaise and Analog Droplet Size Distribution

The variation in the oil droplet size distribution of all mayonnaise and mayonnaise analog samples during storage at 4 °C for 28 days is presented in Figure 5 (I, II, and III). The size distribution for mayonnaise C was 3–15 μm (80% distribution), without significant changes over 28 days. The droplets in mayonnaise analog B were larger and more broadly distributed (20–180 μm) than in mayonnaise C. The analog sample A demonstrated a bimodal droplet distribution with the largest droplet size and range (33–750 μm). No noticeable changes in the droplet size distribution were observed after storing mayonnaise and analogs at 4 °C for 28 days, confirming excellent stability in all samples. Previous work demonstrated differences in droplet size distribution depending on the homogenization qualifications, oil/aqueous phase composition, ingredient viscosity, and emulsion concentration and type [9]. In this study, only the type of emulsifiers was changed. The particle size range of analog B was similar to those in previous reports by Liu et al. (2007) and Di Mattia et al. (2015) [32,33] but was also markedly higher than those obtained by Laca et al. (2010), Nikzade et al. (2012), and Raikos et al. (2020) [4,8,34].

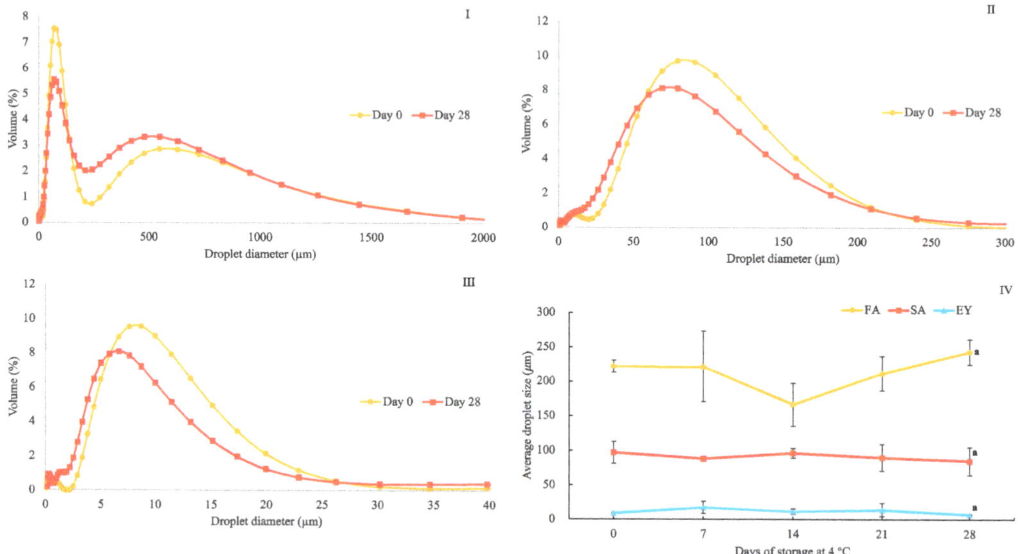

Figure 5. Droplet size distribution of mayonnaise and mayonnaise analog made with freeze-dried aquafaba (**I**), spray-dried aquafaba (**II**) and egg yolk (**III**); and volume average mean droplet diameter (d_{43}, μm) (**IV**) of mayonnaise and mayonnaise analog as a function of time. The letter (a) refers to no significant difference among storage times according to Tukey's test ($p < 0.05$).

The average droplet size (d_{43}) of analog B, 97.3 ± 16 μm, was smaller than that of analog A, 222 ± 9 μm, for all times in the 28-day trial ($p < 0.05$). However, both analogs A and B had significantly larger particle sizes than mayonnaise C (9.02 ± 1 μm). This was also visually evident in the CLSM, suggesting a lower capacity of aquafaba to emulsify and stabilize the mixture compared with egg yolk-based mayonnaise. This result indicates a wider variety of aquafaba mayonnaise analog microstructures depending on the emulsifier condition and the composition [32]. Droplet size is a crucial parameter in mayonnaise and analog evaluation, as it affects rheology, stability, storage life, texture, and taste [35]. In general, oil droplets with smaller particle sizes help in decreasing the movement of droplets and inhibiting coalescence, sedimentation, and other instabilities within the emulsion system, thereby increasing mayonnaise and analog viscosity and stability [9].

The difference in droplet size for analogs A and B suggests an effect of the drying process on aquafaba emulsifying properties. Since the dry matter of aquafaba consists primarily of carbohydrates and proteins, some chemical reactions (e.g., the Maillard reaction) can occur under high-temperature conditions during the spray drying process, promoting protein glycation and the formation of covalent conjugates between proteins and polysaccharides. These interactions can influence aquafaba emulsifying properties, causing a difference between the droplet sizes of analogs A and B [36–39]. Xu and Zhao [38] stated that these protein–polysaccharide conjugates can be easily absorbed in oil-water interfaces to form a thick and stable film and improve colloidal stability.

Different emulsifying components of aquafaba and egg yolk can explain why aquafaba mayonnaise analogs exhibited larger droplet sizes. Stantiall et al. (2018) reported that aquafaba dry matter mainly consisted of insoluble polysaccharides (46%), water-soluble carbohydrates (24%), and protein (19%) [13]. Polysaccharides have remarkable WHC and thickening properties conferred from their hydrophilicity and high molecular weight.

These properties can provide high viscosity of the aqueous phase and restrain the formation of fine, evenly distributed droplets in vegan mayonnaise analogs. In addition, because of the high concentration and high molecular weight of insoluble polysaccharides, a thick gel layer can be produced around the oil droplets to further enlarge their size [9]. Comparatively, the size of emulsifier molecules in egg is much smaller (mainly lecithin, lipoprotein, livetin, and phosvitin) [34,40]. Generally, small molecular size emulsifiers can generate small fat droplets and promote superior stability in an emulsion system [41,42].

4. Conclusions

The impacts of different cooking and drying methods on aquafaba emulsion capacity and stability were investigated. Dried aquafaba powder was used as an emulsifier in a vegan mayonnaise analog, and the physicochemical properties and stability were compared to traditional mayonnaise made from egg yolk. Aquafaba prepared by soaking chickpea seed in 4 °C water for 16 h and cooking for 30 min achieved the highest emulsion capacity and stability. When compared with other drying methods, freeze drying and spray drying produced powdered aquafaba with superior emulsion properties. Spray drying displayed a shorter drying time. Both freeze-dried and spray-dried aquafaba can be used to produce vegan mayonnaise analogs with comparable pH, colour, and stability to egg yolk mayonnaise. However, the droplet size in mayonnaise analogs made from freeze-dried and spray-dried aquafaba were much larger than in egg-yolk mayonnaise. This reduced the long-term stability compared with egg yolk mayonnaise. This study lays the foundation for commercial production of aquafaba powder for long-term storage and transportation and demonstrates the use of dried aquafaba as an egg replacement and emulsifier in the formulation of vegan mayonnaise analogs. Further studies will evaluate the effect of processing methods and investigate the effects of dried aquafaba concentration on the microstructure and stability of aquafaba mayonnaise analogs.

Author Contributions: Conceptualization, R.M. and Y.H.; validation, V.M. and B.T.; investigation, Y.H.; writing-original draft preparation, Y.H.; writing-review and editing, Y.H., S.K.P., T.J.T., V.M., R.M. and M.J.T.R.; supervision, V.M. and M.J.T.R.; funding acquisition, M.J.T.R. All authors have read and agreed to the published version of the manuscript.

Funding: This research was funded by the Saskatchewan Ministry of Agriculture's Agriculture Development Fund (ADF-20180091) and by the Department of Chemical and Biological Engineering, BLE Devolved Graduate Scholarship (University of Saskatchewan, Canada).

Institutional Review Board Statement: Not applicable.

Informed Consent Statement: Not applicable.

Data Availability Statement: The datasets generated for this study are available on request to the corresponding author.

Conflicts of Interest: The authors declare no conflict of interest.

References

1. Raymundo, A.; Franco, J.M.; Empis, J.; Sousa, I. Optimization of the composition of low-fat oil-in-water emulsions stabilized by white lupin protein. *J. Am. Oil Chem. Soc.* **2002**, *79*, 783–790. [CrossRef]
2. Ghoush, M.A.; Samhouri, M.; Al-Holy, M.; Herald, T. Formulation and fuzzy modeling of emulsion stability and viscosity of a gum–protein emulsifier in a model mayonnaise system. *J. Food Eng.* **2008**, *84*, 348–357. [CrossRef]
3. Campos, D.C.P.; Antoniassi, R.; Deliza, R.; de Freitas, S.C.; Felberg, I. Molho cremoso à base de extrato de soja: Estabilidade, propriedades reológicas, valor nutricional e aceitabilidade do consumidor. *Ciênc. Tecnol. Aliment.* **2009**, *29*, 919–926. [CrossRef]
4. Nikzade, V.; Tehrani, M.M.; Saadatmand-Tarzjan, M. Optimization of low-cholesterol-low-fat mayonnaise formulation: Effect of using soy milk and some stabilizer by a mixture design approach. *Food Hydrocoll.* **2012**, *28*, 344–352. [CrossRef]
5. Rahmati, K.; Mazaheri Tehrani, M.; Daneshvar, K. Soy milk as an emulsifier in mayonnaise: Physico-chemical, stability and sensory evaluation. *J. Food Sci. Technol.* **2014**, *51*, 3341–3347. [CrossRef]
6. Ghazaei, S.; Mizani, M.; Piravi-Vanak, Z.; Alimi, M. Particle size and cholesterol content of a mayonnaise formulated by OSA-modified potato starch. *Food Sci. Technol.* **2015**, *35*, 150–156. [CrossRef]

7. He, Y.; Meda, V.; Reaney, M.J.T.; Mustafa, R. Aquafaba, a new plant-based rheological additive for food applications. *Trends Food Sci. Technol.* **2021**, *111*, 27–42. [CrossRef]

8. Raikos, V.; Hayes, H.; Ni, H. Aquafaba from commercially canned chickpeas as potential egg replacer for the development of vegan mayonnaise: Recipe optimisation and storage stability. *Int. J. Food Sci. Technol.* **2020**, *55*, 1935–1942. [CrossRef]

9. Yildirim, M.; Sumnu, G.; Sahin, S. Rheology, particle-size distribution, and stability of low-fat mayonnaise produced via double emulsions. *Food Sci. Biotechnol.* **2016**, *25*, 1613–1618. [CrossRef]

10. Friberg, J.; Mandell, L. Influence of phase equilibria on properties of emulsions. *J. Pharm. Sci.* **1970**, *59*, 1001–1004. [CrossRef]

11. Buhl, T.F.; Christensen, C.H.; Hammershøj, M. Aquafaba as an egg white substitute in food foams and emulsions: Protein composition and functional behavior. *Food Hydrocoll.* **2019**, *96*, 354–364. [CrossRef]

12. Damian, J.J.; Huo, S.; Serventi, L. Phytochemical content and emulsifying ability of pulses cooking water. *Eur. Food Res. Technol.* **2018**, *244*, 1647–1655. [CrossRef]

13. Stantiall, S.E.; Dale, K.J.; Calizo, F.S.; Serventi, L. Application of pulses cooking water as functional ingredients: The foaming and gelling abilities. *Eur. Food Res. Technol.* **2018**, *244*, 97–104. [CrossRef]

14. Serventi, L.; Wang, S.; Zhu, J.; Liu, S.; Fei, F. Cooking water of yellow soybeans as emulsifier in gluten-free crackers. *Eur. Food Res. Technol.* **2018**, *244*, 2141–2148. [CrossRef]

15. Serventi, L. Cooking water composition. In *Upcycling Legume Water: From Wastewater to Food Ingredients*; Springer Nature: Cham, Switzerland, 2020; pp. 73–85. [CrossRef]

16. He, Y.; Shim, Y.Y.; Mustafa, R.; Meda, V.; Reaney, M.J.T. Chickpea cultivar selection to produce aquafaba with superior emulsion properties. *Foods* **2019**, *8*, 685. [CrossRef]

17. Shim, Y.Y.; Mustafa, R.; Shen, J.; Ratanapariyanuch, K.; Reaney, M.J.T. Composition and properties of aquafaba: Water recovered from commercially canned chickpeas. *J. Vis. Exp.* **2018**, *2018*, 56305. [CrossRef] [PubMed]

18. Lafarga, T.; Villaró, S.; Bobo, G.; Aguiló-Aguayo, I. Optimisation of the pH and boiling conditions needed to obtain improved foaming and emulsifying properties of chickpea aquafaba using a response surface methodology. *Int. J. Gastron. Food Sci.* **2019**, *18*, 100177. [CrossRef]

19. Alsalman, F.B.; Tulbek, M.; Nickerson, M.; Ramaswamy, H.S. Evaluation and optimization of functional and antinutritional properties of aquafaba. *Legum. Sci.* **2020**, *2*, 1–15. [CrossRef]

20. Pastor, A.; Compson, Z.G.; Dijkstra, P.; Riera, J.L.; Martí, E.; Sabater, F.; Hungate, B.A.; Marks, J.C. Stream carbon and nitrogen supplements during leaf litter decomposition: Contrasting patterns for two foundation species. *Oecologia* **2014**, *176*, 1111–1121. [CrossRef]

21. AACC Approved Methods of Analysis. Method 44-15.02. Moisture-Air-Oven Methods. In *AACC International Approved Methods*, 11th ed.; Cereals & Grains Association: St. Paul, MN, USA, 2013. [CrossRef]

22. Luo, Z.; Murray, B.S.; Yusoff, A.; Morgan, M.R.A.; Povey, M.J.W.; Day, A.J. Particle-Stabilizing Effects of Flavonoids at the Oil−Water Interface. *J. Agric. Food Chem.* **2011**, *59*, 2636–2645. [CrossRef]

23. Vega, C.; Grover, M.K. Physicochemical Properties of Acidified Skim Milk Gels Containing Cocoa Flavanols. *J. Agric. Food Chem.* **2011**, *59*, 6740–6747. [CrossRef] [PubMed]

24. Xu, Y.; Obielodan, M.; Sismour, E.; Arnett, A.; Alzahrani, S.; Zhang, B. Physicochemical, functional, thermal and structural properties of isolated Kabuli chickpea proteins as affected by processing approaches. *Int. J. Food Sci. Technol.* **2017**, *52*, 1147–1154. [CrossRef]

25. Mun, S.; Kim, Y.L.; Kang, C.G.; Park, K.H.; Shim, J.Y.; Kim, Y.R. Development of reduced-fat mayonnaise using 4αGTase-modified rice starch and xanthan gum. *Int. J. Biol. Macromol* **2009**, *44*, 400–407. [CrossRef] [PubMed]

26. McClements, D.J.; Demetriades, K. An Integrated Approach to the Development of Reduced-Fat Food Emulsions. *Crit. Rev. Food Sci. Nutr.* **1998**, *38*, 511–536. [CrossRef]

27. Worrasinchai, S.; Suphantharika, M.; Pinjai, S.; Jamnong, P. β-Glucan prepared from spent brewer's yeast as a fat replacer in mayonnaise. *Food Hydrocoll.* **2006**, *20*, 68–78. [CrossRef]

28. Depree, J.; Savage, G. Physical and flavour stability of mayonnaise. *Trends Food Sci. Technol.* **2001**, *12*, 157–163. [CrossRef]

29. Jaynes, E.N. Applications in the food industry: II. In *Encyclopedia of Emulsion Technology, Vol. 2: Applications*; Becher, P., Ed.; Marcel Dekker: New York, NY, USA, 1985; pp. 367–369.

30. Langton, M.; Jordansson, E.; Altskär, A.; Sørensen, C.; Hermansson, A.M. Microstructure and image analysis of mayonnaises. *Food Hydrocoll.* **1999**, *13*, 113–125. [CrossRef]

31. Mustafa, R.; He, Y.; Shim, Y.Y.; Reaney, M.J.T. Aquafaba, wastewater from chickpea canning, functions as an egg replacer in sponge cake. *Int. J. Food Sci. Technol.* **2018**, *53*, 2247–2255. [CrossRef]

32. Di Mattia, C.; Balestra, F.; Sacchetti, G.; Neri, L.; Mastrocola, D.; Pittia, P. Physical and structural properties of extra-virgin olive oil based mayonnaise. *LWT Food Sci. Technol.* **2015**, *62*, 764–770. [CrossRef]

33. Liu, H.; Xu, X.M.; Guo, S.D. Rheological, texture and sensory properties of low-fat mayonnaise with different fat mimetics. *LWT Food Sci. Technol.* **2007**, *40*, 946–954. [CrossRef]

34. Laca, A.; Sáenz, M.C.; Paredes, B.; Díaz, M. Rheological properties, stability and sensory evaluation of low-cholesterol mayonnaises prepared using egg yolk granules as emulsifying agent. *J. Food Eng.* **2010**, *97*, 243–252. [CrossRef]

35. McClements, D.J. *Food Emulsions: Principles, Practices, and Techniques*, 3rd ed.; CRC Press: Boca Raton, FL, USA, 2015; pp. 245–373. [CrossRef]

36. Shi, C.; Lin, Z.; Xiao, X.; Zhai, X.; Ma, C.W.; Ren, J. Comparisons of Processing Stability and Antioxidant Activity of the Silkworm Pupae Protein Hydrolysates by Spray-dry and Freeze-dry. *Int. J. Food Eng.* **2018**, *14*, 1–11. [CrossRef]
37. Kasran, M.; Cui, S.W.; Goff, H.D. Covalent attachment of fenugreek gum to soy whey protein isolate through natural Maillard reaction for improved emulsion stability. *Food Hydrocoll.* **2013**, *30*, 552–558. [CrossRef]
38. Xu, W.; Zhao, X.-H. Structure and property changes of the soy protein isolate glycated with maltose in an ionic liquid through the Maillard reaction. *Food Funct.* **2019**, *10*, 1948–1957. [CrossRef] [PubMed]
39. Yang, Y.; Cui, S.; Gong, J.; Miller, S.S.; Wang, Q.; Hua, Y. Stability of citral in oil-in-water emulsions protected by a soy protein–polysaccharide Maillard reaction product. *Food Res. Int.* **2015**, *69*, 357–363. [CrossRef]
40. Moros, J.E.; Franco, J.M.; Gallegos, C. Rheological properties of cholesterol-reduced, yolk-stabilized mayonnaise. *J. Am. Oil Chem. Soc.* **2002**, *79*, 837–843. [CrossRef]
41. Mangino, M.E. Protein interactions in emulsions: Protein-lipid interactions. In *Protein Functionality in Food Systems*; Hettiarachchy, N.S., Ziegler, G.R., Eds.; Marcel Dekker: New York, NY, USA, 1994; pp. 147–180.
42. Euston, S.R.; Hirst, R.L. Comparison of the concentration-dependent emulsifying properties of protein products containing aggregated and non-aggregated milk protein. *Int. Dairy J.* **1999**, *9*, 693–701. [CrossRef]

 *foods*

Article

Effect of Vegetable Juice, Puree, and Pomace on Chemical and Technological Quality of Fresh Pasta

Jinghong Wang [1,2], Margaret Anne Brennan [1], Charles Stephen Brennan [1,2,3] and Luca Serventi [1,*]

[1] Department of Wine, Food and Molecular Biosciences, Lincoln University, P.O. Box 85084, Lincoln 7647, New Zealand; Jinghong.wang@lincolnuni.ac.nz (J.W.); Margaret.Brennan@lincoln.ac.nz (M.A.B.); charles.brennan@rmit.edu.au (C.S.B.)
[2] Riddet Institute, Palmerston North 4442, New Zealand
[3] School of Science, RMIT University, P.O. Box 2474, Melbourne, VIC 3001, Australia
[*] Correspondence: Luca.Serventi@lincoln.ac.nz; Tel.: +64-3-423-0860

Abstract: Vegetable pasta is a premium product, and its consumption may deliver health benefits by increasing vegetable intake. This study investigated the replacement of semolina with juice, puree, and pomace of spinach and red cabbage. The effect of replacement on chemical composition, cooking performance (cooking loss, swelling index, water absorption), texture quality (elasticity, firmness), and colour was evaluated. The cooking loss of pasta made with spinach juice and spinach puree at 1 g/100 g substitution was the same as the control, while all other samples had a higher cooking loss. Spinach pasta had a higher breaking force but lower breaking distance in the tensile test than the control, while red cabbage pasta had a lower breaking force and breaking distance. Spinach juice fortified pasta was firmer than the control. Red cabbage juice pasta was less firm than other forms of fortified pasta at 1 g/100 g substitution level. Spinach and red cabbage juice are better colorants than puree or pomace as they change the colour of the pasta more dramatically at the same substitution level. Cooking performance and texture quality of spinach juice pasta were better than other samples, which indicates a premium pasta product for the food industry.

Keywords: quality; texture; physicochemical; vegetable pasta; colour

Citation: Wang, J.; Brennan, M.A.; Brennan, C.S.; Serventi, L. Effect of Vegetable Juice, Puree, and Pomace on Chemical and Technological Quality of Fresh Pasta. *Foods* **2021**, *10*, 1931. https://doi.org/10.3390/foods10081931

Academic Editor: Emanuele Zannini

Received: 23 July 2021
Accepted: 16 August 2021
Published: 20 August 2021

Publisher's Note: MDPI stays neutral with regard to jurisdictional claims in published maps and institutional affiliations.

1. Introduction

Pasta is a staple cereal food worldwide and it is a good vehicle for delivering functional ingredients [1,2]. Vegetables contain many health-promoting phytochemicals that traditional pasta lacks [3]. Those phytochemicals include dietary fibre, vitamins, polyphenols, carotenoids, glucosinolates, and minerals. Even though consumers are aware of the health benefits of consuming vegetables, their ingrained eating habits prevent them from a sufficient vegetable intake [4]. Hence, incorporating vegetables in staple foods such as pasta or bread may be a good option.

Vegetable pasta has been studied by many researchers [5,6]. An inferior cooking and sensory quality of vegetable pasta have been frequently reported compared to the traditional product [7–10]. Quality changes typically include increased cooking loss (CL), decreased firmness, and elasticity [1]. Authors substituted semolina for vegetable powder [7,11,12]. The powder is not only involved in nutrition loss due to air oven-drying [13], but also related to solid particle-size associated quality impact [14]. Hence, pasta enriched with other forms of vegetables, such as puree, juice, and pomace, were investigated in this study.

There are a few studies that have used other forms of vegetables to enhance pasta. For example, Gull et al. [15] added carrot pomace powder to a pasta formula, while Simonato et al. [16] used 5–10% olive pomace to fortify pasta. Carini et al. [17] added carrot juice to pasta and found that carrot juice pasta had similar extensibility and CL compared to the control, while carrot flour enriched pasta had a very high CL (more than 8 g/100 g) and lower extensibility, which indicates inferior quality. This study's limitation is that the actual substitution level (based on the dry matter) of carrot flour is much higher than carrot juice so that it is not a like for like comparison. Rakhesh et al. [18] made use of carrot, spinach, tomato, and beetroot puree to fortify pasta and found a decreased CL and improved texture of the resultant pasta. This study lacks comparison with the powder form and the description of puree-semolina mixing procedure is unclear. Juice and puree addition also have limitations when combined with pasta. Juice contains a very low level of solids, mostly around 5–15% for fresh vegetable juice [19]. Thus, achieving a high substitution level based on dry matter may be impossible for juice and puree due to excessive hydration, which can cause large lump formation, resulting in difficulties in successful extrusion [20,21]. The water content of juice and puree makes them more difficult to store and transport and may cause an increased cost for the food industry.

This project investigated the optimum method of vegetable fortification to produce vegetable-enriched pasta with better texture and cooking quality. The aim was to compare the key chemical composition, cooking performance, and texture quality of vegetable fortified pasta produced using vegetable juice, puree, and pomace. Our preliminary study showed that vegetable powder shows no significant difference with vegetable puree when added to pasta in key technical tests such as elasticity, firmness, and cooking loss. However, the nutritional quality (e.g., antioxidant ability) of powder enriched pasta was lower than puree enriched pasta. This may be due to the oven drying used to produce the vegetable powder. Therefore, the powder was altered to puree in our study. Two kinds of leafy vegetables, spinach (*Spinacia oleracea* L.) and red cabbage (Brassica oleracea convar. capitata var. capitata f. rubra), were selected in this study. Spinach is cheap and widely available. It is considered to have antioxidant and antidiabetic effects [22]. Spinach is also widely accepted by the food industry to produce commercially available green pasta. Red cabbage is nutritious as it is high in fibre and antioxidant phytochemicals [23]. Red cabbage materials in this study created a novel purple coloured pasta.

2. Materials and Methods

2.1. Raw Material

Semolina (Sun Valley Foods Ltd., Hamilton, New Zealand, labeled protein = 10.7 g/100 g, Fibre = 2.1 g/100 g, sodium = 10 mg/100 g), fresh spinach, and red cabbage were brought from the local market (New World Supermarket, Lincoln, New Zealand).

2.2. Vegetable Preparation

Spinach and red cabbage were washed thoroughly, their roots were removed with a sharp knife, the stem and leaf were put into the juicer (Model: Oscar Neo DA 1000; NATURE'S WONDERLAND Ltd., Brisbane, Australia), and the pomace and juice were collected separately. The vegetable juice was placed into a separate glass jar with a cap and stored at −18 °C until use. The pomace was spread in a backed tray and put into an oven to dry at 60 °C for 7 h. The dried pomace was then ground to a powder with a coffee grinder for 10 s twice, and the resultant pomace powder was stored in a Ziplock plastic bag at room temperature. The spinach puree and red cabbage puree were produced by mixing juice and fresh pomace together using a blender (Nutri-bullet NBO7200-1210DG; Capitalbrands Ltd., Boston, MA, USA), then the spinach puree and red cabbage were collected in a glass jar with a cap and stored at −18 °C. Before use, the puree was defrosted and put into the blender again to homogenise.

2.3. Pasta Preparation

Pasta was prepared using a lab-scale pasta machine (Model: MPF15N235M; Firmer., Ravenna, Italy) with 20 holes of 2.25 mm diameter. The vegetable pomace fortified pasta was prepared by mixing pomace with semolina in a pasta machine, and then 40 °C water was added according to the manual of the device to extrude the pasta. The formula is shown in Table 1. The puree and juice were defrosted and warmed to 40 °C in a water bath. Then, the puree, or juice, and 40 °C water were added to semolina in the pasta machine to extrude the pasta according to Table 1. The substitution level of juice and puree pasta is based on dry matter, according to the solid content measurement of the raw material. 1% substitution of juice, and 2% substitution of puree is the substitution level that can be both achieved by spinach and red cabbage material.

Table 1. Pasta formula to produce every 130 g pasta.

Pasta Type	Semolina g	Water g	Vegetable Amount g	Water from Vegetable g	Dry Matter from Vegetable g	Substitution Level %
C	100	30	0	0	0	0
SJ1	99	9.51	21.49	20.49	1	1
SPU1	99	11.63	19.37	18.37	1	1
SPU2	98	8.73	23.27	21.27	2	2
SPO1	99	30	1	0	1 *	1
SPO2	98	30	2	0	2 *	2
SPO10	90	30	10	0	10 *	10
RCJ1	99	6.93	24.07	23.07	1	1
RCPU1	99	19.43	11.57	10.57	1	1
RCPU2	98	8.85	23.15	21.15	2	2
RCPO1	99	30	1	0	1 *	1
RCPO2	98	30	2	0	2 *	2
RCPO10	90	30	10	0	10 *	10

SJ, SPU, and SPO represent spinach juice pasta, spinach puree pasta, and spinach pomace pasta, respectively; RCJ, RCPU, and RCPO represent red cabbage juice, red cabbage puree, and red cabbage pomace, respectively; 1, 2, and 10 is the substitution level (g/100 g) based on the dry weight. C: control sample. * the water content of pomace (see Table 2) is neglected in this study because it is less than 14%, which is close to that of semolina.

2.4. Proximate Analysis

Solid and moisture content was measured using the oven-dry method (105 °C), according to AACC [24]. The protein content of raw material and pasta was determined by Dumas total N methods (Elemental analyser Vario MAX CN, Frankfurt, Germany), and a conversion factor of 6.25 was applied to both pasta and raw materials to convert N to protein %. It should be noted that the conversion factor for different vegetables might vary as total N included non-amnio acid N like nitrate and N from nucleic acids [25]. Thus, the protein results are proximate, especially for the raw material. Total starch was measured by AOAC official method 966.11 using Megazyme total starch kit. Ash content of raw materials and pasta samples was measured according to AACC [24].

2.5. Cooking Performance

2.5.1. Optimal Cooking Time

Optimal cooking time (OCT) was measured according to AACC [24]. A total of 20 units of 5 cm pasta strands were put into 300 mL boiling water. The OCT was evaluated by taking a strand every 30 s and squeezing it between two transparent glass slides until the white core disappeared.

2.5.2. Pasta Cooking Procedure

Aliquots of 10 g of pasta were cooked in 600 mL of boiling water at OCT, then rinsed with 100 mL of cold water and stained for 30 s to measure cooking loss (CL), swelling index (SI), and water absorption index (WAI).

2.5.3. Cooking Loss

The CL was measured according to AACC [24]. Cooking water was collected by a stainless-steel vessel and dried in an air oven at 105 °C until a constant weight was reached. The residue was weighed and reported as gram residue per 100 g raw material.

2.5.4. Swelling Index and Water Absorption Index

SI and WAI were evaluated according to Desai et al. [26] with slight modification. 10 g of pasta was cooked to OCT and weighed after wash and stain, recorded as P_c. Then, the cooked pasta was dried at 105 °C until a constant weight was reached, recorded as P_{cd}. SI and WAI can be calculated with the following Equations (1) and (2):

$$SI = (P_c - P_{cd})/P_{cd} \tag{1}$$

$$WAI = (P_c - P_u)/P_u * 100 \tag{2}$$

P_u is the weight of uncooked pasta, P_c is the weight of cooked pasta, P_{cd} is the weight of dried, cooked pasta.

2.6. Texture Measurement

The firmness, breaking distance, and breaking force was measured by a Texture Analyser (TA.XT2; Stable Micro systems, Godalming, UK) with a 5 kg load cell. The pasta was cooked to OCT as described above before test. Firmness test is according to Approved Method 66-50 [24], with some modifications. Five strands of cooked pasta were placed on a flat metal plate. A noodle blade was used to compress the cooked pasta strands. The test parameters were set as test speed = 0.2 mm/s, post test speed = 10 mm/s, and distance of 5 mm. Tension test setting was according to [27]. The A/SPR spaghetti/noodle rig (Settings: Pre-test speed, 3 mm/s; test speed, 3 mm/s; initial distance, 10 mm; Final Distance 120 mm) was used in the test. Data are represented as the mean of nine measurements from triplicate cooking batches.

2.7. Colour Measurement

A portable colour meter (Minolta Chroma Meter CR210; Minolta Camera Co., Japan) was used to measure cooked (to OCT) and uncooked pasta. Each pasta was measured nine times from the triplicate cooking batches, and the result was expressed as L * (brightness range from 100 to 0), a * (redness–greenness range from 128 to −128), b * (yellowness–blueness range from 128 to −128). The instrument was calibrated using a standard white tile (L * = 98.03, a * = −0.23, b * = 0.25).

2.8. Statistical Analysis

All experiments were performed in triplicate except for what has been mentioned. All data were statistical analysed by one-way ANOVA, and the difference was evaluated by the Duncan test. SPSS (version 16) was used to perform data and figures.

3. Results and Discussion

3.1. Proximate Composition of Vegetable Pasta

The protein, moisture, and ash content of spinach and red cabbage pasta are shown in Table 2. The protein content in pasta is essential as it is the key of pasta structure. In pasta, gluten protein can be described as the backbone, with starch granules trapped in it playing a crucial function in pasta structure [28]. It is widely accepted that this structure is mainly maintained by disulfide bonds with the help of other non-covalent interactions like hydrogen bonds and ionic bonds [29–31]. It is suggested that protein-rich material may result in protein–protein interaction and form a more cohesive structure, thus helping the gluten form a homogeneous pasta structure [32]. Table 2 (a) shows that spinach raw material is rich in protein, ranked by protein content the spinach pasta is as follows SJ > SPU > SPO based on the dry weight. As a result, all the uncooked spinach pasta shows a significantly higher ($p < 0.05$) protein content than the control. Lisiewska et al. [33] reported that raw spinach contains 36 ± 12 mg/100 g cysteine content, around 1.5% of its total amino acid composition. Cysteine can provide sulfhydryl groups to form disulphide bonds during dough formation [34]. It indicates that protein from spinach may positively impact the formation of a gluten network and pasta quality. For cooked pasta, higher protein content was observed compared to raw. A similar trend was found by Manthey and Hall III [35], who use buckwheat bran flour to enrich pasta. It is possibly due to the leaching of starch into the cooking water, increasing the proportion of protein content in the pasta. Although for SJ1, the uncooked pasta shows significantly higher protein content than the control, the cooked one shows no difference ($p < 0.05$). This may be because the juice sample contains more soluble protein that may be lost during cooking. The cooked red cabbage pasta has a significantly lower protein content than control except for RCPU2, possibly because red cabbage raw material has a lower protein content (RCPO contains 11.06 g/100 g compared to SPO of 23.91 g/100 g).

Table 2. Proximate chemical analysis of vegetable pasta.

(a) Spinach pasta and spinach raw material							
Protein g/100 g dry matter		**Total Starch g/100 g dry matter**		**Moisture g/100 g Material**		**Ash g/100 g dry matter**	
Uncooked	Cooked	Uncooked	Cooked	Uncooked	Cooked	Uncooked	Cooked
Raw material							
Semolina 12.58 ± 0.11	N/A	71.42 ± 0.51	N/A	10.95 ± 0.10	N/A	0.95 ± 0.04	N/A
SJ 38.56 ± 0.06	N/A	N/A	N/A	95.35 ± 0.04	N/A	21.82 ± 0.02	N/A
SPU 31.17 ± 0.07	N/A	N/A	N/A	91.41 ± 0.07	N/A	17.40 ± 0.05	N/A
SPO 23.91 ± 0.13	N/A	N/A	N/A	12.95 ± 0.02	N/A	13.29 ± 0.01	N/A
Spinach pasta							
C 12.49 ± 0.01 d,B	12.96 ± 0.01 de,A	70.35 ± 0.39 a,A	70.24 ± 0.26 a,A	36.16 ± 0.70 ab	64.58 ± 0.23 a	0.68 ± 0.01 g,A	0.44 ± 0.01 f,B
SJ1 12.76 ± 0.01 c,B	13.00 ± 0.06 cd,A	69.60 ± 0.62 b,A	69.46 ± 0.50 b,A	35.71 ± 0.06 b	64.40 ± 0.24 a	0.89 ± 0.00 d,A	0.55 ± 0.01 d,B
SPU1 12.75 ± 0.01 c,B	13.04 ± 0.01 c,A	69.72 ± 0.60 b,A	69.48 ± 0.46 b,A	35.36 ± 0.25 b	64.47 ± 0.15 a	0.84 ± 0.00 e,A	0.55 ± 0.00 d,B
SPU2 12.96 ± 0.01 b,B	13.24 ± 0.06 b,A	68.06 ± 0.41 c,A	68.41 ± 0.45 c,A	35.67 ± 0.22 b	64.50 ± 0.26 a	1.11 ± 0.01 b,A	0.65 ± 0.01 b,B
SPO1 12.78 ± 0.01 c,B	12.89 ± 0.01 e,A	69.44 ± 0.48 b,A	69.29 ± 0.37 b,A	36.71 ± 0.68 ab	65.00 ± 0.16 a	0.81 ± 0.00 f,A	0.58 ± 0.00 c,B
SPO2 12.77 ± 0.01 c,B	13.06 ± 0.01 c,A	68.13 ± 0.62 c,A	68.57 ± 0.52 c,A	35.62 ± 0.25 b	64.61 ± 0.05 a	0.93 ± 0.01 c,A	0.48 ± 0.00 e,B
SPO10 14.13 ± 0.03 a,B	14.44 ± 0.01 a,A	61.39 ± 0.43 d,A	60.93 ± 0.36 d,A	37.27 ± 0.64 a	64.53 ± 0.90 a	1.94 ± 0.01 a,A	1.14 ± 0.01 a,B
(b) Red cabbage pasta and red cabbage raw material							
Raw material							
RCJ 19.23 ± 0.01	N/A	N/A	N/A	95.85 ± 0.00	N/A	10.18 ± 0.01	N/A
RCPU 16.23 ± 0.08	N/A	N/A	N/A	91.36 ± 0.02	N/A	8.26 ± 0.02	N/A
RCPO 11.06 ± 0.01	N/A	N/A	N/A	13.14 ± 0.07	N/A	5.50 ± 0.04	N/A
Red cabbage pasta							
C 12.49 ± 0.01 b,B	12.96 ± 0.01 a,A	70.35 ± 0.39 a,A	70.24 ± 0.26 a,A	36.16 ± 0.70 c	64.58 ± 0.23 b	0.68 ± 0.01 f,A	0.44 ± 0.01 e,B
RCJ1 12.46 ± 0.03 b,B	12.82 ± 0.04 b,A	68.97 ± 0.56 b,A	68.40 ± 0.51 b,A	36.41 ± 0.41 bc	64.68 ± 0.53 ab	0.78 ± 0.01 c,A	0.46 ± 0.00 d,B
RCPU1 12.41 ± 0.01 c,B	12.63 ± 0.03 d,A	68.88 ± 0.21 b,A	68.46 ± 0.42 b,A	37.45 ± 0.13 abc	64.79 ± 0.37 b	0.76 ± 0.00 d,A	0.50 ± 0.01 c,B
RCPU2 12.56 ± 0.01 a,B	12.91 ± 0.01 a,A	67.50 ± 0.58 c,A	67.32 ± 0.57 c,A	36.63 ± 0.75 abc	64.60 ± 0.72 b	0.84 ± 0.02 b,A	0.54 ± 0.02 b,B
RCPO1 12.30 ± 0.01 d,B	12.59 ± 0.01 d,A	68.74 ± 0.47 b,A	68.41 ± 0.42 b,A	37.77 ± 0.33 ab	65.83 ± 0.04 a	0.73 ± 0.01 e,A	0.50 ± 0.01 c,B
RCPO2 12.56 ± 0.05 a,B	12.72 ± 0.01 c,A	67.45 ± 0.57 c,A	67.30 ± 0.40 c,A	37.30 ± 0.36 abc	65.42 ± 0.46 ab	0.78 ± 0.01 c,A	0.55 ± 0.00 b,B
RCPO10 12.41 ± 0.01 c,B	12.77 ± 0.03 bc,A	60.34 ± 0.58 d,A	59.96 ± 0.50 d,A	37.94 ± 0.45 a	64.30 ± 0.14 b	1.16 ± 0.01 a,A	0.95 ± 0.01 a,B

SJ, SPU, and SPO represent spinach juice, spinach puree, and spinach pomace, respectively; RCJ, RCPU, and RCPO represent red cabbage juice, red cabbage puree, and red cabbage pomace, respectively; N/A means not tested. 1, 2, and 10 is the substitution level (g/100 g) based on the dry weight. C: control sample. Results expressed as Mean ± standard deviation calculated from triplicate measurements. Protein starch and ash results are based on a dry weight basis. Values within a column in the same sub-table followed by the same superscripted letters are not significantly different from each other ($p > 0.05$), values followed by the same superscripted capital letter are not significantly different between cooked and uncooked samples according to the ANOVA-Duncan test.

The total starch content of vegetable enriched pasta decreased with increased vegetable substitution. Cooking did not show any significant difference in total starch composition ($p < 0.05$) of vegetable enriched pasta. The ash content of foods is mainly inorganic metal compounds [36]. Perssini, Sensidoni, Pollini, and De Cindio [30] found that sodium chloride content increases the strength and solid-like semolina-flour dough behaviour via optimization of ionic strength. McCann and Day [37] found that salt delays the formation of the gluten network by reducing the rate of gluten hydration. Tang et al. [38] found that salt content can increase the strength of the disulfide bond in flour Raman gluten dough as less free SH groups are detected. Thus, the ash content may influence pasta quality. Table 2 shows spinach raw material characteristics with higher ash content in every form compared with red cabbage raw material. The addition of vegetable material increased the ash content of vegetable pasta significantly ($p < 0.05$) in every sample. This result is similar to Prabhasankar et al. [39], who used Japanese seaweed to fortify pasta. The ash content indicates a higher mineral content in those samples. Cooking causes a decreased ash content of vegetable pasta. It may be because some metal in ash is present in a water-soluble form and is lost during cooking. Desai, Brennan, and Brennan [26] found similar trends showing that cooked fish powder fortified pasta had a lower ash content than before cooking.

3.2. Cooking Quality of Vegetable Pasta

Optimal cooking time (OCT), cooking loss (CL), swelling index (SI), and water absorption index (WAI) are crucial cooking quality attributes of pasta [20]. Those attributes are strongly influenced by the protein–starch matrix formed during cold extrusion [31]. A good quality pasta has a compact protein–starch matrix, which slows the diffusion of the water to the starch core and inhibits amylose leaching into cooking water, giving a longer OCT and a decreased CL [40]. Table 3 shows that red cabbage content increases the cooking loss significantly. The CL of red cabbage pasta ranged from 4.767 to 6.163 g/100 g, compared to 4.399 g/100 g of the control sample. The CL values of the spinach juice pasta and spinach puree pasta shows no significant difference at 1 g/100 g substitution level versus the control. Other spinach pasta samples show increased cooking loss (from 4.447 to 5.920 g/100 g) compared to control. The increased CL indicates a weaker gluten matrix, which may be caused by fibre disruption, competition for water between gluten protein and other compounds (such as water-soluble fibre and soluble salt), and a dilution of gluten, which is caused by the substitution of semolina with vegetable material. The pasta with red cabbage pomace had a higher cooking loss than puree or juice sample at the same substitution level (RCPO1 > RCPU1, RCPO1 > RCJ1, RCPO2 > RCPU2). This may be because it contained less protein compared to pasta made with puree or juice, as shown in Table 2. The protein content and their properties can influence the gluten network formation and pasta structure [20]. The higher protein content in juice and puree can potentially interact with gluten, hence decreasing the disruptive effect caused by fibre and diluted gluten. Carini, Curti, Spotti, and Vittadini [17] found that carrot juice pasta has a much lower CL than pasta with carrot flour. However, the substitution level of carrot juice and carrot flour in that research was not standardised. Kowalczewski et al. [41] report the CL of fresh potato juice fortified pasta is lower than that fortified by spray-dried potato juice. All the vegetable pasta in this study had a CL lower than 8 g/100 g, which is a widely agreed maximum value for consumer acceptability [20,42].

The OCT is not changed in the vegetable pasta of all 1 g/100 g, 2 g/100 g samples, possibly because at low substitution levels the gluten network is not significantly changed to create a measurable impact. However, at a substitution level of 10 g/100 g, SPO10 and RCPO10 have a shorter optimal cooking time (Table 3). The decreased OCT may be caused by decreased water absorption (from 81.27 g/100 g of control to 74.11 g/100 g of SPO10 and 73:80 g/100 g of RCPO10). Similar results were found by Aravind et al. [43] using inulin (soluble fibre) to enrich pasta, and a lower OCT was reported. Cárdenas-Hernández et al. [44] also found OCT was decreased when amaranth flour and amaranth leaves and carboxymethylcellulose were added to semolina to produce pasta. In contrast, Foschia et al. [45] found an increased OCT when using 15 g/100 g dietary fibre (such as long-chain inulin, psyllium, or Glucagel) to substitute semolina.

The swelling index (SI) and water absorption index (WAI) reflect the amount of water absorbed at OCT. Table 3 shows that all spinach pasta samples show the same swelling index compared to control. Similar results were reported by Yadav et al. [46], which shows spinach pasta has no significant difference in water absorption versus control. Red cabbage pomace samples (RCPO1, RCPO10) show a lower WAI ($p < 0.05$) compared to control, while red cabbage juice and puree pasta show no significant SI and WAI difference versus control. It is potentially because the components of the red cabbage pomace have less affinity for water than the components of the red cabbage juice or puree. The results of

WAI and SI of RCPO pasta are consistent with Sun-Waterhouse, Jin, and Waterhouse [6], who found that elderberry juice pasta absorbs less water than other samples. In contrast, water absorption increase was observed in turnip pasta, tomato pasta, and carrot pasta [46], as well as broad bean flour fortified pasta [47]. This study may indicate that the SI and WAI of vegetable pasta is dependent on the intactness or strength of the gluten network and the water-binding capacity of vegetable components.

Table 3. Cooking performance of vegetable pasta.

	Optimal Cooking Time (Mins: Second)	Cooking Loss (g/100 g)	Swelling Index (g Water/g Dry Pasta)	Water Absorption Index (g/100 g)
		Spinach Pasta		
C	7:00	4.399 ± 0.063 [de]	1.863 ± 0.065 [a]	81.27 ± 1.42 [a]
SJ1	7:00	4.367 ± 0.065 [e]	1.801 ± 0.019 [a]	80.62 ± 1.17 [a]
SPU1	7:00	4.447 ± 0.092 [de]	1.814 ± 0.012 [a]	81.92 ± 1.22 [a]
SPU2	7:00	4.800 ± 0.026 [c]	1.817 ± 0.015 [a]	82.09 ± 0.28 [a]
SPO1	7:00	4.503 ± 0.015 [d]	1.858 ± 0.013 [a]	80.86 ± 2.18 [a]
SPO2	7:00	5.001 ± 0.062 [b]	1.826 ± 0.004 [a]	81.93 ± 0.90 [a]
SPO10	6:30	5.920 ± 0.781 [a]	1.821 ± 0.073 [a]	74.11 ± 2.67 [b]
		Red Cabbage Pasta		
C	7:00	4.399 ± 0.063 [e]	1.863 ± 0.065 [ab]	81.27 ± 1.42 [ab]
RCJ1	7:00	4.767 ± 0.021 [d]	1.927 ± 0.004 [a]	82.16 ± 0.78 [a]
RCPU1	7:00	4.803 ± 0.070 [d]	1.832 ± 0.043 [b]	80.07 ± 1.59 [abc]
RCPU2	7:00	4.943 ± 0.068 [c]	1.878 ± 0.025 [ab]	81.29 ± 0.34 [ab]
RCPO1	7:00	5.083 ± 0.379 [b]	1.840 ± 0.030 [b]	77.67 ± 1.54 [c]
RCPO2	7:00	5.067 ± 0.076 [b]	1.826 ± 0.057 [b]	79.03 ± 1.85 [bc]
RCPO10	6:15	6.163 ± 0.067 [a]	1.801 ± 0.012 [b]	73.80 ± 0.96 [d]

SJ, SPU, and SPO represent spinach juice, spinach puree, and spinach pomace, respectively; RCJ, RCPU, and RCPO represent red cabbage juice, red cabbage puree, and red cabbage pomace, respectively; 1, 2, and 10 is the substitution level (g/100 g) based on the dry weight. C: control sample. Results expressed as Mean ± standard deviation calculated from triplicate measurements. Values within a column of the same kind of pasta followed by the same superscripted letter are not significantly different from each other ($p > 0.05$) according to the ANOVA-Duncan test.

3.3. Texture and Colour of Vegetable Pasta

Pasta texture plays an essential role in overall quality and consumer acceptance [15,26]. Elasticity is an important texture profile that is considered to be conferred by gliadins that interact non-covalently with high modular weight glutenin subunits [31]. Elasticity (breaking distance and breaking force) of spinach pasta and red cabbage pasta is shown in Figure 1a,b, respectively. Spinach pasta has a higher breaking force ($p > 0.05$, except SPO1 and SPO2 insignificantly higher) than control. Meanwhile, red cabbage addition shows no significant influence on the breaking force of RCJ1, RCPU1, RCPU2, and RCPO1 and decreased breaking force was observed for RCPO2 and RCPO10. SJ1 and SPU1 have the same breaking distance as control while other spinach pasta and all red cabbage pasta were characterized by lower breaking distance. Juice fortified pasta shows a higher breaking distance compared to puree and pomace fortified pasta (SJ1 > SPO1, SPU2, SPO2, SPO10 significantly, RCJ1 > RCPU1, RCPO1, RCPO2, RCPU2 & RCPO10). At a higher substitution level of 10 g/100 g. The breaking distance of SPO10 and RCPO10 decreased dramatically. The decreased breaking distance indicates a weakened structure. Lu et al. [48] found a lower breaking force compared to control when adding white button mushroom powder (5–15%) and porcini mushroom powder (10–15%) to pasta, respectively. The same authors reported no significant change in breaking force when incorporating 5–15% shiitake mushroom powder to durum wheat to produce pasta. Foschia, Peressini, Sensidoni, Brennan, and Brennan [27] found that breaking force was decreased when durum wheat was substituted with 15 g/100 g dietary fibre (inulin, psyllium and oat material).

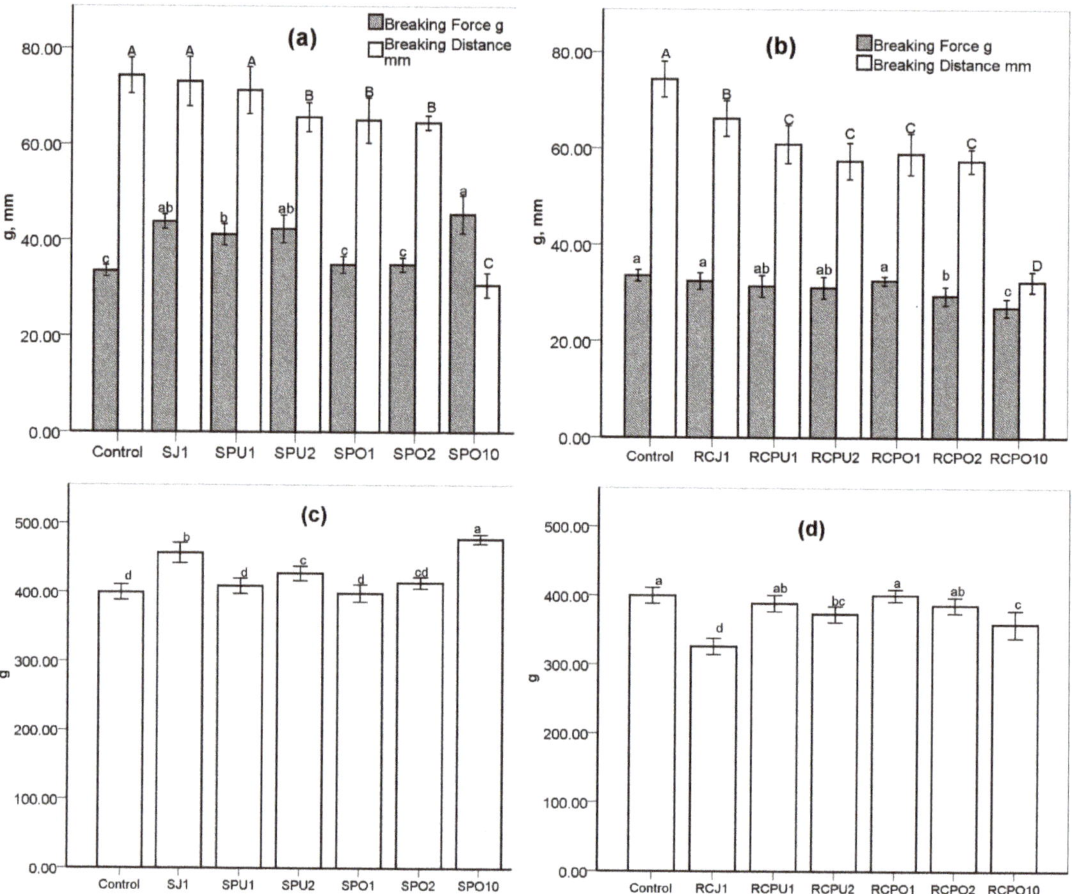

Figure 1. (**a**) Elasticity of spinach pasta, (**b**) Elasticity of red cabbage pasta, (**c**) Firmness of spinach pasta, (**d**) Firmness of red cabbage pasta. SJ, SPU, SPO represent spinach juice, spinach puree, spinach pomace, respectively; RCJ, RCPU, RCPO represent red cabbage juice, red cabbage puree, red cabbage pomace, respectively; 1, 2, and 10 is the substitution level (g/100 g) based on the dry weight. C: control sample. Error bars present the standard deviation of replicates. The same letter mean values are not significantly different from each other ($p > 0.05$).

Firmness is a measure of the force needed to compress pasta strands between teeth, and is an indicator of protein matrix integrity after cooking, which is dependent on the quality of gluten fraction [20]. Figure 1c,d show the firmness of spinach pasta and red cabbage pasta, respectively. The spinach pasta has a greater firmness than the control (except SPU1, SPO1, and SPO2). At the same substitution level, SJ1 has a greater firmness than SPU1 and SPO1. One possible reason for this is that SJ1 has fewer solid components. Those components may form discontinuities or cracks inside the pasta and result in a weakened structure. Red cabbage pasta firmness was equal to or lower than the control, while spinach pasta firmness was equal to or higher than the control. This is possibly because spinach pasta has a higher protein content than red cabbage pasta (as shown in Table 2). The higher protein content may contribute to a stronger protein structure, thus mitigating the disruptive effect of dietary fibre on the gluten network. This assumption is consistent with Petitot, Boyer, Minier, and Micard [8], who substituted 35% of semolina with split pea or faba bean and reported a significantly firmer pasta with a higher protein content. Jayawardena, Morton, Brennan, and Bekhit [32] used 10–25% protein-rich beef lung powder added to durum wheat, and the resultant pasta had a significantly higher firmness and breaking force. The firmness of RCJ1 is the lowest of all tested samples, possibly because of more water swelling (see Table 3 swelling index) by the starch granules, which in turn created a softer texture. Foschia, Peressini, Sensidoni, Brennan, and Brennan [27] found that incorporating short-chain inulin leads to a dramatic decrease in pasta firmness and increased water absorption. Gull, Prasad, and Kumar [15] reported a significantly lower firmness than control when 2–10% carrot pomace was added to the pasta formula.

It may be assumed that the texture profile of vegetable pasta is dependent on the vegetable components. Some components such as fibre and sugar may adversely affect the overall texture and cooking quality as they influence the water absorption, thus causing a change in the hydration process of the starch granules and the gluten network. Fibre particles dilute the gluten and therefore also contribute to gluten network disruption and potentially weaken the structure. Other components, such as protein, may generally have some beneficial effects such as strengthening the gluten network and other interactions to enhance the structure, such as increasing the firmness and breaking force. The overall texture and cooking quality are dependent on the balance of such adverse and beneficial effects from vegetable components. SJ1 in this study provides outstanding cooking and texture quality, with the identical cooking loss, water absorption, and breaking distance compared with control. It also has higher firmness and breaking force than control, thus produces al dente products with a firm, elastic texture. A low substitution level (1 g/100 g according to dry matter), juice form (lower solid particles), and higher protein content than durum wheat may contribute to its distinctive texture quality.

Colour results of vegetable pasta are shown in Table 4. The colour of vegetable pasta is strongly influenced by vegetable addition. Red cabbage pasta has lower brightness and yellowness (less L * and b * value) and higher redness (increased a * value) compared to the control. When comparing the different forms of vegetables, the juice's dye effect is stronger than puree or pomace as RCJ1 (both raw and cooked) has lower brightness and yellowness and more redness than RCPU1 and RCPO1. After cooking, the red cabbage pasta tends to be brighter but less red and yellow. Possibly due to the fact that the phytochemicals that provide the colour are water-soluble and leach into the cooking water. Chigurupati et al. [49] found that red cabbage colour is water-soluble and sensitive to pH change. It was found that the red cabbage colour changed from purple to deep blue when pH changed from acid to neutral. This could explain why cooked red cabbage pasta tends to be bluer (lower b * value except for RCPU1) and the cooking water presents a slightly blue colour as the water boils, leading to acid evaporation [50]; thus, the pH of cooking water tends to be neutral. Spinach addition leads to the decreased brightness, redness, and yellowness (L *, a *, b *, respectively) of the resultant pasta. Cooking procedure decreases the lightness, greenness, and yellowness of spinach pasta. Interestingly, the yellowness decrease of cooked spinach pasta is much lower than control (from 29.58 to 13.74 of control vs. 13.05 to 11.03 of SJ1, for example), indicating that spinach reduces the yellowness decrease during cooking. This is consistent with Nisha et al. [51], who found that thermal treatment causes a decrease in lightness and greenness but improved yellowness of spinach puree.

Table 4. Colour characteristics of cooked and uncooked pasta enriched with spinach and red cabbage.

	Uncooked				Cooked			
	L	a	b	Colour Example	L	a	b	Colour Example
			Spinach Pasta					
C	65.38 ± 0.40 [a]	−0.36 ± 0.07 [a]	29.58 ± 0.18 [a]		61.68 ± 0.30 [a]	−0.66 ± 0.02 [a]	13.74 ± 0.02 [a]	
SJ1	43.04 ± 0.21 [f]	−9.75 ± 0.09 [g]	13.05 ± 0.18 [c]		40.16 ± 0.19 [e]	−7.71 ± 0.15 [g]	11.03 ± 0.06 [b]	
SPU1	49.56 ± 0.11 [c]	−9.59 ± 0.02 [f]	14.79 ± 0.03 [b]		43.63 ± 0.04 [c]	−6.85 ± 0.07 [f]	10.39 ± 0.26 [c]	
SPU2	45.66 ± 0.09 [e]	−7.30 ± 0.03 [e]	10.68 ± 0.03 [e]		40.06 ± 0.41 [e]	−6.25 ± 0.13 [e]	9.27 ± 0.16 [d]	
SPO1	51.02 ± 0.46 [b]	−5.12 ± 0.20 [b]	12.69 ± 0.11 [d]		46.49 ± 0.31 [b]	−4.79 ± 0.19 [d]	7.93 ± 0.38 [e]	
SPO2	48.54 ± 0.44 [d]	−7.07 ± 0.09 [d]	10.23 ± 0.29 [f]		41.43 ± 0.26 [d]	−4.54 ± 0.27 [c]	7.36 ± 0.23 [f]	
SPO10	38.74 ± 0.12 [g]	−5.78 ± 0.06 [c]	7.60 ± 0.17 [g]		29.83 ± 0.08 [f]	−3.45 ± 0.09 [b]	4.95 ± 0.04 [g]	
			Red Cabbage Pasta					
C	65.38 ± 0.40 [a]	−0.36 ± 0.07 [g]	29.58 ± 0.18 [a]		61.68 ± 0.30 [a]	−0.66 ± 0.02 [e]	13.74 ± 0.02 [a]	
RCJ1	41.93 ± 0.74 [f]	8.85 ± 0.05 [a]	−7.17 ± 0.10 [g]		46.46 ± 0.58 [g]	3.12 ± 0.12 [c]	−8.61 ± 0.16 [e]	
RCPU1	47.54 ± 0.39 [c]	5.91 ± 0.02 [d]	−2.22 ± 0.02 [e]		54.26 ± 0.33 [c]	−1.26 ± 0.03 [f]	−1.38 ± 0.38 [b]	
RCPU2	44.55 ± 0.35 [e]	6.30 ± 0.02 [c]	−5.49 ± 0.02 [f]		50.07 ± 0.85 [e]	4.47 ± 0.18 [b]	−8.68 ± 0.28 [e]	
RCPO1	49.59 ± 0.38 [b]	3.25 ± 0.01 [f]	3.59 ± 0.04 [b]		58.02 ± 0.09 [b]	−1.69 ± 0.07 [g]	−2.31 ± 0.40 [c]	
RCPO2	45.29 ± 0.32 [d]	4.01 ± 0.01 [e]	2.05 ± 0.02 [c]		53.23 ± 0.11 [d]	2.05 ± 0.04 [d]	−5.31 ± 0.27 [d]	
RCPO10	36.55 ± 0.12 [g]	6.82 ± 0.03 [b]	−1.59 ± 0.50 [d]		47.12 ± 0.06 [f]	5.75 ± 0.06 [a]	−5.04 ± 0.04 [d]	

SJ, SPU, and SPO represent spinach juice pasta, spinach puree pasta, and spinach pomace pasta, respectively; RCJ, RCPU, and RCPO represent red cabbage juice, red cabbage puree, and red cabbage pomace, respectively; 1, 2, and 10 is the substitution level (g/100 g) based on the dry weight. C: control sample. Results expressed as Mean ± standard deviation calculated from ten measurements. L a b colour is converted to R G B colour through https://www.nixsensor.com/free-color-converter/ (accessed on 22 July 2021) and colour was output through EXCEL. While the colour convertor can only input integer colour number, the generated example colour is proximate. Values within a column from the same kind of pasta followed by the same superscripted letter are not significantly different from each other ($p > 0.05$), according to the ANOVA- Duncan test.

4. Conclusions

The results show that the juice, puree, and pomace of vegetables behave differently when incorporated into a pasta formulation. Those differences are plausible due to heterogeneous compositions in the varied forms of vegetables. At a low substitution level (1−2 g/100 g), juice, puree, and pomace can all be used to produce pasta with acceptable cooking performance and texture quality. Juice fortified pasta has lower cooking losses and better elasticity compared to puree and pomace fortified pastas. Among all pasta samples in this study, the cooking performance and texture quality of spinach juice pasta were better than other vegetable pastas and comparable or even better than control. This is probably due to its higher protein (cysteine-rich) composition and low substitution level (less gluten dilution and structure interruption). The study may indicate that vegetable juice with high protein content, such as spinach juice, can be used to produce premium pasta products for the food industry.

Author Contributions: Conceptualization, J.W. and C.S.B.; methodology, M.A.B.; software, J.W.; validation, L.S., C.S.B. and M.A.B.; formal analysis, J.W.; investigation, J.W.; resources, L.S.; data curation, J.W.; writing—original draft preparation, J.W.; writing—review and editing, J.W., L.S., C.S.B. and M.A.B.; visualization, J.W.; supervision, L.S., C.S.B. and M.A.B.; project administration, L.S.; funding acquisition, L.S. and C.S.B. All authors have read and agreed to the published version of the manuscript.

Funding: This research received no external funding. The research fund was offered by Lincoln University.

Institutional Review Board Statement: Not applicable.

Informed Consent Statement: Not applicable.

Data Availability Statement: The data presented in this study are available in the Tables 2–4 and Figure 1 within the article.

Acknowledgments: The authors thank Letitia Stipkovits for the laboratory training and are grateful to other colleagues for their technical support.

Conflicts of Interest: The authors declare no conflict of interest.

References

1. Oliviero, T.; Fogliano, V. Food design strategies to increase vegetable intake: The case of vegetable enriched pasta. *Trends Food Sci. Technol.* **2016**, *51*, 58–64. [CrossRef]
2. Peressini, D.; Cavarape, A.; Brennan, M.A.; Gao, J.; Brennan, C.S. Viscoelastic properties of durum wheat doughs enriched with soluble dietary fibres in relation to pasta-making performance and glycaemic response of spaghetti. *Food Hydrocoll.* **2020**, *102*, 105613. [CrossRef]
3. Marinelli, V.; Padalino, L.; Conte, A.; Del Nobile, M.A.; Briviba, K. Red Grape Marc Flour as Food Ingredient in Durum Wheat Spaghetti: Nutritional Evaluation and Bioaccessibility of Bioactive Compounds. *Food Sci. Technol. Res.* **2018**, *24*, 1093–1100. [CrossRef]
4. Sobota, A.; Wirkijowska, A.; Zarzycki, P. Application of vegetable concentrates and powders in coloured pasta production. *Int. J. Food Sci. Technol.* **2020**, *55*, 2677–2687. [CrossRef]
5. Mridula, D.; Gupta, R.K.; Bhadwal, S.; Khaira, H. Optimization of Groundnut Meal and Capsicum Juice for Protein and Antioxidant Rich Pasta. *Agric. Res.* **2016**, *5*, 293–304. [CrossRef]
6. Sun-Waterhouse, D.; Jin, D.; Waterhouse, G.I.N. Effect of adding elderberry juice concentrate on the quality attributes, polyphenol contents and antioxidant activity of three fibre-enriched pastas. *Food Res. Int.* **2013**, *54*, 781–789. [CrossRef]
7. Jalgaonkar, K.; Jha, S.K.; Mahawar, M.K. Influence of incorporating defatted soy flour, carrot powder, mango peel powder, and moringa leaves powder on quality characteristics of wheat semolina-pearl millet pasta. *J. Food Process. Preserv.* **2018**, *42*, e13575. [CrossRef]
8. Petitot, M.; Boyer, L.; Minier, C.; Micard, V. Fortification of pasta with split pea and faba bean flours: Pasta processing and quality evaluation. *Food Res. Int.* **2010**, *43*, 634–641. [CrossRef]
9. Rayas-Duarte, P.; Mock, C.; Satterlee, L. Quality of spaghetti containing buckwheat, amaranth, and lupin flours. *Cereal Chem.* **1996**, *73*, 381–387.
10. Ahmad, N.; Ur-Rehman, S.; Shabbir, M.A.; Abdullahl, M.; Shehzad, M.A.; Ud-Din, Z.; Roberts, T.H. Fortification of durum wheat semolina with detoxified matri (*Lathyrus sativus*) flour to improve the nutritional properties of pasta. *J. Food Sci. Technol.* **2018**, *55*, 2114–2121. [CrossRef]
11. Sahni, P.; Shere, D. Physico-chemical and sensory characteristics of beet root pomace powder incorporated fibre rich cookies. *Int. J. Food Ferment. Technol.* **2016**, *6*, 309. [CrossRef]
12. Sant'Anna, V.; Christiano, F.D.P.; Marczak, L.D.F.; Tessaro, I.C.; Thys, R.C.S. The effect of the incorporation of grape marc powder in fettuccini pasta properties. *LWT Food Sci. Technol.* **2014**, *58*, 497–501. [CrossRef]
13. Karam, M.C.; Petit, J.; Zimmer, D.; Baudelaire Djantou, E.; Scher, J. Effects of drying and grinding in production of fruit and vegetable powders: A review. *J. Food Eng.* **2016**, *188*, 32–49. [CrossRef]
14. Padalino, L.; Conte, A.; Lecce, A.; Likyova, D.; Sicari, V.; Pellicano, T.; Poiana, M.; Del Nobile, M. Durum wheat whole-meal spaghetti with tomato peels: How by-product particles size can affect final quality of pasta. *J. Food Process. Technol.* **2015**, *6*, 1.
15. Gull, A.; Prasad, K.; Kumar, P. Effect of millet flours and carrot pomace on cooking qualities, color and texture of developed pasta. *LWT Food Sci. Technol.* **2015**, *63*, 470–474. [CrossRef]
16. Simonato, B.; Trevisan, S.; Tolve, R.; Favati, F.; Pasini, G. Pasta fortification with olive pomace: Effects on the technological characteristics and nutritional properties. *LWT* **2019**, *114*, 108368. [CrossRef]
17. Carini, E.; Curti, E.; Spotti, E.; Vittadini, E. Effect of Formulation on Physicochemical Properties and Water Status of Nutritionally Enriched Fresh Pasta. *Food Bioprocess Technol.* **2012**, *5*, 1642–1652. [CrossRef]
18. Rakhesh, N.; Fellows, C.M.; Sissons, M. Evaluation of the technological and sensory properties of durum wheat spaghetti enriched with different dietary fibres. *J. Sci. Food Agric.* **2015**, *95*, 2–11. [CrossRef]
19. Wu, J.S.; Shen, S.-C.; Sinha, N.; Hui, Y. Processing of vegetable juice and blends. In *Handbook of Vegetables and Vegetable Processing*; Wiley Online Library: Hoboken, NJ, USA, 2011; pp. 335–350.
20. Bustos, M.; Perez, G.; Leon, A. Structure and quality of pasta enriched with functional ingredients. *Rsc Adv.* **2015**, *5*, 30780–30792. [CrossRef]
21. González, A.; Bordón, M.G.; Bustos, M.C.; Córdova Salazar, K.L.; Ribotta, P.D.; Martínez, M.L. Study of the incorporation of native and microencapsulated chia seed oil on pasta properties. *Int. J. Food Sci. Technol.* **2021**, *56*, 233–241. [CrossRef]
22. Vyas, M. Nutritional profile of spinach and its antioxidant & antidiabetic evaluation. *Int. J. Green Pharm. (IJGP)* **2017**, *11*, 192–197. [CrossRef]
23. Iborra-Bernad, C.; Tárrega, A.; García-Segovia, P.; Martínez-Monzó, J. Advantages of sous-vide cooked red cabbage: Structural, nutritional and sensory aspects. *LWT Food Sci. Technol.* **2014**, *56*, 451–460. [CrossRef]
24. AACC. *Approved Methods of the AACC*, 10th ed.; Cereals & Grains Association: Saint Paul, MN, USA, 2000.
25. Simonne, A.H.; Simonne, E.H.; Eitenmiller, R.R.; Mills, H.A.; Cresman, C.P., III. Could the Dumas Method Replace the Kjeldahl Digestion for Nitrogen and Crude Protein Determinations in Foods? *J. Sci. Food Agric.* **1997**, *73*, 39–45. [CrossRef]
26. Desai, A.; Brennan, M.A.; Brennan, C.S. The effect of semolina replacement with protein powder from fish (*Pseudophycis bachus*) on the physicochemical characteristics of pasta. *LWT* **2018**, *89*, 52–57. [CrossRef]
27. Foschia, M.; Peressini, D.; Sensidoni, A.; Brennan, M.A.; Brennan, C.S. How combinations of dietary fibres can affect physico-chemical characteristics of pasta. *LWT Food Sci. Technol.* **2015**, *61*, 41–46. [CrossRef]

28. Bonomi, F.; D'Egidio, M.G.; Iametti, S.; Marengo, M.; Marti, A.; Pagani, M.A.; Ragg, E.M. Structure–quality relationship in commercial pasta: A molecular glimpse. *Food Chem.* **2012**, *135*, 348–355. [CrossRef]

29. Ooms, N.; Delcour, J.A. How to impact gluten protein network formation during wheat flour dough making. *Curr. Opin. Food Sci.* **2019**, *25*, 88–97. [CrossRef]

30. Perssini, D.; Sensidoni, A.; Pollini, C.M.; De Cindio, B. Rheology of wheat doughs for fresh pasta production: Influence of semolina-flour blends and salt content. *J. Texture Stud.* **2000**, *31*, 163–182. [CrossRef]

31. Sicignano, A.; Di Monaco, R.; Masi, P.; Cavella, S. From raw material to dish: Pasta quality step by step. *J. Sci. Food Agric.* **2015**, *95*, 2579–2587. [CrossRef]

32. Jayawardena, S.R.; Morton, J.D.; Brennan, C.S.; Bekhit, A.E.-D.A. Utilisation of beef lung protein powder as a functional ingredient to enhance protein and iron content of fresh pasta. *Int. J. Food Sci. Technol.* **2019**, *54*, 610–618. [CrossRef]

33. Lisiewska, Z.; Kmiecik, W.; Gębczyński, P.; Sobczyńska, L. Amino acid profile of raw and as-eaten products of spinach (*Spinacia oleracea* L.). *Food Chem.* **2011**, *126*, 460–465. [CrossRef]

34. Filip, S.; Vidrih, R. Amino acid composition of protein-enriched dried pasta: Is it suitable for a low-carbohydrate diet? *Food Technol. Biotechnol.* **2015**, *53*, 298–306. [CrossRef]

35. Manthey, F.A.; Hall, C.A., III. Effect of processing and cooking on the content of minerals and protein in pasta containing buckwheat bran flour. *J. Sci. Food Agric.* **2007**, *87*, 2026–2033. [CrossRef]

36. Reilly, C. *Metal Contamination of Food*; Wiley Online Library: Hoboken, NJ, USA, 1980.

37. McCann, T.H.; Day, L. Effect of sodium chloride on gluten network formation, dough microstructure and rheology in relation to breadmaking. *J. Cereal Sci.* **2013**, *57*, 444–452. [CrossRef]

38. Tang, Y.; Yang, Y.; Wang, Q.; Tang, Y.; Li, F.; Zhao, J.; Zhang, Y.; Ming, J. Combined effect of carboxymethylcellulose and salt on structural properties of wheat gluten proteins. *Food Hydrocoll.* **2019**, *97*, 105189. [CrossRef]

39. Prabhasankar, P.; Ganesan, P.; Bhaskar, N.; Hirose, A.; Stephen, N.; Gowda, L.R.; Hosokawa, M.; Miyashita, K. Edible Japanese seaweed, wakame (*Undaria pinnatifida*) as an ingredient in pasta: Chemical, functional and structural evaluation. *Food Chem.* **2009**, *115*, 501–508. [CrossRef]

40. Diantom, A.; Curti, E.; Carini, E.; Boukid, F.; Mattarozzi, M.; Vodovotz, Y.; Careri, M.; Vittadini, E. A multi-scale approach for pasta quality features assessment. *LWT Food Sci. Technol.* **2019**, *101*, 285–292. [CrossRef]

41. Kowalczewski, P.; Lewandowicz, G.; Makowska, A.; Knoll, I.; Błaszczak, W.; Białas, W.; Kubiak, P. Pasta Fortified with Potato Juice: Structure, Quality, and Consumer Acceptance. *J. Food Sci.* **2015**, *80*, S1377–S1382. [CrossRef]

42. Sissons, M. Role of durum wheat composition on the quality of pasta and bread. *Food* **2008**, *2*, 75–90.

43. Aravind, N.; Sissons, M.; Egan, N.; Fellows, C. Effect of insoluble dietary fibre addition on technological, sensory, and structural properties of durum wheat spaghetti. *Food Chem.* **2012**, *130*, 299–309. [CrossRef]

44. Cárdenas-Hernández, A.; Beta, T.; Loarca-Piña, G.; Castaño-Tostado, E.; Nieto-Barrera, J.O.; Mendoza, S. Improved functional properties of pasta: Enrichment with amaranth seed flour and dried amaranth leaves. *J. Cereal Sci.* **2016**, *72*, 84–90. [CrossRef]

45. Foschia, M.; Peressini, D.; Sensidoni, A.; Brennan, M.A.; Brennan, C.S. Synergistic effect of different dietary fibres in pasta on in vitro starch digestion? *Food Chem.* **2015**, *172*, 245–250. [CrossRef]

46. Yadav, D.N.; Sharma, M.; Chikara, N.; Anand, T.; Bansal, S. Quality Characteristics of Vegetable-Blended Wheat–Pearl Millet Composite Pasta. *Agric. Res.* **2014**, *3*, 263–270. [CrossRef]

47. Tazrart, K.; Zaidi, F.; Lamacchia, C.; Haros, M. Effect of durum wheat semolina substitution with broad bean flour (*Vicia faba*) on the *Maccheronccini* pasta quality. *Eur. Food Res. Technol.* **2016**, *242*, 477–485. [CrossRef]

48. Lu, X.; Brennan, M.A.; Serventi, L.; Mason, S.; Brennan, C.S. How the inclusion of mushroom powder can affect the physicochemical characteristics of pasta. *Int. J. Food Sci. Technol.* **2016**, *51*, 2433–2439. [CrossRef]

49. Chigurupati, N.; Saiki, L.; Gayser, C.; Dash, A.K. Evaluation of red cabbage dye as a potential natural color for pharmaceutical use. *Int. J. Pharm.* **2002**, *241*, 293–299. [CrossRef]

50. Hanschen, F.S. Domestic boiling and salad preparation habits affect glucosinolate degradation in red cabbage (*Brassica oleracea var. capitata f. rubra*). *Food Chem.* **2020**, *321*, 126694. [CrossRef] [PubMed]

51. Nisha, P.; Singhal, R.S.; Pandit, A.B. A study on the degradation kinetics of visual green colour in spinach (*Spinacea oleracea* L.) and the effect of salt therein. *J. Food Eng.* **2004**, *64*, 135–142. [CrossRef]

 *foods*

Article

Delivery of Phenolic Compounds, Peptides and β-Glucan to the Gastrointestinal Tract by Incorporating Dietary Fibre-Rich Mushrooms into Sorghum Biscuits

Juncai Tu [1,2], Margaret Anne Brennan [1], Gang Wu [1,2], Weidong Bai [3], Ping Cheng [3], Bin Tian [1] and Charles Stephen Brennan [1,2,4,*]

1   Department of Wine, Food and Molecular Biosciences, Lincoln University, Christchurch 7647, New Zealand; Juncai.Tu@lincolnuni.ac.nz (J.T.); Margaret.Brennan@lincoln.ac.nz (M.A.B.); Gang.Wu@lincolnuni.ac.nz (G.W.); Bin.Tian@lincoln.ac.nz (B.T.)
2   Riddet Institute, Palmerston North 4474, New Zealand
3   College of Light Industry and Food Sciences, Zhongkai University of Agriculture and Engineering, Guangzhou 510225, China; whitebai2001@163.com (W.B.); nkpcheng@163.com (P.C.)
4   School of Science, RMIT University, G.P.O. Box 2474, Melbourne, VIC 3001, Australia
*   Correspondence: Charles.Brennan@rmit.edu.au; Tel.: +61-(3)-9925-7460

Abstract: Sorghum biscuits were enriched with mushroom powders (*Lentinula edodes*, *Auricularia auricula* and *Tremella fuciformis*) at 5%, 10% and 15% substitution levels. An in vitro gastrointestinal digestion was used to evaluate the effect of this enrichment on the phenolic content and soluble peptide content as well as antioxidant activities of the gastric or intestinal supernatants (bio-accessible fractions), and the remaining portions of phenolic compounds, antioxidants and β-glucan in the undigested residue (non-digestible fraction). The phenolic content of the gastric and intestinal supernatants obtained from digested mushroom-enriched biscuits was found to be higher than that of control biscuit, and the phenolic content was positively correlated to the antioxidant activities in each fraction ($p < 0.001$). *L. edodes* and *T. fuciformis* enrichment increased the soluble protein content (small peptide) of sorghum biscuits after in vitro digestion. All mushroom enrichment increased the total phenolic content and β-glucan content of the undigested residue and they were positively correlated ($p < 0.001$). The insoluble dietary fibre of biscuits was positively correlated with β-glucan content ($p < 0.001$) of undigested residue. These findings suggested that enriching food with mushroom derived dietary fibre increases the bioavailability of the non-digestible β-glucan and phenolic compounds.

Keywords: *Lentinula edodes*; *Auricularia auricula*; *Tremella fuciformis*; phenolic compounds; β-glucan

Citation: Tu, J.; Brennan, M.A.; Wu, G.; Bai, W.; Cheng, P.; Tian, B.; Brennan, C.S. Delivery of Phenolic Compounds, Peptides and β-Glucan to the Gastrointestinal Tract by Incorporating Dietary Fibre-Rich Mushrooms into Sorghum Biscuits. *Foods* **2021**, *10*, 1812. https://doi.org/10.3390/foods10081812

Academic Editor: Alessandra Marti

Received: 28 June 2021
Accepted: 3 August 2021
Published: 5 August 2021

1. Introduction

A diet rich in biologically active ingredients (such as polyphenols and dietary fibre) can help lower the risk of chronic diseases, such as obesity, bowel inflammation and cancer and helps to regulate gut microbiota. Sorghum is rich in phenolic compounds, including phenolic acids, tannins and flavonoids, and the amount and diversity of the major polyphenols in sorghum are higher than wheat, maize and rice [1]. Previous research into the prevention of chronic disease using sorghum has concentrated on the bioactive polyphenols in relation to their effects on antioxidant capacity, oxidative stress reduction, metabolism of glucose and lipid, inflammatory activity and regulation of the gut microbiota [2]. In addition to this, sorghum is gluten-free which makes it suitable for those suffering from coeliac disease [3]. Sorghum is not a commonly-consumed cereal, but it has been reported that cereal products that claim they are a source of fibre and show potential to reduce the risk of diabetes and cardiovascular disease can gain an increase in consumer liking [4]. Therefore, sorghum has the potential to be an alternative to wheat flour traditionally used in cereal-based foods.

Sorghum has a low protein digestibility, which contributes to the hydrophobic nature of kafirins, and the way in which the proteins bind with starch granules and phenolic compounds [5–8]. Cooking increases the protein digestibility of sorghum [7], which means that biscuits could be a suitable food to deliver bioactive compounds from sorghum to the gastrointestinal tract. In common with other cereals, sorghum is deficient in lysine and is considered to have poor quality protein from a nutritional point of view [9]. Mushrooms are a good source of lysine and can be incorporated into sorghum flour to improve the protein quality [10]. They have been recognised as the only non-animal food source that can provide vitamin D, mainly in the form of D2 and D3, both of which exhibit anti-inflammation, anti-tumour and anti-cancer properties [11]. Mushrooms have a high dietary fibre content and show immunomodulatory and anti-cancer activities [12]. Mushroom β-glucan can be digested by the colonic microbiota to produce short-chain fatty acids (SCFAs) these can help to regulate blood pressure, appetite, glucose homeostasis and improve gut integrity [13]. The insoluble dietary fibre, present in mushrooms, is also fermented to produce SCFAs, however, an in vivo study using pigs showed that β-glucan had a higher fermentation rate and SCFAs production rate than other insoluble fibres [14].

It is of great interest to incorporate mushrooms into cereal products to improve the nutritional quality and functionalities of products in recent studies [15–18]. However, mushroom dietary fibre might negatively affect the bioaccessibility of the phenolic compounds and the digestion of other nutrients when they are enriched in products. They can create associations with polyphenols before and during gastrointestinal digestion [19]. Some of the phenolic compounds are available in the stomach and upper intestine to reduce the free radicals present. The remaining phenolic compounds pass into the colon and are available to be bio-transformed to metabolites via fermentation by colonic microbiota [20,21]. Research is needed to analyse the portions of the phenolic compounds and other nutrients that can reach to the upper or lower part of gastrointestinal tract and how nutrients are delivered when products are enriched by fibre-rich mushrooms.

The main aim of this study was to evaluate the effect of mushroom powder on the digestion of the developed mushroom enriched sorghum biscuits and the release of proteins and phenolic compounds in relation to the antioxidant properties. The effect of mushroom incorporation on the β-glucan content and the colonic bioavailability of phenolic compounds were evaluated.

2. Materials and Methods

2.1. Materials

Sorghum flour (Davis Trading, New Zealand) and dried shiitake (*Lentinula edodes*), black ear (*Auricularia auricula*) and silver ear (*Tremella fuciformis*) mushrooms (Jade Phoenix, Guangzhou, China) were used in this study.

2.2. Preparation of Biscuits

Dried mushrooms were crushed with a Coffee Grinder (Breville, Sydney, Australia) and were further processed to powder using a Laboratory Mill 3310 (PerkinElmer, Waltham, MA, USA). The dough of sorghum biscuits was prepared by mixing sorghum flour (225 g) with 65 g sugar, 64 g vegetable shortening, 2.1 g salt, 2.5 g sodium bicarbonate and 50 g distilled water with a stand mixer (Breville, Australia). The dough was rolled and cut (6 mm thickness and 57 mm diameter) before putting into an oven and bake for 15 min at 160 °C. The mushroom-enriched biscuits had 5%, 10% and 15% sorghum flour replaced with mushroom powder (As shown in Table S1).

2.3. In Vitro Gastrointestinal Digestion

An in vitro gastrointestinal digestion, including gastric and intestinal stage, was simulated according to the method of Wu, et al. [22]. The biscuits (2 g) were dispersed into gastric (pepsin) solution and incubated at 37 °C for 2 h. For the intestinal stage, the pH was adjusted by 2 mL of 1 mol/L $NaHCO_3$ and 5 mL of 0.1 mol/L sodium maleate buffer (pH 6). After the pH adjustment, 0.1 mL of α-amyloglucosidase (3000 U/mL) was added, following by adding 5 mL of freshly prepared pancreatin-bile solution and incubated at 37 °C for 2 h.

The samples obtained after gastric and intestinal digestion were centrifuged at 13,000× *g* for 10 min (4 °C) to separate the bio-accessible fraction (supernatant) and the undigested residue. The supernatants were used for the determination of phenolic compound content and antioxidant activity. The undigested residue was freeze-dried and ground into a powder before further evaluation. The bio-accessibility index (%) is a measure of how available the phenolic compounds for absorption [20], and it was calculated according to the equation,

$$\text{BI } (\%) = \frac{Phenolic\ content\ of\ supernatant}{Total\ phenolic\ content\ of\ biscuit} \times 100.$$

The phenolic content in the equation was determined as described in Section 2.6.1. Total phenolic content was the sum of free and bound phenolic content.

2.4. BCA Assay and Protein Profile

After in vitro gastric and intestinal digestion described in Section 2.3, aliquots (0.5 mL) of digesta were taken and heated at 95 °C for 5 min. The samples were stood for 1 h at room temperature followed by centrifugation (13,000× *g*, 10 min), and the soluble protein content of the supernatant (bio-accessible fraction) was measured by Pierce™ BCA Protein Assay Kit (Thermo Fisher Scientific). The soluble protein content (SPC) was calculated according to the equation,

$$\text{SPC } (\text{mg/g dw}) = \frac{Protein\ weight\ in\ supernatant}{Total\ dry\ weight\ of\ biscuit} \times 100.$$

An SDS-PAGE assay was carried out according to the method used by Gong, et al. [23] with NuPAGE™ 4–12% Bis-Tris electrophoresis gels (Bio-Rad, Richmond, CA, USA). The marker for molecular weight (10–250 kDa) was used as a reference to the protein bands. The proteins in biscuits and digesta supernatants were extracted with the NuPAGE™ LDS sample buffer and reducing agent (×1) followed by a heat treatment (100 °C for 5 min). After centrifuging, 10 μL of the maker and 20 μL of sample extracts were loaded into the gel and run at 170 V for 40 min. The gel was stained with Commassie blue G-250 for 1 h and then was destained overnight.

2.5. Extraction of Phenolic Compounds

Free phenolic compounds were extracted with methanol as reported by Wang, et al. [24]. Samples (1 g) were stirred with 30 mL of 70% methanol (*v/v*) on a magnetic multi-stirrer overnight at ambient temperature and centrifuged for 10 min (9000× *g*, 4 °C). The supernatants were transferred to a volumetric flask (50 mL). The resulting pellets were further extracted twice with 10 mL of 70% methanol for 30 s over a vortex and the mixtures were centrifuged. The supernatants were combined and transferred quantitatively to a volumetric flask for the determination of phenolic content.

To extract the bound phenolic compounds, the method according to Li, et al. [25] was followed. The residues obtained after methanol extractions were subjected to alkaline hydrolysis by adding 20 mL of 4 mol/L NaOH. The samples were stirred at ambient temperature for 4 h before centrifuging at 9000× *g* for 10 min. The hydrolysed samples were acidified with 5 mol/L HCl to pH 2 and then extracted with ethyl acetone four times and centrifuged. Supernatants were collected and the organic phase was evaporated under reduced pressure at 30 °C. The samples were re-dissolved in 70% methanol. All extractions were performed in triplicate and samples were kept in the dark at −20 °C prior to the determination of phenolic content.

2.6. Determination of Phenolic Content and Antioxidant Activity

2.6.1. Phenolic Content Determination

Phenolic compound content was determined by the Folin-Ciocalteu method according to Polat, et al. [26]. The methanol extracts (free phenolic content), alkaline hydrolysed supernatants (bound phenolic content) and supernatants from the in vitro gastrointestinal digestion (bio-accessible phenolic content) were all analysed using this method. Results were expressed as milligrams of gallic acid equivalents (GAE) per gram of dry weight products.

2.6.2. Antioxidant Activity

The antioxidant activity of the methanol extracts of biscuits and supernatants from the in vitro gastrointestinal digestion was determined using 2,2-diphenyl-1-picrylhydrazyl (DPPH) radical scavenging activity and ferric reducing antioxidant power (FRAP) assays according to the methods described by Wu, et al. [27]. The results were expressed as micromoles of Trolox equivalents per gram dry weight (μmol TE/g dry weight), and μmol Fe^{2+} equivalents (Fe^{2+} E)/g dry weight of samples, respectively.

2.7. β-Glucan Determination

The β-glucan content of the mushroom powders, biscuits and dried undigested residues were determined using the Yeast-mushroom β-glucan assay kit (Megazyme, International Ireland Ltd., Wicklow, Ireland) according to McCleary and Draga [28]. The principle of this method was to determine the total glucan and α-glucan. The β-glucan content was calculated by subtracting α-glucan content from the total glucan content.

2.8. Nutritional Analysis

Crude protein content was measured using the Dumas method with the conversion factor of 6.25 for biscuits. The contents of insoluble dietary fibre (IDF), soluble dietary fibre (SDF) and total dietary fibre (TDF) were evaluated using commercial Megazyme assay kits (Megazyme International Ireland Ltd., Wicklow, Ireland) based on the method of Leon Prosky, et al. [29].

2.9. Statistical Analysis

All samples were analysed in triplicate and recorded by mean values $\pm$ standard deviation. Significant differences between multiple mean values were analysed by the One-way ANOVA and Tukey test ($p < 0.05$) using Minitab$^{®}$ (vision 19). Pearson's correlation coefficients were performed using Minitab$^{®}$ (vision 19) to assess the correlations between observed values ($p < 0.001$). Principal component analysis was conducted using Graphpad Prism 9.0 (GraphPad, CA, USA) to evaluate the effects of mushrooms substitutions on the biscuits variances.

3. Results and discussion

3.1. Phenolic Content

The free, bound and total phenolic content of the sorghum flour, the mushroom powders and the mushroom enriched sorghum biscuits are shown in Table 1. The sorghum flour contained a total 2.98 mg GAE/g dry weight (dw) of free phenolic content (methanol extracts), and this value was significantly higher than both *A. auricula* (1.37 mg GAE/g dw) and *T. fuciformis* (1.23 mg GAE/g dw) mushrooms, but lower than *L. edodes* (7.16 mg GAE/g dw). A similar trend was observed in the total phenolic content (TPC) which was the sum of the free and bound phenolic content. The bound phenolic content of *A. auricula*, *T. fuciformis* and sorghum flour was much higher than their free phenolic content. The differences in the phenolic content of the mushrooms affected the phenolic content of the mushroom enriched sorghum biscuits.

Table 1. The phenolic content of sorghum flour, mushrooms and mushroom-enriched sorghum biscuits. Phenolic content of the digesta supernatant after gastric and intestinal in vitro digestion, and the bio-accessibility index of the phenolic compounds in those digesta. Different uppercase letters represent significant difference of values between the ingredients ($p < 0.05$), while the lowercase letters represent significant difference of values between the biscuits ($p < 0.05$).

Samples	FPC	BPC	TPC	Gastric Fractions	Intestinal Fractions	Bio-Accessibility Index (%)	
						BI$_G$	BI$_I$
Ingredients							
Sorghum	2.98 ± 0.18 [B]	7.49 ± 0.14 [A]	10.47 ± 0.30 [B]	3.89 ± 0.01 [C]	6.45 ± 0.06 [C]	37.23 ± 1.11 [C]	61.69 ± 2.14 [C]
L. edodes	7.16 ± 0.11 [A]	7.47 ± 0.22 [A]	14.63 ± 0.24 [A]	9.55 ± 0.08 [A]	13.36 ± 0.38 [A]	65.26 ± 0.64 [A]	91.33 ± 1.30 [A]
A. auricula	1.37 ± 0.07 [C]	7.30 ± 0.17 [A]	8.67 ± 0.10 [C]	4.80 ± 0.22 [B]	6.93 ± 0.28 [BC]	55.29 ± 2.47 [B]	79.87 ± 4.00 [B]
T. fuciformis	1.23 ± 0.07 [C]	6.65 ± 0.03 [B]	7.89 ± 0.06 [D]	4.58 ± 0.11 [B]	7.50 ± 0.12 [B]	58.08 ± 1.75 [B]	95.08 ± 1.19 [A]
Biscuits							
Control	1.78 ± 0.01 [c]	3.48 ± 0.04 [e]	5.26 ± 0.03 [e]	2.31 ± 0.12 [e]	3.36 ± 0.10 [e]	43.99 ± 2.51 [d]	63.84 ± 2.27 [e]
5% LEB	1.79 ± 0.01 [c]	3.68 ± 0.05 [bcd]	5.48 ± 0.04 [d]	2.44 ± 0.05 [de]	4.09 ± 0.18 [c]	44.58 ± 0.69 [d]	74.71 ± 2.91 [bc]
10% LEB	1.94 ± 0.03 [b]	3.74 ± 0.03 [bc]	5.68 ± 0.06 [bc]	2.98 ± 0.06 [ab]	4.48 ± 0.07 [b]	52.42 ± 1.08 [abc]	78.82 ± 0.67 [b]
15% LEB	2.08 ± 0.03 [a]	3.82 ± 0.02 [ab]	5.90 ± 0.04 [a]	3.10 ± 0.05 [a]	4.53 ± 0.09 [ab]	52.65 ± 1.11 [ab]	76.88 ± 1.20 [b]
5% AAB	1.75 ± 0.02 [cd]	3.95 ± 0.02 [a]	5.70 ± 0.04 [b]	2.54 ± 0.10 [cde]	3.87 ± 0.03 [cd]	44.50 ± 1.93 [d]	67.83 ± 0.57 [de]
10% AAB	1.67 ± 0.02 [ef]	3.81 ± 0.03 [ab]	5.48 ± 0.05 [d]	2.60 ± 0.05 [cd]	3.61 ± 0.06 [de]	47.51 ± 0.92 [cd]	65.87 ± 1.58 [de]
15% AAB	1.62 ± 0.02 [f]	3.85 ± 0.15 [ab]	5.47 ± 0.16 [d]	2.91 ± 0.13 [ab]	3.85 ± 0.12 [cd]	53.17 ± 2.28 [a]	70.28 ± 2.44 [cd]
5% TFB	1.79 ± 0.02 [c]	3.70 ± 0.04 [bcd]	5.49 ± 0.04 [cd]	2.78 ± 0.03 [bc]	3.46 ± 0.08 [e]	50.57 ± 0.80 [abc]	62.96 ± 1.71 [e]
10% TFB	1.74 ± 0.03 [cd]	3.61 ± 0.04 [cde]	5.35 ± 0.04 [de]	2.55 ± 0.13 [cde]	4.83 ± 0.06 [a]	47.67 ± 2.58 [cd]	90.32 ± 1.27 [a]
15% TFB	1.70 ± 0.02 [de]	3.56 ± 0.07 [de]	5.26 ± 0.06 [e]	2.52 ± 0.12 [de]	4.70 ± 0.21 [ab]	47.78 ± 1.73 [bcd]	89.37 ± 2.94 [a]

Values = means ± standard deviation ($n = 3$). FPC—free phenolic content (methanol extraction); BPC—bound phenolic content (alkaline hydrolysis); TPC—total phenolic content; TPC = FPC + BPC. Values in the same column for ingredients with different uppercase letters are significantly different ($p < 0.05$). Values in the same column for biscuits with different lowercase letters are significantly different ($p < 0.05$). Abbreviations: LEB- *L. edodes* biscuit; AAB—*A. auricula* biscuit; TFB—*T. fuciformis* biscuit; BI (bio-accessibility index) = phenolic content of gastric (or intestinal) supernatant/total phenolic content of biscuit.

The replacement (5–15%) of sorghum flour by mushroom powder significantly changed the phenolic content of the biscuits ($p < 0.05$). Free phenolic content of the biscuits was ($p < 0.05$) increased by the inclusion of *L. edodes* mushroom with the substitution levels, and slightly decreased by the substitution with *A. auricula* and *T. fuciformis*. The bound phenolic content of mushroom-enriched biscuits was ranged from 3.56 to 3.93 mg GAE/g dw, and the values were higher than the control biscuits (3.48 mg GAE/g dw). Biscuits enriched with *L. edodes* and *A. auricula* had an increased total phenolic content. Enrichment with *T. fuciformis* at the 5% substitution level increased the total phenolic content, but there was no significant difference at 10% and 15% levels. That means the sorghum total phenolic content was not significantly diluted by *A. auricula* and *T. fuciformis* enrichment. In bakery products, many inner physiochemical reactions related to phenolic compounds can occur upon the thermal treatment, such as the liberation of bound phenolic compounds, degradation and oxidation [30]. Previous studies reported that roasting sorghum grains at high temperatures (150 and 180 °C) led to the degradation and loss of phenolic compounds (such as gallic acid, chlorogenic acid, ellagic acid, luteolin and quercetin) [31]. The increase in bound and total phenolic content of mushroom biscuits could be due to the mushroom dietary fibres inhibiting the release of the bound phenolic compounds and the loss of free phenolic compounds during baking (160 °C). The sorghum biscuit phenolic compounds may have become attached to the mushroom dietary fibres during the biscuit making process (mixing, agitation and rolling) through non-covalent bonding [19]. This interaction could increase the amount of bound phenolic compounds in the biscuits.

In vitro simulated gastrointestinal digestion was performed to evaluate the bio-accessibility of the phenolic compounds in the biscuits. Even though the in vitro model of digestion cannot represent the real digestion in human gastrointestinal tract with limitations to mimic the morphology and anatomical structure of digestion tract and peristaltic movement, it can be a simple and rapid method with no ethical restrictions to be used to analyse how foods being digested by enzymes and the effects of interactions of food ingredients on the release of nutrients. After the gastric stage of in vitro digestion, the bio-accessible fractions had a significantly higher phenolic content than the methanol extracts. The gastric digestion of biscuits partially released phenolic compounds into the supernatant, which had a significantly lower phenolic content (2.31–3.10 mg GAE/g dw) than the biscuit total phenolic content (5.26–5.90 mg GAE/g dw). Digestion with pepsin broke down the protein and disrupted the physical texture of the food, indicating a release of protein bound phenolic compounds or those entrapped in the food macro-structure. The increase in phenolic content after gastric digestion has been reported in many studies, such as wheat-shiitake noodles [24] and *Moringa oleifera* leaf-wheat pasta [32]. Compared with the control the phenolic content of the gastric supernatant, obtained after digestion of all *L. edodes* enriched biscuits (as well as 10–15% *A. auricula* and 5% *T. fuciformis*), was significantly ($p < 0.05$) increased.

The supernatants from the in vitro gastrointestinal digesta had 20–47% greater phenolic content compared to the gastric digesta. During this stage the starch was digested by α-amylase, releasing the phenolic compounds that had been bound to the macromolecules. The inclusion of mushrooms in most substitution levels significantly increased the phenolic content compared with the control biscuit after gastrointestinal digestion ($p < 0.05$), except for 5% *T. fuciformis*. Zieliński, et al. [33] observed a 4-fold increase in phenolic compound content after gastrointestinal digestion of buckwheat biscuits compared with the biscuits before digestion. Phenolic compounds can bind onto starch, protein and dietary fibres, and they are freed from the food matrix under gastrointestinal digestion due to the change of pH (from 2 to 7) and enzymatic hydrolysis of nanoparticles [19,34]. The hydrolysis of those macromolecules and their interactions with the phenolic compounds may positively affect the release of phenolic compounds during digestion. The increase of phenolic content in gastric and intestinal fractions for mushroom enriched biscuits could be that mushroom phenolic compounds are more digestible than sorghum phenolic compounds.

The phenolic bio-accessibility index after simulated gastric (BI_G) and intestinal (BI_I) digestion are shown in Table 1, the BI_G of biscuits was 43.99–53.17%, and the BI_I values increased to 62.96–90.32%. Compared with the control biscuit the BI_G values were significantly increased in several samples, including 10%–15% *L. edodes*, 15% *A. auricula* and 5% *T. fuciformis* enriched biscuits. An increase of BI_I value was observed in all levels of *L. edodes* incorporated biscuits, and 15% *A. auricula* and 10–15% *T. fuciformis* enriched biscuits. During the digestion process digestive enzymes, bile salts and pH change all act on the phenolic compounds via processes such as oxidation and hydrolysis, this affects their structure and stability altering their form and thus influencing their bio-accessibility [20,35]. Meng, et al. [36] and Quan, et al. [37] reported that alkaline conditions in the intestinal tract can degrade the phenolic compounds of fruit during in vitro digestion. The BI_I values of *T. fuciformis* biscuits (10% and 15%) almost reached 100%, however, there was an abundance of phenolic compounds retained in the undigested residues (Table 2). It should be noted that the bio-accessibility index could be affected by the limitations of the Folin-Ciocalteu assay. The Folin reagent may react with some fatty acids, Fe^{2+} ions, free amino acids and peptides released from the food matrix [33,38] which would result in the overestimation of the phenolic content of the bioavailable fractions. The alkaline extraction of bound phenolic compounds, however, (as described in 2.5) may destroy some phenolic compounds, resulting in their loss and subsequent under estimation. Insoluble dietary fibre (Table S1) can be resistant to the release of phenolic compounds even under alkaline hydrolysis which could affect the mushroom-enriched biscuits as they have a high insoluble dietary fibre content. Some previous studies determined the content of phenolic compounds by the Folin-Ciocalteu assay and calculated bio-accessibility after in vitro gastrointestinal digestion of cereal products (Ketnawa, Suwannachot, & Ogawa, 2020; Wang et al., 2020; Zieliński, Szawara-Nowak, & Wronkowska, 2020). However, their calculations for the bio-accessibility were based on methanol extraction only and did not take into consideration the bound phenolic compounds. For instance, Zieliński, Szawara-Nowak and Wronkowska [33] found more than three times of increase of bioavailable phenolic content than the methanol extracts of buckwheat biscuits, and they reported a bio-accessibility index of over 300%. In contrast, Blanco Canalis, Baroni, Leon and Ribotta [21] found a lower phenolic content in the in vitro digested bio-accessible fractions than in the acetone-water extracts of peach puree enriched wheat cookies.

Table 2. Phenolic content and potential antioxidant activity of the undigested residue after in vitro gastrointestinal digestion. Different letters represent significant difference of values between the biscuits ($p < 0.05$).

Samples	Phenolic Content (mg GAE/g dw)			FRAP (μmol Fe^{2+} E/g dw)			DPPH (μmol TE/g dw)		
	Free	Bound	Total	Free	Bound	Total	Free	Bound	Total
Control biscuit	0.58 ± 0.01 [e]	0.56 ± 0.01 [f]	1.15 ± 0.01 [f]	6.34 ± 0.41 [d]	8.41 ± 0.17 [d]	14.75 ± 0.46 [e]	0.95 ± 0.00 [e]	0.91 ± 0.06 [c]	1.85 ± 0.06 [e]
5% LEB	0.78 ± 0.01 [b]	0.90 ± 0.02 [ab]	1.67 ± 0.03 [b]	7.04 ± 0.17 [ab]	12.59 ± 0.38 [a]	19.62 ± 0.45 [a]	1.07 ± 0.01 [ab]	1.05 ± 0.03 [ab]	2.12 ± 0.03 [abcd]
10% LEB	0.84 ± 0.01 [a]	0.94 ± 0.05 [a]	1.78 ± 0.05 [a]	7.05 ± 0.20 [ab]	13.19 ± 0.41 [a]	20.24 ± 0.55 [a]	1.09 ± 0.02 [a]	1.14 ± 0.06 [a]	2.23 ± 0.06 [ab]
15% LEB	0.85 ± 0.02 [a]	0.96 ± 0.03 [a]	1.77 ± 0.06 [a]	7.31 ± 0.07 [a]	12.91 ± 0.10 [a]	20.22 ± 0.04 [a]	1.09 ± 0.00 [a]	1.15 ± 0.02 [a]	2.24 ± 0.02 [a]
5% AAB	0.60 ± 0.01 [e]	0.77 ± 0.02 [cde]	1.37 ± 0.02 [de]	5.27 ± 0.13 [e]	9.92 ± 0.26 [c]	15.20 ± 0.37 [de]	0.93 ± 0.02 [e]	1.06 ± 0.01 [ab]	1.99 ± 0.02 [de]
10% AAB	0.65 ± 0.01 [d]	0.79 ± 0.02 [cd]	1.45 ± 0.01 [d]	5.30 ± 0.09 [e]	10.18 ± 0.15 [bc]	15.49 ± 0.24 [cde]	1.04 ± 0.01 [bc]	1.10 ± 0.04 [ab]	2.13 ± 0.04 [abc]
15% AAB	0.72 ± 0.02 [c]	0.84 ± 0.03 [bc]	1.56 ± 0.03 [c]	5.64 ± 0.08 [e]	10.99 ± 0.16 [b]	16.63 ± 0.24 [bc]	1.04 ± 0.03 [bc]	1.10 ± 0.05 [ab]	2.14 ± 0.07 [abc]
5% TFB	0.59 ± 0.01 [e]	0.71 ± 0.01 [e]	1.30 ± 0.01 [e]	6.41 ± 0.06 [cd]	9.61 ± 0.69 [c]	16.02 ± 0.75 [bcd]	1.00 ± 0.01 [cd]	1.10 ± 0.02 [ab]	2.10 ± 0.03 [bcd]
10% TFB	0.67 ± 0.01 [d]	0.72 ± 0.03 [de]	1.39 ± 0.02 [de]	6.70 ± 0.07 [bcd]	10.00 ± 0.17 [bc]	16.70 ± 0.24 [b]	0.99 ± 0.01 [d]	1.00 ± 0.01 [bc]	2.00 ± 0.01 [d]
15% TFB	0.70 ± 0.04 [cd]	0.75 ± 0.03 [de]	1.45 ± 0.04 [d]	6.85 ± 0.12 [abc]	10.07 ± 0.47 [bc]	16.91 ± 0.39 [b]	0.99 ± 0.01 [d]	1.04 ± 0.06 [ab]	2.02 ± 0.07 [cd]

Values = means ± standard deviation ($n = 3$). Values in the same column with different letters are significantly different ($p < 0.05$). Total = Free + Bound for phenolic content, FRAP and DPPH. LEB—*L. edodes* biscuit; AAB—*A. auricula* biscuit; TFB—*T. fuciformis* biscuit.

3.2. Protein Profile and Soluble Protein Content after Digestion

The incorporation of *L. edodes* mushroom (5–15%) and *A. auricula* (15%) mushroom significantly increased the protein content of sorghum biscuits (Table S1). However, there was no significant difference between *T. fuciformis* enriched biscuits and the control biscuits. The protein profile distribution of the biscuits, with 15% mushroom enrichment and their digests were analysed using SDS-page under reducing conditions (Figure 1a,b). Before digestion, the pattern of proteins in biscuits displayed major bands with a molecular weight between 18–28 kDa, corresponding to kafirins [5]. The non-kafirin fractions were storage proteins globulin-1 (~65 kDa) and granule-bound starch synthase 1 (~50 kDa) [39]. Sorghum kafirins can be classified into three main fractions according to their molecular weight in α-kafirins (25 and 23 kDa), β-kafirins (20, 18 and 16 kDa) and γ-kafirins (28 kDa) [7]. The sorghum biscuits showed a high-intensity band at ~23 kDa (α-kafirins). The protein profile of 15% *T. fuciformis* had a small band at ~12 kDa, showing an increase in the small *Mw* protein fractions. After the complete gastrointestinal digestion, an overall decline or even disappearance of the band was observed due to enzymatic hydrolysis and proteolysis of the proteins, that produced small peptides (<10 kDa), oligopeptides and free amino acids which are not retained in the gels. The distinct bands (~25–50 kDa) that appeared on the gels of the digesta correspond to pancreatin and pepsin, this was also found in previous studies [40,41].

L. edodes digesta supernatant had a higher ($p < 0.05$) soluble protein content than the supernatant of sorghum flour in both gastric and intestinal fractions, while *A. auricula* and *T. fuciformis* had a lower soluble protein content (Figure S2). The soluble protein content of the gastric and intestinal supernatant of the biscuits is shown in Figure 1c. The initial pepsin hydrolysis of sorghum and mushroom proteins in the gastric stage produced large peptides and few of any small peptides or free amino acids. The intestinal stage was crucial for producing oligopeptides and free amino acids [42]. The BCA protein assay used in this study identified the soluble peptides and proteins with three or more amino acid residues. The gastric fractions had a higher soluble protein content than the intestinal fractions, as large polypeptides are hydrolysed, by the enzymes in the pancreatin, into free amino acids or dipeptides and these are not detected by the BCA reagent. The soluble proteins in the intestinal supernatants are not present in the SDS-gels, indicating that these hydrolysed protein fractions are small peptides (<10 kDa) and oligopeptides (short-chain peptides). These small oligopeptides show antioxidant, anti-inflammatory, anticancer, hypocholesterolemic and antihypertensive activities and are likely to be readily absorbed by the intestinal wall [43,44]. Further studies are needed to identify the structure-related absorption and function of the oligopeptides derived from the biscuits after in vitro gastrointestinal digestion.

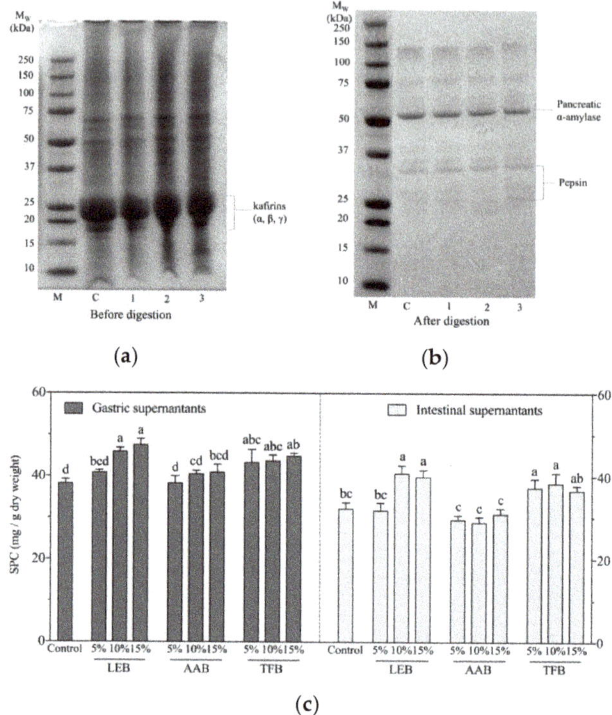

Figure 1. SDS-page analysis of sorghum biscuits enriched with mushroom powder (15% substitution) (**a**) before and (**b**) after gastrointestinal digestion. (**c**) Soluble protein content (SPC) of biscuits after in vitro gastrointestinal digestion. (M—marker; C—control biscuit; 1—15% *L. edodes* biscuit; 2—15% *A. auricula* biscuit; 3—15% *T. fuciformis* biscuit). Error bars represent standard deviation ($n = 3$). Columns with different letters are significantly different within the same chart ($p < 0.05$). LEB—*L. edodes* biscuit; AAB—*A. auricula* biscuit; TFB—*T. fuciformis* biscuit.

The soluble protein content of gastric and intestinal supernatant of *L. edodes* and *T. fuciformis* enriched biscuits was higher than the control biscuit ($p < 0.05$), except for 5% *L. edodes* biscuit. The soluble protein content of 10% and 15% *A. auricula* biscuit was higher than the control in gastric fraction, but in intestinal supernatant it was not significantly different to the control. Kafirins, the main protein in sorghum, are proline-rich chains with low water solubility and low enzyme accessibility [5]. While peptide bonds that contain proline cannot be hydrolysed by pancreatic enzymes [42], their digestion and the release of proteins can be influenced by other components in the food matrix such as fibre. Mushrooms are rich in fibre and adding powdered mushrooms may improve protein hydrolysis via the effects of fibre on the physical structure of the biscuit. Ashwath Kumar, et al. [45] discovered that fibre enriched wheat biscuits (TDF, 9.09%) had a higher rate of protein hydrolysis. Sciarini, et al. [46] found that the addition of oat bran fibre and resistant starch into a gluten-free bread (rice flour) increased the percentage of protein hydrolysis and suggested that this was achieved by disrupting the crumb structure. Fibre can also act as a physical barrier to some enzymes and delay the hydrolysis of proteins or polypeptides [18]. The solubility and digestion of protein could also be related to the molecular weight of the protein, and the 15% *T. fuciformis* biscuit had a low *Mw* of protein fraction and a higher soluble protein content compared to the control. Phenolic compounds have been reported to inhibit various digestion enzymes [31,47], by interacting with hydrolysis enzymes. *L. edodes* and *T. fuciformis* biscuits at both 10% and 15% levels had higher intestinal digesta phenolic content than the control, and this is consistent with the soluble protein content in that fraction. The results indicated that the phenolic compounds released after digestion could impede the protein digestion, resulting in an increase of small peptides and oligopeptides retained in the bioaccessible fractions rather than being digested to free amino acids.

3.3. In Vitro Antioxidant Activity after Digestion

The antioxidant activities (FRAP and DPPH) of *L. edodes*, *A. auricula* and *T. fuciformis* and sorghum flour are shown in Figure S2. The *L. edodes* had a significantly higher reducing capacity (FRAP) and free radical scavenging ability (DPPH) than the other mushrooms and the flour both before and after digestion ($p < 0.05$), which was consistent with the phenolic content in each fraction.

The FRAP and DPPH of the sorghum biscuits before and after digestion are shown in Figure 2a,b. It can be seen that *L. edodes* biscuits had a higher reducing capacity (FRAP) and free radical scavenging ability (DPPH) than other biscuits both before and after digestion ($p < 0.05$), which is consistent with the phenolic content of each fraction. The antioxidant activity of the digesta supernatants was significantly higher than the methanol extract of the biscuits, and the activity after intestinal digestion was increased almost three-fold. After in vitro digestion, the physical structure and inter and intra-molecular bonds in the biscuits are hydrolysed by enzymes and the nutrients and antioxidants are released [24]. *A. auricula* and *T. fuciformis* enriched biscuits had a higher content of soluble dietary fibre and this macromolecule could bind to antioxidants in stomach and intestine, delivering the antioxidants to colon. Baczek, et al. [48] found an increase of antioxidant properties (ABTS and FRAP) in the soluble fractions after in vitro digestion of oat-buckwheat bread.

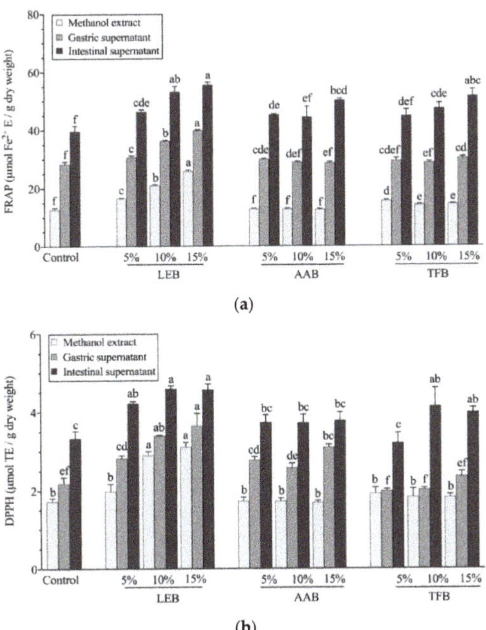

Figure 2. In vitro bio-accessible antioxidant activity of mushroom-enriched sorghum biscuits assessed by Ferrous reducing antioxidant power (FRAP, **a**) and Diphenyl-2-picrylhydrazyl radicals scavenging ability (DPPH, **b**). Values are means ± standard deviation ($n = 3$). The statistical analysis of significance was performed between 10 of the biscuit samples for each fraction (methanol extract, gastric supernatant and intestinal supernatant). Products with different letters are significantly different ($p < 0.05$). LEB—*L. edodes* biscuit; AAB—*A. auricula* biscuit; TFB—*T. fuciformis* biscuit.

The digestion process, whereby food is exposed to digestive enzymes and variations in pH, is crucial to cause the release of phenolic compounds from the molecules that have bound them [49] The released phenolic compounds are the main contributor to the antioxidant properties [50], however, other molecules associated with the binding or trapping of phenolic compounds can also affect the antioxidant activity. Compared with the control, sorghum biscuits enriched with *L. edodes* had increased the antioxidant activity (FRAP and DPPH) in both gastric and intestinal fractions. *T. fuciformis* 10% and 15% enriched biscuits also had increased ($p < 0.05$) antioxidant activities after digestion. *A. auricula* 15% enriched biscuits had higher FRAP values in the upper gastrointestinal tract than the control samples, but the addition of *A. auricula* did not increase DPPH values. The FRAP assay is based on the reduction of ions from Fe^{3+} to Fe^{2+}, and the DPPH reagent can receive hydrogen atoms from antioxidants [20]. That means that the antioxidant compounds assessed by the DPPH are not the same as that of FRAP. The antioxidant activity of food after gastrointestinal digestion is vital for health. For example, antioxidants can scavenge and suppress the excess reactive oxygen species (ROS) in the organisms and prevent oxidative-related diseases [51]. Otherwise, excessive ROS causes inflammation which leads to diseases, such as inflammatory bowel disease.

3.4. Phenolic and Antioxidants Content in Undigested Residue

Both the free and bound phenolic content of undigested residue increased with the incorporation of mushrooms, as can be seen in Table 2. Mushroom enrichment in the sorghum biscuits increased the dietary fibre content of biscuits and total phenolic content of the undigested residue meaning that there were more phenolic compounds available to transit into the colon. The phenolic compounds in vegetables or mushrooms are normally conjugated with dietary fibres [15], which explains the increased phenolic content in the undigested residue. Phenolic compounds in the pellets can undergo biological metabolism by colonic microbiota and microbial enzymes [52]. Dong, et al. [53] demonstrated that the bound polyphenols in carrot dietary fibre were liberated during in vitro fermentation, and the fermented polyphenols could promote the growth of specific beneficial flora and suppress the harmful bacterial flora. Fermentation by microbiota can promote the bioavailability and absorption of phenolic compounds. The bio-absorption of polyphenolic metabolites fermented by gut microbiota may reach the liver via the hepatic portal vein after absorption, and undergo further degradation and enter into systematic circulation before reaching targeted tissues and cells [54].

The free phenolic content of the undigested residue and the FRAP activity was decreased in the *A. auricula* biscuits compared to the control and was increased in the *L. edodes* and *T. fuciformis* biscuits. However, the total reducing capacity (FRAP) of the undigested residue increased when the biscuits were enriched with mushroom powder, due to the increase in antioxidant activity of the bound fraction of the undigested residue. The DPPH activity of the undigested residue of mushroom enriched biscuits showed that they had a higher ($p < 0.05$) total radical scavenging capacity than the control biscuits.

3.5. β-Glucan Potential Colon-Bioavailability

β-glucan is abundant in both cereals and mushrooms, and it can bring various disease prevention properties, such as reducing postprandial blood glucose and lowering LDL cholesterol. The sorghum flour contained a total of 6.41 g/100g dw of β-glucan, which was much lower than mushroom samples, as can be seen in Figure 3a. The *L. edodes* had a high content of β-glucan (27.78 g/100g dw), *A. auricula* had slightly less (21.55 g/100g dw) and *T. fuciformis* had least (17.44 g/100g dw), and the results are similar to the β-glucan content of several other mushroom cultivars (9 to 27 g/100g dw) [55]. Cereal β-glucan has a fibrous structure with a combination of 1-3 β-glycosidic and 1-4 β-glycosidic linkages, while mushroom β-glucan mainly consists of 1-3 β-glycosidic with 1-6 β-glycosidic branches. The different sources of β-glucan have diverse molecular structures, such as molecular weight, conformation and branching degree, influencing their solubility, viscosity and rheological characteristics [56]. For example, the chemical structure and molecular weight of β-glucan are two main factors that determine the solubility of β-glucan [57]. The sorghum flour, *L. edodes* and *A. auricula* all had a much lower soluble fibre content than β-glucan content, especially *L. edodes* samples indicating that more β-glucan is water-insoluble in *L. edodes*. Morales, et al. [58] found that *Lentinula edodes* had a low yield (4.2%) of hot water-soluble extract and a high content (27%) of β-glucan, which was consistent with *L. edodes* mushroom used in this study. They further determined the β-glucan of the hot water-soluble extract and the resulting insoluble fraction after extraction, finding that the insoluble fraction had a higher content of β-glucan (38%) than the soluble extract (13.2%). The solubility of β-glucan largely depends on the percentage of 1-3 β-glycosidic linkages, because this type of linkage leads to twists in the straight-chain polymer allowing water molecules to enter into the chains [59]. Lentinan is a water-soluble polysaccharide from *L. edodes* with a primary structure of the (1-3)-β-D-glucan [60]. The solubility might also be influenced by interactions with other macronutrients in mushrooms, such as insoluble fibre and proteins [61]. β-glucan could covalently connect with the chitin (insoluble fibre) [62], leading to a high content of insoluble β-glucan content in mushroom samples. Alahmed and Simsek [57] found a decrease in the solubility of oat β-glucan due to the increase of molecular weight because of a rise in cohesive energy density. The molecular weight of β-glucan also relates to its viscosity and thus affects its functionalities (such as hypoglycaemic and hypocholesterolemic properties) [56].

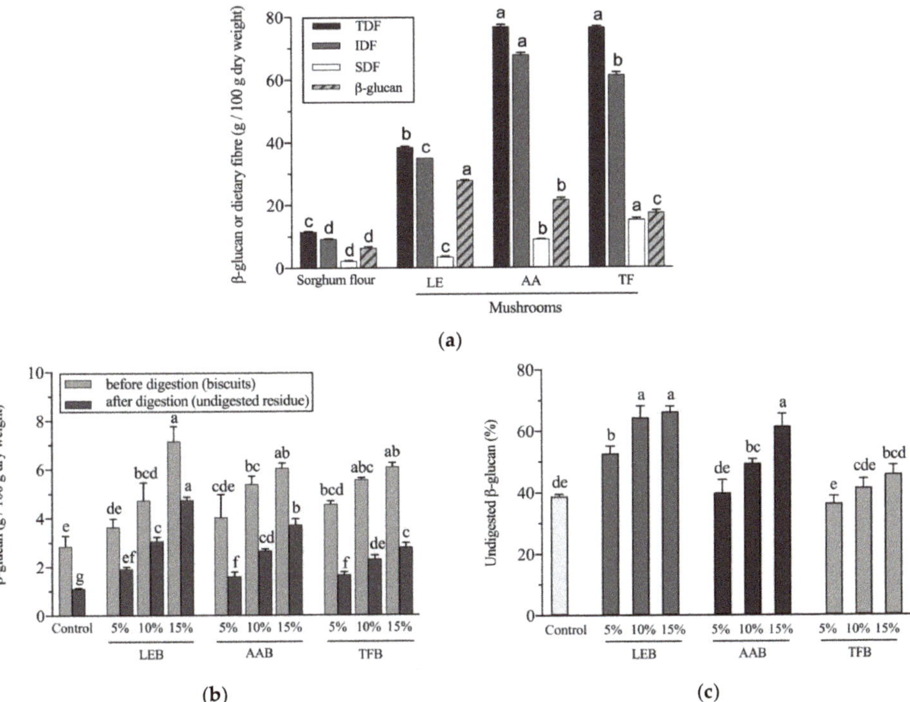

Figure 3. β-glucan and dietary fibre content of sorghum flour and mushrooms (**a**); and β-glucan content of biscuits and undigested residue (g/100 g dry weight biscuit) (**b**); and the undigested β-glucan (%) after gastrointestinal digestion (**c**). Undigested β-glucan (%) was calculated by the following formula: Undigested β-glucan (%) = (gram of β-glucan in the undigested residue/gram of β-glucan in the initial biscuits) × 100. Values = means ± standard deviation (*n* = 3). The statistical analysis of significance was performed between four of the materials for figure (**a**) and between 10 of the biscuit samples for figures (**b**,**c**). Products with different letters are significantly different (*p* < 0.05). LEB—*L. edodes* biscuit; AAB—*A. auricula* biscuit; TFB—*T. fuciformis* biscuit.

The addition of powder of any of the three mushroom species used in this study significantly ($p < 0.05$) increased the β-glucan contents of sorghum biscuits and their undigested residues (Figure 3b). After in vitro digestion, β-glucan content was higher in the undigested residue than in the original biscuits showing that it is mostly indigestible, especially in *L. edodes*-enriched samples. The dietary fibre (non-digestible polysaccharides) and the nutrients, phenols and other bioactive compounds that are bound to them that remain after gastric and intestinal digestion, this remaining portion is a prediction of composition of the digesta that would pass into the colon. The solubility of β-glucan during the in vitro digestion process affects the precipitation of β-glucan, and the intestinal digestion condition (pH 7) may cause the β-glucan to aggregate [52]. The undigested residue of *L. edodes* enriched biscuits had an increased β-glucan (1.92–4.74 g/100g dw biscuit) compared to the control (1.10 g/100g dw biscuit). Significantly higher content was also found in the undigested residues of the *A. auricula* enriched biscuits (1.61–3.71 g/100g dw biscuit) and *T. fuciformis* enriched biscuits (1.66–2.79 g/100g dw biscuit). The expression "undigested β-glucan" (%) was used to indicate the percentage of β-glucan from the biscuits that would potentially be available for fermentation by the colonic microbiota. As shown in Figure 3c, the undigested β-glucan (%) increased significantly in mushroom enriched biscuits after in vitro gastrointestinal digestion, except for 5% *A. auricula* and 5% *T. fuciformis*. The control biscuit had a 38.75% of undigested β-glucan, and the percentage increased to 66.36% in 15% *L. edodes* biscuit as well as 61.44% in 15% *A. auricula* biscuit. There was only a 7.11% increase in 15% *T. fuciformis* biscuit.

The β-glucan that is undigested by the stomach or intestine acts as a conveyer of phenolic compounds to the colon, it does this by forming a gel network, which limits the solubilisation of phenolic compounds in stomach and intestine [52]. Another important function of β-glucan in the colon is to be fermented and utilised by gut microbiota. In vivo studies have indicated that consuming mushroom β-glucan or polysaccharides causes an increase in the production of short-chain fatty acids (SCFAs) as a result of fermentation by colonic microbiota; this in return, modulates the gut flora community and regulates inflammatory bowel diseases [61,63,64].

3.6. Principal Component Analysis and Correlations

The principal component 1 (PC1) explained 53.15% of the total variance, while the PC2 accounts for the subsequent 21.42% of the total variance. In this case, Figure 4 explains 74.57% of the variability, which shows that PC1 and PC2 both had a large contribution to explain the response variables. The control biscuit was loaded in the negative axis of PC1 and the positive axis of PC2 (Figure 4a). The control biscuit was positively related to starch content, and more negatively related to insoluble dietary fibre, total dietary fibre, β-glucan, undigested residue β-glucan, bound phenolic content and intestinal digesta phenolic content. PC1 highly discriminated the *L. edodes* enriched biscuits from the control biscuit in terms of the increased substitution levels. *L. edodes* incorporation was more positively related to the free phenolic content, protein, soluble protein content of intestinal supernatant, and antioxidant activity of the bio-accessible fractions. The score of the *A. auricula* and *T. fuciformis* enriched biscuits moved downward with increasing substitution levels. That means that the enrichment with *A. auricula* and *T. fuciformis* was mainly characterised by high dietary fibre, β-glucan and β-glucan of undigested residue content and a high bound phenolic content. Overall, the results analysed by the PCA model illustrated that the enrichment with mushroom powders had a significant effect on the parameters of the sorghum biscuits that were analysed.

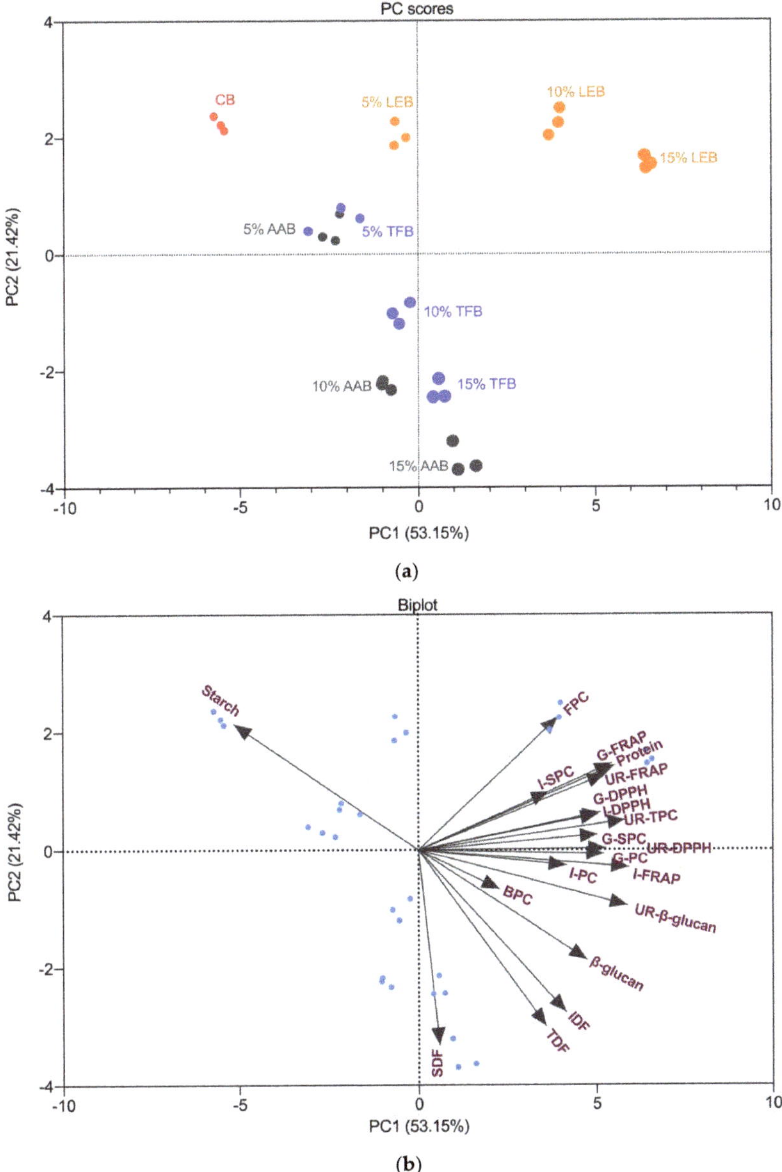

Figure 4. Principal component analysis of principal component scores biplot. PC scores distribution of different biscuit samples incorporated with dried mushrooms (**a**) and Biplot component loading (**b**) as obtained from the principal component analysis. Abbreviations: CB—control biscuits; LEB—*L. edodes* biscuit; AAB—*A. auricula* biscuit; TFB—*T. fuciformis* biscuit; FPC—free phenolic content biscuits; BPC—bound phenolic content of biscuits; G-PC, G-FRAP and G-DPPH represent the phenolic content and antioxidant activity of gastric supernatant; I-PC, I-FRAP and I-DPPH, represented the phenolic content and antioxidant activity of intestinal supernatant; G-SPC—soluble protein content of gastric supernatant; I-SPC—soluble protein content of intestinal supernatant; TDF—total dietary fibre; SDF—soluble dietary fibre; IDF—insoluble dietary fibre; UR-TPC, UR-FRAP and UR-DPPH represented the total phenolic content and total antioxidant activity of the undigested residue; UR-β-glucan—β-glucan content of undigested residue.

Pearson's correlation was conducted to evaluate the correlations between the nutritional composition, bio-accessible phenolic content, peptides, antioxidants and β-glucan remaining after in vitro digestion. The protein content of biscuits was positively correlated with the soluble protein content of gastric supernatant ($r = 0.662$, $p < 0.001$) and intestinal supernatant ($r = 0.464$, $p < 0.01$) (Table S2). This means that the biscuits with a higher protein content had a higher soluble peptide content in the in vitro digested supernatant. The phenolic content of both gastric and intestinal supernatant was positively correlated ($p < 0.001$) with the antioxidant activities (FRAP and DPPH). There was a positive correlation between soluble protein content and phenolic content of gastric supernatant ($r = 0.601$, $p < 0.001$). A similar correlation was observed in the intestinal supernatant ($r = 0.594$, $p = 0.001$). These findings were added of interest and suggested that the soluble peptides in the digest supernatants might interact with phenolic compounds and potentially help transport the bioactive compounds for further intestinal uptake.

Insoluble dietary fibre content was positively correlated with the phenolic content of both gastric and intestinal supernatant and antioxidant activity of the intestinal supernatant. The results showed that there were positive correlations between biscuit insoluble dietary fibre and gastric digesta phenolic content ($r = 0.524$, $p < 0.01$), intestinal digesta phenolic content ($r = 0.463$, $p = 0.01$), intestinal digesta FRAP ($r = 0.634$, $p < 0.001$) and intestinal digesta DPPH ($r = 0.362$, $p < 0.05$). The soluble dietary fibre did not show significant correlations with the intestinal digesta phenolic and antioxidant content. The explanation could be that insoluble dietary fibre has a higher phenolic and antioxidant content than soluble dietary fibre [15].

Mushroom enrichment improved the total insoluble dietary fibre content of sorghum biscuits. Part of the phenolic compounds and antioxidants contained in insoluble dietary fibre were released during the gastrointestinal digestion of biscuits. Phenolic content of gastric and intestinal supernatant from mushroom enriched biscuits were increased. Phenolic compounds trapped in mushroom dietary fibre could be easier to be released than sorghum insoluble fibre during digestion. It might be due to the difference in the structure of insoluble fibre between mushroom (chitin-β-glucan) and sorghum (cellulose). One recent study compared the release of phenolic compounds from different cereal sources of insoluble dietary fibre (wheat, barley, quinoa and triticale) under in vitro simulated gastrointestinal digestion and found that insoluble fibre from quinoa and triticale had a significantly higher phenolic content in both gastric and intestinal fractions than wheat and barley [65].

Dietary fibres were reported to have negative effects on the release and absorption of phenolic compounds by their molecular interactions [66]. A positive correlation was found between β-glucan content of the undigested residue and its total phenolic content ($r = 0.754$, $p < 0.001$), FRAP ($r = 0.588$, $p = 0.001$) and DPPH ($r = 0.706$, $p < 0.001$). Insoluble dietary fibre had a lower correlation coefficient ($r = 0.456$, $p < 0.05$) with total phenolic content of the undigested residue. The β-glucan remaining might be the main insoluble fibre that contributes to the potential delivery of bioactive compounds to the colon. The enriched mushroom insoluble dietary fibre might have a higher content of bound phenolic compounds than sorghum insoluble fibre. Insoluble dietary fibre also favoured the accumulation of β-glucan in the colon as a positive correlation was found between insoluble dietary fibre and β-glucan of undigested residue ($r = 0.788$, $p < 0.001$).

3.7. Nutritional Value

Industrial production of biscuits would require the information of the functional ingredients in the manipulation of the quality and nutritional value of the final products. The optical addition of mushroom powder into cereal products were between 5% and 15% with minor or no negative effects on the sensory acceptability and slight changes in physical characteristics [16]. In the nutritional view of point, enrichment of sorghum biscuits with mushrooms enhanced the dietary fibre and β-glucan content. Previous study has illustrated a close association between an increase of dietary fibre consumption and a lower incidence of obesity and type-2 diabetes in relation to gut microbiota [67]. Daily consumption of 3 g of β-glucan is recommended to have a healthy indicate of cholesterol-lowering [68]. This means that a serving of approximately 50 g of biscuits enriched with 15% mushroom can satisfy this recommendation. Glycaemic carbohydrate is normally high in biscuits, but mushroom fortification reduced the in vitro glycaemic glucose value (Table S2), especially for 15% *A. auricula* and *T. fuciformis* enriched sample with a lower than 30 g glucose per 100 g dry weight. The high fibre content and low glucose level of biscuits could be considered to be diabetic-friendly. The bioavailable small peptides and antioxidants was improved by mushroom enrichment, which can be linked to help relief of chronic diseases. Apart from being nutrient-rich, sorghum biscuits are gluten-free that offers significant opportunities for those population with coeliac disease. These properties added the interest of developing functional products in the future.

4. Conclusions

Enrichment of sorghum biscuits with mushroom powders improved their nutritional quality. Most of mushroom-containing biscuits had a higher content of phenolic compounds quantified in their in vitro digested supernatants and better antioxidant activity than the control biscuit, indicating that mushroom fortification enhanced the bioavailable phenolic content. The digested *L. edodes* and *T. fuciformis* biscuits contained more soluble peptides than the control and *A. auricula* biscuits. The soluble peptides had a small molecular weight that cannot be identified by the SDS-page gel. The undigested residue of mushroom enriched biscuits had a higher remaining portions of phenolic compound and β-glucan than the control, which was related to the insoluble dietary fibre. These findings provide an understanding of the nutritional and functional benefits of mushroom enriched biscuits under the in vitro gastrointestinal digestion and support the idea that mushroom enrichment can potentially increase the bio-accessible phenolic content and the proportions of phenolic compounds and β-glucan delivered to colon. This means the developed biscuits would have a real application in the future with a potential to attenuate chronic diseases. Future experiments are required to evaluate the functionalities of the biscuits ingredients released or remained during digestion through modelled cell line and in vivo studies.

Supplementary Materials: The following are available online at https://www.mdpi.com/article/10.3390/foods10081812/s1, Table S1: Formula of sorghum biscuits enriched with dried mushroom powders, Table S2: Starch, protein and dietary fibre contents and in vitro glycaemic response of sorghum biscuits enriched with dried mushroom powders, Figure S1: Soluble protein content (SPC) of mushrooms and sorghum flour after in vitro gastrointestinal digestion., Figure S2: In vitro bioaccessible antioxidant activity of mushrooms and sorghum flour assessed by FRAP (a) and DPPH (b). Values are means ± standard deviation ($n = 3$). Products with different letters are significantly different ($p < 0.05$). Figure S3: Pearson's correlations between the observed biscuits parameters before and after in vitro digestion.

Author Contributions: J.T.: methodology, formal analysis, investigation, writing—original draft preparation, data curation. M.A.B.: writing—review and editing. G.W.: writing—review and editing. W.B.: supervision. P.C.: supervision. B.T.: supervision. C.S.B.: Conceptualisation, supervision, writing—review and editing. All authors have read and agreed to the published version of the manuscript.

Funding: This research received no external funding.

Acknowledgments: The authors would like to thank Xiaodan Hui and Letitia Stipkovits for the technical support.

Conflicts of Interest: The authors declare no conflict of interest.

References

1. Espitia-Hernandez, P.; Chavez Gonzalez, M.L.; Ascacio-Valdes, J.A.; Davila-Medina, D.; Flores-Naveda, A.; Silva, T.; Ruelas Chacon, X.; Sepulveda, L. Sorghum (*Sorghum bicolor* L.) as a potential source of bioactive substances and their biological properties. *Crit. Rev. Food Sci. Nutr.* **2020**, 1–12. [CrossRef]
2. Girard, A.L.; Awika, J.M. Sorghum polyphenols and other bioactive components as functional and health promoting food ingredients. *J. Cereal Sci.* **2018**, *84*, 112–124. [CrossRef]
3. Acquisgrana, M.d.R.; Gomez Pamies, L.C.; Martinez Amezaga, N.M.J.; Quiroga, F.M.; Ribotta, P.D.; Benítez, E.I. Impact of moisture and grinding on yield, physical, chemical and thermal properties of wholegrain flour obtained from hydrothermally treated sorghum grains. *Int. J. Food Sci. Technol.* **2020**, *55*, 2901–2909. [CrossRef]
4. Gosine, L.; McSweeney, M.B. Consumers' attitudes towards alternative grains: A conjoint analysis study. *Int. J. Food Sci. Technol.* **2019**, *54*, 1588–1596. [CrossRef]
5. Espinosa-Ramírez, J.; Serna-Saldívar, S.O. Functionality and characterization of kafirin-rich protein extracts from different whole and decorticated sorghum genotypes. *J. Cereal Sci.* **2016**, *70*, 57–65. [CrossRef]
6. Marengo, M.; Bonomi, F.; Marti, A.; Pagani, M.A.; Elkhalifa, A.E.O.; Iametti, S. Molecular features of fermented and sprouted sorghum flours relate to their suitability as components of enriched gluten-free pasta. *LWT Food Sci. Technol.* **2015**, *63*, 511–518. [CrossRef]
7. Cabrera-Ramirez, A.H.; Luzardo-Ocampo, I.; Ramirez-Jimenez, A.K.; Morales-Sanchez, E.; Campos-Vega, R.; Gaytan-Martinez, M. Effect of the nixtamalization process on the protein bioaccessibility of white and red sorghum flours during in vitro gastrointestinal digestion. *Food Res. Int.* **2020**, *134*, 109234. [CrossRef] [PubMed]
8. De Morais Cardoso, L.; Pinheiro, S.S.; Martino, H.S.; Pinheiro-Sant'Ana, H.M. Sorghum (*Sorghum bicolor* L.): Nutrients, bioactive compounds, and potential impact on human health. *Crit. Rev. Food Sci. Nutr.* **2017**, *57*, 372–390. [CrossRef]
9. Taylor, J.R.N.; Belton, P.S.; Beta, T.; Duodu, K.G. Increasing the utilisation of sorghum, millets and pseudocereals: Developments in the science of their phenolic phytochemicals, biofortification and protein functionality. *J. Cereal Sci.* **2014**, *59*, 257–275. [CrossRef]
10. Lu, X.; Brennan, M.A.; Serventi, L.; Brennan, C.S. Incorporation of mushroom powder into bread dough-effects on dough rheology and bread properties. *Cereal Chem.* **2018**, *95*, 418–427. [CrossRef]
11. Radzki, W.; Ziaja-Sołtys, M.; Nowak, J.; Topolska, J.; Bogucka-Kocka, A.; Sławińska, A.; Michalak-Majewska, M.; Jabłońska-Ryś, E.; Kuczumow, A. Impact of processing on polysaccharides obtained from button mushroom (*Agaricus bisporus*). *Int. J. Food Sci. Technol.* **2019**, *54*, 1405–1412. [CrossRef]
12. Khongdetch, J.; Laohakunjit, N.; Kaprasob, R. King Boletus mushroom-derived bioactive protein hydrolysate: Characterisation, antioxidant, ACE inhibitory and cytotoxic activities. *Int. J. Food Sci. Technol.* **2021**. [CrossRef]
13. Colosimo, R.; Warren, F.J.; Edwards, C.H.; Ryden, P.; Dyer, P.S.; Finnigan, T.J.A.; Wilde, P.J. Comparison of the behavior of fungal and plant cell wall during gastrointestinal digestion and resulting health effects: A review. *Trends Food Sci. Technol.* **2021**, *110*, 132–141. [CrossRef]
14. Williams, B.A.; Mikkelsen, D.; le Paih, L.; Gidley, M.J. In vitro fermentation kinetics and end-products of cereal arabinoxylans and (1,3;1,4)-β-glucans by porcine faeces. *J. Cereal Sci.* **2011**, *53*, 53–58. [CrossRef]
15. Wang, L.; Brennan, M.A.; Guan, W.; Liu, J.; Zhao, H.; Brennan, C.S. Edible mushrooms dietary fibre and antioxidants: Effects on glycaemic load manipulation and their correlations pre-and post-simulated in vitro digestion. *Food Chem.* **2021**, *351*, 129320. [CrossRef]
16. Tu, J.; Brennan, M.; Brennan, C. An insight into the mechanism of interactions between mushroom polysaccharides and starch. *Curr. Opin. Food Sci.* **2021**, *37*, 17–25. [CrossRef]
17. Biao, Y.; Chen, X.; Wang, S.; Chen, G.; McClements, D.J.; Zhao, L. Impact of mushroom (*Pleurotus eryngii*) flour upon quality attributes of wheat dough and functional cookies-baked products. *Food Sci. Nutr.* **2020**, *8*, 361–370. [CrossRef]
18. Gonzalez, A.; Cruz, M.; Losoya, C.; Nobre, C.; Loredo, A.; Rodriguez, R.; Contreras, J.; Belmares, R. Edible mushrooms as a novel protein source for functional foods. *Food Funct.* **2020**, *11*, 7400–7414. [CrossRef]
19. Jakobek, L.; Matić, P. Non-covalent dietary fiber-Polyphenol interactions and their influence on polyphenol bioaccessibility. *Trends Food Sci. Technol.* **2019**, *83*, 235–247. [CrossRef]
20. Barros, R.G.C.; Pereira, U.C.; Andrade, J.K.S.; de Oliveira, C.S.; Vasconcelos, S.V.; Narain, N. In vitro gastrointestinal digestion and probiotics fermentation impact on bioaccessbility of phenolics compounds and antioxidant capacity of some native and exotic fruit residues with potential antidiabetic effects. *Food Res. Int.* **2020**, *136*, 109614. [CrossRef]
21. Blanco Canalis, M.S.; Baroni, M.V.; Leon, A.E.; Ribotta, P.D. Effect of peach puree incorportion on cookie quality and on simulated digestion of polyphenols and antioxidant properties. *Food Chem.* **2020**, *333*, 127464. [CrossRef]
22. Wu, G.; Hui, X.; Mu, J.; Gong, X.; Stipkovits, L.; Brennan, M.A.; Brennan, C.S. Functionalization of sodium caseinate fortified with blackcurrant concentrate via spray-drying and freeze-drying techniques: The nutritional properties of the fortified particles. *LWT Food Sci. Technol.* **2021**, *142*, 111051. [CrossRef]
23. Gong, X.; Morton, J.D.; Bhat, Z.F.; Mason, S.L.; Bekhit, A.E.D.A. Comparative efficacy of actinidin from green and gold kiwi fruit extract onin vitrosimulated protein digestion of beef Semitendinosusand its myofibrillar protein fraction. *Int. J. Food Sci. Technol.* **2019**, *55*, 742–750. [CrossRef]
24. Wang, L.; Zhao, H.; Brennan, M.; Guan, W.; Liu, J.; Wang, M.; Wen, X.; He, J.; Brennan, C. In vitro gastric digestion antioxidant and cellular radical scavenging activities of wheat-shiitake noodles. *Food Chem.* **2020**, *330*, 127214. [CrossRef]

25. Li, Q.; Yang, S.; Li, Y.; Huang, Y.; Zhang, J. Antioxidant activity of free and hydrolyzed phenolic compounds in soluble and insoluble dietary fibres derived from hulless barley. *LWT Food Sci. Technol.* **2019**, *111*, 534–540. [CrossRef]
26. Polat, H.; Dursun Capar, T.; Inanir, C.; Ekici, L.; Yalcin, H. Formulation of functional crackers enriched with germinated lentil extract: A Response Surface Methodology Box-Behnken Design. *LWT Food Sci. Technol.* **2020**, *123*, 109065. [CrossRef]
27. Wu, G.; Hui, X.; Stipkovits, L.; Rachman, A.; Tu, J.; Brennan, M.A.; Brennan, C.S. Whey protein-blackcurrant concentrate particles obtained by spray-drying and freeze-drying for delivering structural and health benefits of cookies. *Innov. Food Sci. Emerg. Technol.* **2021**, *68*, 102606. [CrossRef]
28. McCleary, B.V.; Draga, A. Measurement of beta-glucan in mushrooms and mycelial products. *J. AOAC Int.* **2016**, *99*, 364–373. [CrossRef]
29. Prosky, L.; Asp, N.-G.; Schweizer, T.F.; Devries, J.W.; Furda, I. Determination of insoluble and soluble dietary fiber in foods and food products: Collaborative study. *J. AOAC Int.* **1992**, *75*, 360–367. [CrossRef]
30. Lang, G.H.; Lindemann, I.D.S.; Ferreira, C.D.; Hoffmann, J.F.; Vanier, N.L.; de Oliveira, M. Effects of drying temperature and long-term storage conditions on black rice phenolic compounds. *Food Chem.* **2019**, *287*, 197–204. [CrossRef] [PubMed]
31. Irondi, E.A.; Adegoke, B.M.; Effion, E.S.; Oyewo, S.O.; Alamu, E.O.; Boligon, A.A. Enzymes inhibitory property, antioxidant activity and phenolics profile of raw and roasted red sorghum grains in vitro. *Food Sci. Hum. Wellness* **2019**, *8*, 142–148. [CrossRef]
32. Rocchetti, G.; Rizzi, C.; Pasini, G.; Lucini, L.; Giuberti, G.; Simonato, B. Effect of moringa oleifera l. leaf powder addition on the phenolic bioaccessibility and on in vitro starch digestibility of durum wheat fresh pasta. *Foods* **2020**, *9*, 628. [CrossRef] [PubMed]
33. Zieliński, H.; Szawara-Nowak, D.; Wronkowska, M. Bioaccessibility of anti-AGEs activity, antioxidant capacity and phenolics from water biscuits prepared from fermented buckwheat flours. *LWT Food Sci. Technol.* **2020**, *123*, 109051. [CrossRef]
34. Liu, C.; Ge, S.; Yang, J.; Xu, Y.; Zhao, M.; Xiong, L.; Sun, Q. Adsorption mechanism of polyphenols onto starch nanoparticles and enhanced antioxidant activity under adverse conditions. *J. Funct. Foods* **2016**, *26*, 632–644. [CrossRef]
35. Dantas, A.M.; Mafaldo, I.M.; Oliveira, P.M.L.; Lima, M.D.S.; Magnani, M.; Borges, G. Bioaccessibility of phenolic compounds in native and exotic frozen pulps explored in Brazil using a digestion model coupled with a simulated intestinal barrier. *Food Chem.* **2019**, *274*, 202–214. [CrossRef] [PubMed]
36. Meng, X.j.; Tan, C.; Feng, Y. Solvent extraction and in vitro simulated gastrointestinal digestion of phenolic compounds from purple sweet potato. *Int. J. Food Sci. Technol.* **2019**, *54*, 2887–2896. [CrossRef]
37. Quan, W.; Tao, Y.; Lu, M.; Yuan, B.; Chen, J.; Zeng, M.; Qin, F.; Guo, F.; He, Z. Stability of the phenolic compounds and antioxidant capacity of five fruit (apple, orange, grape, pomelo and kiwi) juices during in vitro-simulated gastrointestinal digestion. *Int. J. Food Sci. Technol.* **2018**, *53*, 1131–1139. [CrossRef]
38. Wu, T.; Taylor, C.; Nebl, T.; Ng, K.; Bennett, L.E. Effects of chemical composition and baking on in vitro digestibility of proteins in breads made from selected gluten-containing and gluten-free flours. *Food Chem.* **2017**, *233*, 514–524. [CrossRef]
39. Sousa, R.; Portmann, R.; Dubois, S.; Recio, I.; Egger, L. Protein digestion of different protein sources using the INFOGEST static digestion model. *Food Res. Int.* **2020**, *130*, 108996. [CrossRef] [PubMed]
40. Shen, C.R.; Liu, C.L.; Lee, H.P.; Chen, J.K. The identification and characterization of chitotriosidase activity in pancreatin from porcine pancreas. *Molecules* **2013**, *18*, 2978–2987. [CrossRef]
41. Vilcacundo, R.; Martínez-Villaluenga, C.; Hernández-Ledesma, B. Release of dipeptidyl peptidase IV, α-amylase and α-glucosidase inhibitory peptides from quinoa (Chenopodium quinoa Willd.) during in vitro simulated gastrointestinal digestion. *J. Funct. Foods* **2017**, *35*, 531–539. [CrossRef]
42. Joye, I. Protein digestibility of cereal products. *Foods* **2019**, *8*, 199. [CrossRef] [PubMed]
43. Phongthai, S.; Rawdkuen, S. Fractionation and characterization of antioxidant peptides from rice bran protein hydrolysates stimulated by in vitro gastrointestinal digestion. *Cereal Chem.* **2020**, *97*, 316–325. [CrossRef]
44. Morales, D.; Miguel, M.; Garces-Rimon, M. Pseudocereals: A novel source of biologically active peptides. *Crit. Rev. Food Sci. Nutr.* **2020**, *9*, 1537–1544. [CrossRef]
45. Ashwath Kumar, K.; Sharma, G.K.; Anilakumar, K.R. Influence of multigrain premix on nutritional, in-vitro and in-vivo protein digestibility of multigrain biscuit. *J. Food Sci. Technol.* **2019**, *56*, 746–753. [CrossRef]
46. Sciarini, L.S.; Bustos, M.C.; Vignola, M.B.; Paesani, C.; Salinas, C.N.; Perez, G.T. A study on fibre addition to gluten free bread: Its effects on bread quality and in vitro digestibility. *J. Food Sci. Technol.* **2017**, *54*, 244–252. [CrossRef]
47. Papoutsis, K.; Zhang, J.; Bowyer, M.C.; Brunton, N.; Gibney, E.R.; Lyng, J. Fruit, vegetables, and mushrooms for the preparation of extracts with alpha-amylase and alpha-glucosidase inhibition properties: A review. *Food Chem.* **2021**, *338*, 128119. [CrossRef] [PubMed]
48. Baczek, N.; Jarmulowicz, A.; Wronkowska, M.; Haros, C.M. Assessment of the glycaemic index, content of bioactive compounds, and their in vitro bioaccessibility in oat-buckwheat breads. *Food Chem.* **2020**, *330*, 127199. [CrossRef] [PubMed]
49. Ketnawa, S.; Suwannachot, J.; Ogawa, Y. In vitro gastrointestinal digestion of crisphead lettuce: Changes in bioactive compounds and antioxidant potential. *Food Chem.* **2020**, *311*, 125885. [CrossRef] [PubMed]
50. Lu, X.; Brennan, M.A.; Narciso, J.; Guan, W.; Zhang, J.; Yuan, L.; Serventi, L.; Brennan, C.S. Correlations between the phenolic and fibre composition of mushrooms and the glycaemic and textural characteristics of mushroom enriched extruded products. *LWT Food Sci. Technol.* **2020**, *118*, 108730. [CrossRef]
51. Sanchez, C. Reactive oxygen species and antioxidant properties from mushrooms. *Synth. Syst. Biotechnol.* **2017**, *2*, 13–22. [CrossRef] [PubMed]

52. Mosele, J.I.; Motilva, M.J.; Ludwig, I.A. Beta-glucan and phenolic compounds: Their concentration and behavior during in vitro gastrointestinal digestion and colonic fermentation of different barley-based food products. *J. Agric. Food Chem.* **2018**, *66*, 8966–8975. [CrossRef] [PubMed]

53. Dong, R.; Liu, S.; Zheng, Y.; Zhang, X.; He, Z.; Wang, Z.; Wang, Y.; Xie, J.; Chen, Y.; Yu, Q. Release and metabolism of bound polyphenols from carrot dietary fiber and their potential activity in in vitro digestion and colonic fermentation. *Food Function* **2020**, *11*, 6652–6665. [CrossRef] [PubMed]

54. Gowd, V.; Karim, N.; Shishir, M.R.I.; Xie, L.; Chen, W. Dietary polyphenols to combat the metabolic diseases via altering gut microbiota. *Trends Food Sci. Technol.* **2019**, *93*, 81–93. [CrossRef]

55. Sari, M.; Prange, A.; Lelley, J.I.; Hambitzer, R. Screening of beta-glucan contents in commercially cultivated and wild growing mushrooms. *Food Chem.* **2017**, *216*, 45–51. [CrossRef]

56. Du, B.; Meenu, M.; Liu, H.; Xu, B. A concise review on the molecular structure and function relationship of β-glucan. *Int. J. Mol. Sci.* **2019**, *20*, 4032. [CrossRef]

57. Alahmed, A.; Simsek, S. Pre-harvest glyphosate application effects on properties of β-glucan from oat groats. *J. Cereal Sci.* **2020**, *96*, 103119. [CrossRef]

58. Morales, D.; Smiderle, F.R.; Piris, A.J.; Soler-Rivas, C.; Prodanov, M. Production of a β-d-glucan-rich extract from Shiitake mushrooms (Lentinula edodes) by an extraction/microfiltration/reverse osmosis (nanofiltration) process. *Innov. Food Sci. Emerg. Technol.* **2019**, *51*, 80–90. [CrossRef]

59. Shoukat, M.; Sorrentino, A. Cereal β-glucan: A promising prebiotic polysaccharide and its impact on the gut health. *Int. J. Food Sci. Technol.* **2021**, *56*, 2088–2097. [CrossRef]

60. Zhuang, H.; Chen, Z.; Feng, T.; Yang, Y.; Zhang, J.; Liu, G.; Li, Z.; Ye, R. Characterization of Lentinus edodes beta-glucan influencing the in vitro starch digestibility of wheat starch gel. *Food Chem.* **2017**, *224*, 294–301. [CrossRef]

61. Vetvicka, V.; Gover, O.; Karpovsky, M.; Hayby, H.; Danay, O.; Ezov, N.; Hadar, Y.; Schwartz, B. Immune-modulating activities of glucans extracted from Pleurotus ostreatus and Pleurotus eryngii. *J. Funct. Foods* **2019**, *54*, 81–91. [CrossRef]

62. Singh, A.; Dutta, P.K. Extraction of chitin-glucan complex from Agaricus bisporus: Characterization and antibacterial activity. *J. Polym. Mater.* **2017**, *34*, 1–9.

63. Wang, X.; Wang, W.; Wang, L.; Yu, C.; Zhang, G.; Zhu, H.; Wang, C.; Zhao, S.; Hu, C.A.; Liu, Y. Lentinan modulates intestinal microbiota and enhances barrier integrity in a piglet model challenged with lipopolysaccharide. *Food Funct.* **2019**, *10*, 479–489. [CrossRef] [PubMed]

64. Zhao, R.; Ji, Y.; Chen, X.; Su, A.; Ma, G.; Chen, G.; Hu, Q.; Zhao, L. Effects of a beta-type glycosidic polysaccharide from Flammulina velutipes on anti-inflammation and gut microbiota modulation in colitis mice. *Food Funct.* **2020**, *11*, 4259–4274. [CrossRef] [PubMed]

65. Liu, M.; Liu, X.; Luo, J.; Bai, T.; Chen, H. Effect of digestion on bound phenolic content, antioxidant activity and hypoglycemic ability of insoluble dietary fibre from four Triticeae crops. *J. Food Biochem.* **2021**, *45*, e13746. [CrossRef] [PubMed]

66. Quirós-Sauceda, A.E.; Palafox-Carlos, H.; Sáyago-Ayerdi, S.G.; Ayala-Zavala, J.F.; Bello, L.A.; Álvarez-Parrilla, E.; de la Rosad, L.A.; González-Córdova, A.F.; González-Aguilar, G.A. Dietary fiber and phenolic compounds as functional ingredients: Interaction and possible effect after ingestion. *Food Funct.* **2014**, *5*, 1063–1072. [CrossRef]

67. Liu, H.; Wu, H.; Wang, Q. Health-promoting effects of dietary polysaccharide extracted from Dendrobium aphyllum on mice colon, including microbiota and immune modulation. *Int. J. Food Sci. Technol.* **2019**, *54*, 1684–1696. [CrossRef]

68. FDA. Food labelling: Health claims; soluble fiber from whole oats and risk of coronary heart disease. *Fed. Regist.* **1997**, *62*, 5343–15344.

Article

Physicochemical Properties and Drivers of Liking and Disliking for Cooked Rice Containing Various Types of Processed Whole Wheat

Da-Been Lee [1,†], Mi-Ran Kim [1,†], Jeong-Ae Heo [2], Yang-Soo Byeon [1,3] and Sang-Sook Kim [1,*]

1   Research Group of Food Processing, Korea Food Research Institute, Wanju-gun 55365, Korea; Lee.Da-been@kfri.re.kr (D.-B.L.); ranni1027@kfri.re.kr (M.-R.K.); 50024@kfri.re.kr (Y.-S.B.)
2   Technical Assistance Center, Korea Food Research Institute, Wanju-gun 55365, Korea; Heo.Jeongae@kfri.re.kr
3   Department of Biotechnology, College of Life Sciences and Biotechnology, Korea University, Seoul 02841, Korea
*   Correspondence: sskim@kfri.re.kr; Tel.: +82-63-219-9042
†   Both authors contributed equally to this work.

Abstract: For utilization of whole wheat (WW) in cooked rice products, WW was processed by four different methods (steeping (S_WW), milling (M_WW), enzymatic treatment (E_WW), and passing through a roll mill (1 mm) (R_WW)). Additionally, the physicochemical properties of cooked rice containing various processed wheat were investigated. The hardness of the cooked rice decreased significantly with R_WW and E_WW compared to WW. As a result of a consumer acceptance test, the cooked rice samples containing M_WW and E_WW with high liking scores frequently included 'chewiness' as a reason for liking, and the cooked rice with WW and S_WW was mentioned as being 'too hard' as a reason for disliking. The cooked rice with R_WW, which had the lowest liking score, was mentioned as having appearance characteristics such as 'husk', 'clumpy appearance', and 'messy appearance' as reasons for disliking. The overall results of this study suggest the inclusion of M_WW or E_WW with cooked rice considering health-related benefits and consumer acceptability.

Keywords: cooked rice; processed whole wheat; physicochemical properties; consumer acceptance; drivers of liking and disliking

Citation: Lee, D.-B.; Kim, M.-R.; Heo, J.-A.; Byeon, Y.-S.; Kim, S.-S. Physicochemical Properties and Drivers of Liking and Disliking for Cooked Rice Containing Various Types of Processed Whole Wheat. *Foods* **2021**, *10*, 1470. https://doi.org/10.3390/foods10071470

Academic Editors: Charles Brennan, Luca Serventi, Rana Mustafa and Mike Sissons

Received: 22 April 2021
Accepted: 21 June 2021
Published: 25 June 2021

Publisher's Note: MDPI stays neutral with regard to jurisdictional claims in published maps and institutional affiliations.

1. Introduction

Rice is a major food grain in Korea, and 95% of it is a commercial staple food consumed in the form of cooked rice [1]. Various cooked rice products have been developed by mixing other grains. In particular, wheat, as the second major grain, is consumed at levels of approximately 32 kg per capita per year in Korea [2]. However, most wheat is imported from the United States, Australia, and Canada, and only approximately 2% of the total amount consumed is produced in Korea [2]. Wheat is rich in starch, protein, dietary fiber, minerals, phenolic compounds, and phytochemicals [3]. As interest in health increases, the demand and interest in whole wheat is increasing. The consumption of these processed products is effective in preventing adult diseases such as hypertension and diabetes. In particular, whole wheat contains fiber, vitamin B, vitamin E, iron, and magnesium, is particularly rich in food fiber and is known to lower the risk of obesity, stroke, heart disease, diabetes and colon cancer [4–6]. Additionally, whole wheat has a low glycemic index (GI), which is good for diabetes management and lowers blood cholesterol, reducing the risk of arteriosclerosis and hypercholesterolemia [7]. The benefits of eating whole grain foods are well known, but their use is limited because of the resulting low sensory quality for processed foods [8].

Food development is a consumer-oriented task, and understanding consumers' preferences has become a key factor in the success of such research [9,10]. Therefore, it is important to understand how consumers perceive products [11]. Thus, obtaining consumer feedback on the sensory description of a product as an alternative to conventional sensory profiling has become of great interest over the past two decades [12]. The conventional approach used to understand consumers' preferences is internal and external preference mapping, which combines descriptive data provided by trained panels with acceptance tests performed by consumers [13]. The preference mapping technique requires consumer testing by consumers for acceptability and quantitative descriptive analysis of sensory properties by trained panels, which is expensive and time-consuming [9,12]. Furthermore, the terms generated by the trained panel may differ from those used by consumers and can be difficult for consumers to understand [10,14]. Therefore, ten Kleij et al. [14] proposed a textual analysis of open-ended questions to complement the preference mapping technique. In many studies, free-comment responses explicitly written by consumers in the form of open-ended questions have been mainly used to reduce respondents' complaints by allowing them to explain their responses to other questionnaire items [15]. Moreover, free-comment responses have a definite advantage of being easy to understand because consumers' opinions are expressed in their own language and do not require deep thinking. In addition, it is possible to obtain rich information, including what the researcher did not predict [16].

The objectives of this study were to evaluate the consumer acceptability of processed whole wheat products for cooked rice with improved texture using various processing methods to find out the potential of entering the market. Additionally, functional activity and physicochemical properties were investigated. In addition to consumer acceptances, consumer perceptions were studied by applying two open-ended questions to identify the drivers of liking and disliking for cooked rice samples of cooked rice samples with various types of processed whole wheat.

2. Materials and Methods

2.1. Physicochemical Property

2.1.1. Materials

The five types of cooked rice were prepared with 60% rice (Sindongjin, Yeonggwang, Jeonlanamdo, Korea, 2019) and 40% whole wheat or various types of processed whole wheat (Beakchal, Yeonggwang, Jeonlanamdo, Korea, 2018). Whole wheat (WW) was processed in four ways: steeping (S_WW), milling (M_WW), enzymatic treatment (E_WW), and passing through a roll mill (1 mm) (R_MM). The S_WW was WW steeped with tap water at 4 °C for 24 h; M_WW was WW milled to remove 5% bran using a pearling machine (2RSB-10FS, Kett, Tokyo, Japan); E_WW was WW treated with 5% viscozyme (Novozyme, Bagværd, Denmark) at 50 °C for 24 h; R_WW was WW passed three times through a roll mill with gaps of 2.0, 1.5 and 1.0 mm, sequentially, after steeping in tap water at 4 °C for 16 h.

2.1.2. Cooked Rice with Whole Wheat

Rice (540 g) and wheat (360 g) were washed with water using a rice cleaner (PR7J, Aiho, Tokyo, Japan). Filltered water was added to the rice and wheat at a weight ratio of 1.6:1 (14% moisture basis) and, then, the mixture was cooked using an electric rice cooker (CRP-LHTR1010FWM, Cuckoo, Yangsan-si, Gyeongsangnam-do, Korea). Cooked rice in a bowl was stirred smoothly five times with a spoon and then cooled for 5 min at room temperature. The stirring and cooling procedures of the cooked rice were repeated twice.

2.1.3. Physical Property Analysis of Cooked Rice with Whole Wheat

The moisture content of cooked rice samples containing processed WW was measured according to American Association of Cereal Chemists (AACC) Method 44-15A (AACC, 2010) using a dry oven (HK-DO1000F, Hankuk S & I Co., Hwaseong, Korea). Texture profile analysis (TPA) of cooked rice was performed based on the modified AACC Method 74-09 (AACC, 2010) with a Texture Analyzer TA-XT plus system (Stable Micro System Ltd., Haslemere, UK). The cooked rice (12 g) was placed into a cylindrical container (4 cm diameter $\times$ 1 cm length) and compressed by approximately 40% using a 20 mm plunger. The return distance was 30 mm, the return speed was 1.7 mm/s, and the contact force was 50 g. Hardness, adhesiveness, springiness, cohesiveness, chewiness, and resilience were determined from the two-cycle curves using Texture Export for Window (Stable Micro Systems, Godalming, UK). All physicochemical analyses were repeated three times.

2.1.4. Starch Hydrolysis Index

The in vitro starch hydrolysis index (HI) of cooked rice samples containing processed WW was determined according to a modified method [17]. Two grams of freeze-dried cooked rice powder was mixed with 100 mL of 0.05 M sodium potassium phosphate buffer (pH 6.9). Then, 110 U pancreatic amylase (Type I-A, Sigma Aldrich, St. Louis, MO, USA), was added, and the mixture was incubated in a shaking incubator (120 rpm) at 37 °C. Next, 2 mL aliquots were added every 0, 30, 60, 90, 120 and 180 min, boiled at 95 °C for 5 min, and cooled on ice. The aliquots were centrifuged at 4000 rpm for 10 min, and the reducing sugar content in the supernatant was measured by a modified method from Somogyi [18]. For each starch hydrolysis curve, the area under the curve (AUC) was calculated using SigmaPlot 13.0 (Systat Software Inc., Point Richmond, CA, USA). Hydrolysis indexes were repeated three times and calculated with the following equation.

$$HI = (AUC \text{ test food} / \text{average AUC reference sample}) \times 100 \tag{1}$$

2.2. Consumer Acceptance Test

2.2.1. Sample Preparation and Presentation

The cooked rice samples used in the consumer acceptance test were prepared in the same way as those used to assess the physicochemical properties (Section 2.1.2). The cooked rice samples (50 g) were placed in a bowl (85 $\times$ 50 mm, diameter $\times$ depth) with a lid using a stainless-steel scoop. The cooked rice samples were coded with three-digit random numbers and served at 50 $\pm$ 5 °C. The samples were stored in a heating cabinet (HA-DB3000, Hains Co., Incheon, Korea) to maintain the temperature throughout the evaluation session. The serving order of the samples was determined by a William Latin square design [19], and samples were presented in sequential monadic order to avoid carryover effects [20]. Filtered water at ambient temperature was provided to the consumers to cleanse the palate between samples.

2.2.2. Cooked Rice with Whole Wheat

A total of 103 consumers were students and research scientists at the Korea Food Research Institute (KFRI) recruited through e-mails. The participants were 36% male and 64% female, with an average age of 34.3 years. Written consent was given by all consumers. The testing was carried out in the sensory laboratory of the Korea Food Research Institute (KFRI) equipped with individual booths. Ten to twelve consumers assessed in the experiment at one time and all consumers completed the experiment on the same day. In the consumer acceptance test, the consumers rated the overall appearance, odor, taste/flavor, and texture liking of the samples on a standard 9-point hedonic scale (1 = dislike extremely, 5 = neither like nor dislike, 9 = like extremely) [21]. After evaluating their levels of liking, the consumers were asked to answer 2 open questions, freely describing the reasons for liking and disliking each sample. The procedure followed the method explained by Symoneaux et al. [10]. Answering was not mandatory. In this way, the participants could express only reasons for liking, only reasons for disliking, both or none for each sample. Data collection was carried out with Compusense 5.0 software (Compusense Inc., Guelph, ON, Canada).

2.3. Statistical Analysis

2.3.1. Physicochemical Properties

The data were analyzed using SPSS statistical software (version 21; SPSS Corp., Chicago, IL, USA). ANOVA followed by Duncan's multiple range test was applied to determine significant differences.

2.3.2. Consumer Acceptance Test

For the liking data, analysis of variance (ANOVA) using a general linear model (GLM) was performed to determine the effects of the sample as a fixed source of variation and panel as a random effect. When the effects were significant, significant differences were calculated using Duncan's multiple range test ($p < 0.05$). Pearson correlation coefficients between liking attributes were also calculated.

The open-ended questions were qualitatively analyzed. The reasons for liking and disliking the samples described by the panelist were written in each column and were divided by semicolons (;). The frequency of descriptors was calculated by textual analysis. Next, terms with similar meanings were combined with representative words. For example, 'hardness', 'firmness', and 'rigidity' were merged to 'hardness'. The synonyms were reviewed by three other native researchers for validation. After refining the descriptors for statistical analysis, attributes that were mentioned by more than 5% of the consumers for at least one sample were used for further analysis. Finally, a contingency table was generated with rows of samples and columns as reasons for liking and disliking. As recommended by Symoneaux et al. [10], chi-square analysis per cell was applied to the contingency table to identify statistically significant components within a matrix. Additionally, correspondence analysis (CA) was performed to visually summarize the relationship between samples and descriptions. According to the suggestion of Lê et al. [22], only the terms with cos2 values above 0.8 were shown in the perceptual map to distinguish the attributes that are highly correlated with dimensions 1 and 2.

The data were analyzed using SPSS statistical software (version 21; SPSS Corp., Chicago, IL, USA) and FactoMineR 2.3 [23] of R Studio 1.4.1103 [24] based on R statistical system 4.0.3 [25].

3. Results

3.1. Characteristics of Cooked Rice with Various Types of Processed Whole Wheat

The water content and TPA results of the cooked rice with various types of processed WW are shown in Table 1. The water content was the highest in the cooked rice with R_WW (62.27%), while it was the lowest in the cooked rice with WW (57.77%). The hardness of the cooked rice decreased significantly with R_WW and E_WW. The cohesiveness of cooked rice with E_WW and R_WW markedly increased. There was no significant difference in the HI values among cooked rice samples with various types of processed WW.

Table 1. Water content, TPA and HI of cooked rice samples with various processed whole wheat.

Sample	Water Content (%) ***1	TPA						HI
		Hardness (g) ***	Adhesiveness	Springiness	Cohesiveness ***	Chewiness (g) ***	Resilience **	
WW	57.8 c2	4693 a	−79.2	0.50	0.35 b	810 a	0.16 ab	526
S_WW	60.6 b	3830 b	−85.8	0.48	0.31 c	563 bc	0.13 c	563
M_WW	60.6 b	3822 b	−87.1	0.43	0.32 bc	532 bc	0.14 bc	593
E_WW	60.8 b	3349 c	−82.9	0.47	0.38 a	600 b	0.17 a	596
R_WW	62.3 a	2469 d	−80.2	0.52	0.39 a	505 c	0.18 a	602

TPA: Texture profile analysis, HI: starch hydrolysis index, WW: cooked rice with whole wheat, S_WW: cooked rice with steeped whole wheat, M_WW: cooked rice with milled whole wheat, E_WW: cooked rice with enzyme treated whole wheat, R_WW: cooked rice with rolled whole wheat. [1] Significance levels are as follows: (***) $p < 0.001$; (**) $p < 0.01$. [2] Mean within a column not sharing a superscript letter are significantly different ($p < 0.05$).

3.2. Consumer Acceptance Test

3.2.1. Consumers' Liking Scores

The ANOVA results showed that the cooked rice samples were different in terms of all liking attributes ($p < 0.05$). The mean scores of liking for cooked rice samples are presented in Table 2. Based on the consumer ratings, the overall liking for the cooked rice with M_WW (5.8) was the highest, while that with the cooked rice with R_WW (4.4) was the lowest. In particular, the appearance liking score of the cooked rice with R_WW (4.0), in which the bran of the wheat kernel surface was destroyed, was significantly lower than that of the other samples. The results of Pearson's correlation analysis showed a high correlation between the overall liking and taste/flavor liking score (r = 0.76) and texture liking score (r = 0.74). The texture liking score clearly indicated the enhanced textural quality of the cooked rice with M_WW and E_WW.

Table 2. Mean liking scores of various processed whole wheat samples.

Sample	Overall ***[1]	Appearance ***	Odor ***	Texture ***	Taste/Flavor ***
r^2	1.00	0.59	0.59	0.74	0.76
WW	4.8 ± 1.9 cd[3]	5.0 ± 1.8 b	5.9 ± 1.6 ab	4.6 ± 2.3 b	5.4 ± 1.8 b
S_WW	5.0 ± 1.8 bc	5.0 ± 1.9 b	5.9 ± 1.5 ab	4.5 ± 2.1 b	5.5 ± 1.7 ab
M_WW	5.8 ± 1.5 a	5.6 ± 1.7 a	6.1 ± 1.5 a	5.7 ± 1.8 a	5.9 ± 1.6 a
E_WW	5.3 ± 1.7 b	5.0 ± 1.7 b	5.5 ± 2.0 b	5.5 ± 1.7 a	5.6 ± 2.0 ab
R_WW	4.4 ± 1.8 d	4.0 ± 1.9 c	4.7 ± 1.9 c	4.6 ± 1.9 b	4.6 ± 2.0 c

WW: cooked rice with whole wheat, S_WW: cooked rice with steeped whole wheat, M_WW: cooked rice with milled whole wheat, E_WW: cooked rice with enzyme treated whole wheat, R_WW: cooked rice with rolled whole wheat. [1] Significance levels are as follows: (***) $p < 0.001$. [2] Correlation coefficients between overall liking and other likings attributes. [3] Mean within a column not sharing a superscript letter are significantly different ($p < 0.05$).

3.2.2. Drivers of Liking and Disliking

Consumers described 117 and 139 reasons for liking and disliking the samples, respectively. Three native researchers merged similar terms to representative words, resulting in 67 and 80 reasons for liking and disliking, respectively. Among these terms, 16 reasons for each like and dislike were mentioned at least 5% in one sample. The reasons for liking the samples were mainly terms such as 'corn odor', 'taste', 'nutty flavor', 'odor', and 'chewiness', which were described 563 times in total. The reasons for disliking the samples were frequently mentioned, including attributes such as 'too watery', 'strong after-effect', 'too hard', 'roughness', and 'appearance', a total of 642 times. In addition, 100 consumers answered no reason for liking, and 82 consumers answered no reason for disliking, regardless of the sample.

The chi-square per cell enables the identification of the more or less used attributes for samples [10]. From Table 3, it can be observed that the cooked rice with S_WW was more often considered 'not watery' than the other samples, but was not described as exhibiting 'softness'. The cooked rice with E_WW received more mentions of 'chewiness' as a reason for liking than the other samples. In contrast, the cooked rice with R_WW had the lowest citation frequency for 'chewiness'. As a result of describing the reasons for disliking samples, the cooked rice with R_WW was cited more often as having negative attributes related to appearance, such as 'husk', 'clumpy appearance', and 'messy appearance'. The cooked rice with WW and S_WW had similar tendencies. In both samples, 'too watery' was mentioned less than expected, and 'too hard' was mentioned more than expected. In addition, the cooked rice with WW was mentioned as 'feels undercooked', and the cooked rice with S_WW had more 'roughness' mentions than other samples. Twenty-eight subjects even responded that there was 'no liking reason' for the cooked rice with WW. The cooked rice with M_WW was characterized by a higher citation frequency for 'too watery' and 'too sticky' and fewer citations for 'roughness' as dislikes. However, the twenty-four consumers responded that there was 'no disliking reason' for cooked rice with M_WW. The cooked rice containing E_WW was significantly less frequently referred to as 'too hard' for reasons of disliking.

Table 3. The frequency of the major reasons of liking and disliking for cooked rice samples by more than 5% of consumer in the free comment as a response to open-ended questions.

Major Driver of (Dis) Liking	Samples					
	WW	S_WW	M_WW	E_WW	R_WW	Total
Reasons of liking						
Appearance	6	4	9	3	1	23
Odor	8	8	4	8	9	37
Corn odor	6	6	4	8	2	26
Nutty odor	5	5	5	2	4	21
Taste	5	4	6	8	3	26
Sweetness	2	6	7	6	3	24
Nutty flavor	6	9	5	3	9	32
Chewiness	16	17	20	25 [(+)1*2]	2 [(−)***]	80
Softness	2	0 [(−)*]	8	5	8	23
Sticky	4	6	5	3	4	22
Not watery	2	5 [(+)*]	1	1	0	9
Less after-effect	3	4	5	1	6	19
Easy to chew	0	1	4	0	10 [(+)***]	15
Looks healthy	6	3	3	2	2	16
Harmonious	0	0	5	3	5	13
No liking reasons	28 [(+)*]	19	14	13	26	100
Reasons of disliking						
Appearance	9	6	4	10	22 [(+)***]	51
Husk	0	2	1	3	7 [(+)***]	13
Clumpy appearance	0	0	1	0	6 [(+)***]	7
Messy appearance	0	0	0	0	6 [(+)***]	6
Odor	2	2	2	3	7	16
Off-odor	0 [(−)***]	3	1	10	7	21
Taste	0	0	0	1	5 [(+)***]	6
Texture	4	2	3	1	6	16
Too watery	8 [(−)***]	5 [(−)**]	29 [(+)***]	10	25	77
Too hard	33 [(+)***]	22 [(+)*]	10	3 [(−)**]	0 [(−)***]	68
Roughness	17	18 [(+)*]	1 [(−)***]	8	10	54
Too sticky	2	1	8 [(+)***]	4	3	18
Chewiness	3	5	1	1	1	11
Strong after-effect	14	20	8	11	15	68
Feels undercooked	10 [(+)***]	1	1	1	0	13
No disliking reasons	13	21	24 [(+)***]	14	10 [(−)***]	82
Driver of (dis)liking						
Total reasons of liking	103	109	128	122	101	563
Total reasons of disliking	145	122	97	115	163	642

[1] (+) or (−) indicate that observed value is higher or lower than the expected value. [2] Significance levels are as follows: (***) $p < 0.001$; (**) $p < 0.01$; (*) $p < 0.05$, p-value obtain from Chi-square per cell.

Figure 1 shows the perceptual map result of correspondence analysis (CA). Dimensions (Dim) 1 and 2 explained 77.4% of the total variation (52.4% and 25.0% of the variation, respectively). The results of CA regarding the reasons for liking and disliking samples were largely divided into three parts. Dim 1 differentiated the cooked rice with R_WW (positive axis) with the lowest overall liking score from the cooked rice with WW and S_WW (negative axis). The subjects commonly mentioned 'taste' and 'odor' as reasons for disliking the cooked rice with R_WW. In addition, the subjects particularly disliked the cooked rice with R_WW due to the appearance, mentioning terms such as 'messy appearance', 'clumpy appearance', and 'husk'. On the other hand, the cooked rice with R_WW was mentioned as being 'easy to chew' as a reason for liking. The cooked rice with

WW and S_WW was described with 'too hard' and 'roughness' as reasons for disliking. The cooked rice with R_WW (positive axis), WW, and S_WW (negative axis), separated by Dim 1, had similar overall liking scores, but there were differences in the reasons for liking and disliking these samples. Subsequently, the cooked rice with E_WW and M_WW, with relatively high overall liking scores, was on the negative axis of Dim 2. These samples were associated with the term 'harmonious' and texture attributes such as 'chewiness' and 'softness'. However, some subjects mentioned 'too sticky' as a reason for disliking these samples. To summarize the finding of the CA results, Dim 1 was mainly defined by reasons of liking and disliking (positive axis: appearance-related attributes, negative axis: texture-related attributes), and Dim 2 was described by overall liking score (positive axis: relatively low, negative axis: relatively high).

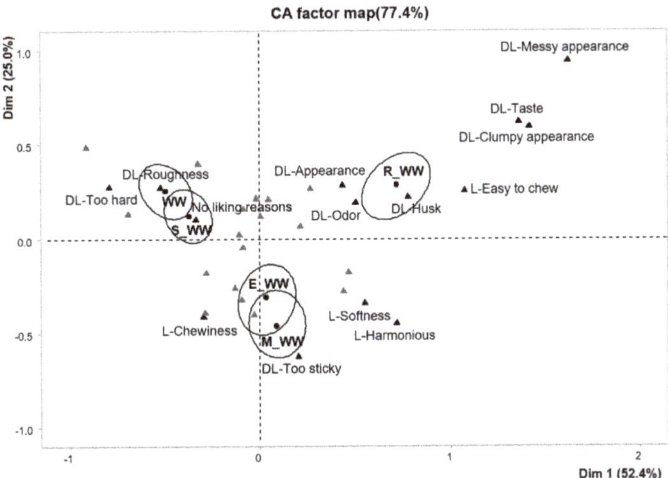

Figure 1. Correspondence analysis plot for reasons of (dis) liking attributes and their corresponding various processed whole wheat sample loadings. Filled circle shapes indicate the sample loadings and filled triangle point up shapes refer to the descriptors.

4. Discussion

The moisture content of a variety of grain kernels such as wheat, brown rice and paddies was maximized by steeping for more than 5 h regardless of temperature [26–28]. In this study, the water content of cooked rice with S_WW and E_WW was increased compared to that with WW since S_WW and E_WW were steeped in water for 24 h. As Park et al. [29] and Gujral et al. [30] reported an increased water content with the degree of milling for brown rice, in this study, the water content of cooked rice with M_WW and R_WW was increased as the bran of the wheat kernel surface was destroyed. Texture is the expression of the structure and surface properties of foods through human senses, which affect consumer perceptions of product acceptability [31]. Various instrumental methods have been developed to evaluate the textural properties of foods [32]. Texture profile analysis (TPA) using a texture analyzer is the most frequently used method of applying compression to imitate the mastication process [33,34]. Cohesiveness is a mechanical texture property related to the deformation of food before breaking. The WW sample showed lower cohesiveness than the E_WW sample because it has a fibrous and firm outer layer. Grains with an outer fibrous brown layer, such as brown rice, can prevent the structure from collapsing after the first compression, and the grain retains its shape after compression, resulting in low cohesiveness [30]. Hardness is the first characteristic perceived during the mastication process and shows a good correlation between sensory data and instrumental data in terms of textural characteristic [35]. Kim et al. [36] showed that texture is the most influential characteristics for the overall quality of cooked rice in the sensory evaluation. Particularly, in Kim et al.'s [37] study, which analyzed the correlation between overall sensory quality and TPA of cooked rice, hardness was a negative correlation with overall sensory quality. Texture influenced overall product quality as well as consumer acceptance. Choi et al. [38] reported that stickiness was important for their overall liking of cooked rice. Park et al. [29] reported that the hardness and chewiness tend to decrease as the degree of milling increases. When classifying cooked rice samples, the factors of hardness and stickiness are more than the taste [39], and it was reported that the change in hardness by processing is an important factor when preparing cooked rice using whole wheat with low preference. Normally, the hardness of cooked rice is 2000 g to 3000 g, which is much lower compared to the cooked rice sample with WW [29,40,41]. As the hardness of M_WW and E_WW decreased, the overall consumer liking score and texture liking score increased. The responses to open-ended questions for the cooked rice samples with M_WW and E_WW confirmed the above results by showing that the mechanical texture characteristic of hardness has a substantial influence on sensory evaluation when it is seen that there are few people who dislike it because it is too hard [40].

In this study, consumers were asked to describe the reasons for liking and disliking samples, as previously proposed by Symoneaux et al. [10]. As a result of textual analysis, the number of likes and dislikes was correlated with the overall liking score, and it was possible to predict the level of overall liking. The cooked rice with M_WW with the highest liking score had 128 likes and 97 dislikes, while the cooked rice with R_WW with the lowest liking score had 101 likes and 163 dislikes (Table 3). The higher the liking score of the sample was, the greater the number of likes and the fewer dislikes, and vice versa [10,11]. In addition, when there were more reasons for liking than disliking, the overall liking score was more than 5 points based on the 9-point hedonic scale, and this trend was observed to have the same pattern in Symoneaux et al. [10] which evaluated samples using a 7-point hedonic scale. However, even though the samples had similar liking scores, this may have been caused by different reasons. The cooked rice with WW, S_WW, and R_WW had similar liking scores, but the reasons for liking and disliking these samples were different. For cooked rice with WW and S_WW, a common reason for disliking was 'too hard', while this description was never mentioned for cooked rice with R_WW. Appearance characteristics such as 'messy appearance', 'clumpy appearance', and 'husk' were frequently described as reasons for disliking cooked rice with R_WW. Often, when consumers answer only one open-ended question, it can be difficult to know whether what they refer to is positive or negative for them. Asking separately for reasons of liking and disliking makes the transcription of terms easier and provides insight to clearly identify the attributes that are positively and negatively correlated with the consumer's preference without further interpretation [11,42]. Furthermore, we can identify which attributes are more important to consumers' preferences, and even the same terms can be liked by some consumers and disliked by others. In conclusion, asking consumers separately for reasons of liking and disliking is a better way to understand their opinion [10].

It is also noteworthy that consumers used more diverse terms to describe the reasons for disliking the samples than liking them. The terms were integrated by three native researchers, obtaining 60 and 80 types of reasons for liking and disliking, respectively. In addition, the chi-square per cell analysis results showed that differences between samples were often larger in terms of dislikes than likes. In previous studies, using open-ended questions [14,43], consumers reportedly cited more likes than dislikes, which was the opposite of this study. The samples surveyed in this study had a relatively lower overall liking score than in previously conducted studies. Alternatively, Letarte et al. [44] reported that dislikes result from more specific and intense sensory experiences than likes. This result may also be due to the positive-negative asymmetry effect in which negative information is weighted more than positive information [45].

Consumers often referred to sensory-related terms rather than holistic or emotional terms. Only three ('easy to chew', 'looks healthy', and 'harmonious') of the 16 reasons for liking and two ('messy appearance' and 'feels undercooked') of the 16 reasons for disliking mentioned by more than 5% of consumers were non-sensory terms. This is because consumers are relatively familiar with the samples. In fact, 73.8% of consumers who participated in this study responded that they ate multigrain cooked rice more than once a week (data not shown).

The main advantage of open-ended questions is that information about attributes that are important to consumers is collected directly in their own language. A limited number of available attributes in conventional profiling can lead to a dumping effect, and if consumers like the product, a halo effect can occur when evaluating attributes [14]. However, the spontaneity given to consumers in free comment methods allows them to freely describe their sensory perception and emphasize the perceived dominant descriptors. In other words, information about consumers' perceptions of characteristics can be provided. Nevertheless, open-ended questions still exhibit labor-intensive issues associated with the preprocessing phase, in which spelling errors are eliminated and synonyms are grouped.

5. Conclusions

This study was performed regarding the physical properties and consumer acceptance for cooked rice with various types of processed whole wheat. The results of TPA showed that all processed whole wheat samples were less hard than cooked rice with WW, indicating a distinct improvement in the texture property. Thus, consumer acceptance test of these processed whole wheat samples was conducted, with M_WW scoring relatively high overall liking for consumers. This result suggested that M_WW was suitable for cooked rice.

On the other side, two open-ended questions, evaluated separately on the reasons of liking and disliking, confirmed that the sensory characteristics associated with texture were identified as important factors in the formation of acceptance in cooked rice containing various types of processed whole wheat. Among the texture properties, 'chewiness' was the driver of liking, while 'hardness' and 'roughness' were the drivers of disliking. In particular, incomplete appearances such as 'husk', 'clumpy appearance', and 'messy appearance' were drivers of disliking. Additionally, the results of this study imply that the level of overall liking for samples could be predicted by including a separate open-ended question for reasons of liking and disliking.

The present study has several limitations in generalizing results. Although the goal was to verify the marketability of relatively new foods, whole wheat products for cooked rice, comparative analysis with multi grain cooked rice that occupies the market was not particularly investigated in the present study. Further investigations on comparative analysis with marketed products may provide useful insights in developing marketing strategies for these food items.

Author Contributions: Conceptualization, D.-B.L. and S.-S.K.; methodology, M.-R.K.; software, M.-R.K.; validation, S.-S.K., D.-B.L. and M.-R.K.; formal analysis, J.-A.H. and Y.-S.B.; investigation, S.-S.K.; resources, S.-S.K.; data curation, D.-B.L. and M.-R.K.; writing—original draft preparation, D.-B.L. and M.-R.K.; writing—review and editing, S.-S.K.; visualization, D.-B.L. and M.-R.K.; supervision, S.-S.K.; project administration, S.-S.K.; funding acquisition, S.-S.K. All authors have read and agreed to the published version of the manuscript.

Funding: This research was funded by Korea Institute of Planning and Evaluation for Technology in Food, Agriculture, Forestry (IPET) through Agri-Bio Industry Technology Development Program (Grant number: 317019-4) and High Value-added Food Technology Program(Grant number: 321039-4), funded by Ministry of Agriculture, Food and Rural Affairs (MAFRA).

Data Availability Statement: The data to support the findings of this study are included within this article.

Conflicts of Interest: The authors declare no conflict of interest.

 *foods*

Article

Effect of Spray-Drying and Freeze-Drying on the Composition, Physical Properties, and Sensory Quality of Pea Processing Water (*Liluva*)

Weijun Chen [1], Hoi Tung Chiu [2], Ziqian Feng [1], Evelyne Maes [3] and Luca Serventi [1,*]

[1] Department of Wine, Food and Molecular Biosciences, Faculty of Agriculture and Life Sciences, Lincoln University, RFH Building, Lincoln P.O. Box 85054, Christchurch 7647, New Zealand; Weijun.Chen@lincolnuni.ac.nz (W.C.); Tyrande.Feng@lincolnuni.ac.nz (Z.F.)

[2] Dry Food NZ Ltd., 12 Ngati Kuia Drive, Port, Havelock 7100, New Zealand; monika@dfnz.co

[3] Proteins & Metabolites Team, AgResearch Limited, Lincoln 7674, New Zealand; Evelyne.maes@agresearch.co.nz

* Correspondence: Luca.Serventi@lincoln.ac.nz

Abstract: Spray-drying and freeze-drying can extend the shelf life and improve the transportability of high-nutritional foods such as *Liluva* (processing water of legumes). Nonetheless, the effects of these processes on nutrition, physiochemical properties, and sensory quality are unknown. In this study, particle sizes, protein profiles, colour, and preliminary sensory profile of pea powder samples were determined by Mastersizer 3000, protein gels, chroma meter, and 9-point hedonic scale, respectively. Results indicated that no significant difference was found in the molecular weight distribution of protein bands in pea water and sensory profile after drying. Fibre content in pea water after spray-drying was higher while soluble carbohydrates and minerals were lower than those after freeze-drying. Spray-drying decreased pea water's lysine content, particle size, redness colour, and yellowness colour, while it increased its light colour; however, freeze-drying showed the opposite results. Overall, spray-drying could be a better drying technology that can be applied to dry pea water. Further experiments are required, however, to determine the influence of drying technologies on emulsifying activity.

Keywords: split yellow peas; soaking water; cooking water; spray-drying; freeze-drying; proximate composition; protein profile; particle size; colour; sensory

Citation: Chen, W.; Chiu, H.T.; Feng, Z.; Maes, E.; Serventi, L. Effect of Spray-Drying and Freeze-Drying on the Composition, Physical Properties, and Sensory Quality of Pea Processing Water (*Liluva*). *Foods* **2021**, *10*, 1401. https://doi.org/10.3390/foods10061401

Academic Editor: Alcina M.M.B. Morais

Received: 17 May 2021
Accepted: 8 June 2021
Published: 17 June 2021

Publisher's Note: MDPI stays neutral with regard to jurisdictional claims in published maps and institutional affiliations.

1. Introduction

Legumes, such as peas, chickpeas, and beans, are low price and high nutritional value foods that are widely consumed by people all around the world. Among legumes, peas are a good source of plant-based protein because the protein content in peas is high at about 22.3 g of protein/100 g [1]. Although the antinutrients (including phytic acid, tannins, and proteolytic inhibitors) found in peas decrease the digestibility of this protein, the soaking, cooking, or baking of peas can reduce the antinutrients and improve the protein bioactivity [1].

In industrial production of legumes, wastewater is generated. *Liluva*, namely the water generated by the soaking, cooking, or canning process of legumes, can be upcycled into functional food ingredients [2]. The wastewater from soaking and cooking of 100 g split yellow peas contained about 1.89 g and 4.4 g solids respectively, including protein, soluble and insoluble carbohydrates, and minerals. In peas' soaking and cooking water, soluble carbohydrates represented 37% and 30% of the dry matter [3,4]. Additionally, the protein content in peas' soaking and cooking water comprises a high percentage of the dry matter as well at around 30% [3,4]. With such high contents of soluble carbohydrates and proteins, pea water has the potential to be a foaming or emulsifier agent [2], which are two important agents in baking products to increase the volume and stabilize the starch–lipid networks of bakery foods [5]. Therefore, *Liluva* is a valuable ingredient that can be applied in the food industry to increase bioactive content and improve the functional properties of food products.

In order to store and transfer pea water more conveniently, and to increase the range of applications, drying methods, such as spray-drying and freeze-drying, can be applied to remove the water and convert the pea water into a powder. Spray-drying is a thermal method widely used in the food industry to produce a dry powder from a liquid [6]. On the other hand, freeze-drying is a nonthermal method that is commonly used for the dehydration of heat-sensitive food through combining freeze and vacuum drying [7]. Even though the shelf life of peas' soaking and cooking water may be extended by drying because of the low water activity [8], its composition and functional properties might change with drying.

The aim of this study was to analyse the effects of spray-drying and freeze-drying on peas' soaking and cooking water, with particular emphasis on spray-drying. Proximate composition, content of free amino acids, protein bands, particle sizes, and colour were investigated in this article. Additionally, preliminary sensory analysis of a food product (sponge cake) containing peas' raw water or reconstituted spray-dried powder was covered.

2. Materials and Methods

2.1. Sample Preparation

Split yellow peas (Cates, New Zealand), wheat flour (Pams, New Zealand), apple vinegar (DYC, New Zealand), icing sugar (Pams, New Zealand), and baking powder (Pams, New Zealand) were used in this study. The soaking water and cooking water of split yellow peas were prepared as described by Serventi [2]. Briefly, split yellow pea soaking water was prepared by soaking split yellow peas in water at a ratio of 1:3.3 (pea to water) for 17 h. After soaking, the soaked split yellow peas were cooked in water at a ratio of 1:1.75 (pea to water) for 90 min. Water samples after soaking and cooking were collected separately. Next, the soaking and cooking water of split yellow peas was spray-dried by Dry Food NZ Ltd. (Havelock, New Zealand) (the processing parameters of spray-drying technique are shown in Appendix A) and freeze-dried (Lincoln University, New Zealand).

2.2. Proximate Composition

The proximate composition of concentrated powders was quantified with the following methods: moisture content (AACC method 44-15A) [9], soluble carbohydrates (Pollock and Jones, Jermyn) [10,11], protein content by total nitrogen (AOAC method 954.01) [12] with a conversion factor of 6.25, fat (AOAC method 920.39) [12], ash (AOAC method 930.05) [13], and insoluble carbohydrates by difference.

2.3. Free Amino Acid Profile

The free amino acid profile was analysed by the Agilent 1100 series HPLC system (Agilent Technologies, Walbronn, Germany) with a 150 mm × 4.6 mm, 3 μm C-18 column (Winlab, Scotland) at 40 °C according to Heems, Luck, Fraudeau, and Verette (1998) and Carducci et al. (1996) [14,15]. The precolumn derivatization was performed on the autosampler. O-phthaldialdehyde (OPA) and 9-fluorenylmethyl chloroformate (FMOC) were used as primary and secondary amino acid derivatization reagents, respectively. The detection was performed using a fluorescence detector with the following settings: 335 nm (excitation) and 440 nm (emission). The detector was switched to second channel at 21 min to detect secondary amino acid proline, and the parameters changed to 260 nm (excitation) and 315 nm (emission). To make solvent A, 0.01 M Na_2HPO_4 was added with 0.8% THF and adjusted to pH = 7.5 with H_3PO_4, while solvent B comprised 50% methanol and 50% acetonitrile. Solvents A and B were used for the separation with the following pump gradients: 0 min, 0% B; 14 min, 40% B; 20 min, 50% B; 24 min, 100% B; 29 min, 100% B; 30 min, 0% B; 36 min, 0% B, with a flow rate of 0.7 mL/min. Sample injection volume was 12 μL.

2.4. Protein Analysis via SDS-PAGE

A sodium dodecyl sulfate polyacrylamide gel electrophoresis (SDS-PAGE) was performed to determine the molecular weight distribution of the protein present in the *Liluva* samples (6 samples in total, including the raw soaking and cooking water, spray-dried soaking and cooking water, and freeze-dried soaking and cooking water of split yellow peas) as described by Buhl et al. (2019) with modifications [16]. Invitrogen™ NuPAGE™ 4–12% Bis-Tris precast gels (Bio-Rad, Richmond, CA, USA) were used to evaluate the protein profile in this experiment. A molecular weight marker (10–250 kDa) was applied to estimate the molecular weight of the protein bands. Prior to heat treatment (100 °C for 5 min), 6 samples (concentration of 0.1%) were mixed 3:1 with the NuPAGE™ LDS Sample Buffer (4X) and Sample Reducing Agent. After heating the mixed solution, 8 µL of the molecular markers and 20 µL of the samples were loaded into the gel. Then, 200 V of a constant current and 45 min of running time were set for the electrophoresis, which was conducted in the running buffer (0.25 M Tris, 0.192 M glycine, 0.1% SDS). Next, the gel was stained with Commassie blue G-250 staining solution for 1 h. Afterwards, the gel was destained with the destain solution (20% Methanol, 10% Acetic acid) overnight.

2.5. Particle Size

The particle size of split yellow pea water and powder samples was measured with a Mastersizer 3000 (Malvern Panalytical) with constant refractive index (1.538) and absorption index (0.01). Modified from the method mentioned by Govoreanu, Saveyn, Van der Meeren, Nopens, and Vanrolleghem (2009) [17], the machines were initially rinsed with pure RO (reversed osmosis) water to stabilize size distribution. Then, samples were diluted into about 400 mL RO water to reach the acceptable obscuration limits (10–20%) in an automated flexible volume wet sample dispersion. After reaching the obscuration limits, samples' particle size distribution was automatically measured by the machine in quintuples.

2.6. Colour

Modified from the method described by Yagiz, Balaban, Kristinsson, Welt, and Marshall (2009) [18], a handheld Konica Minolta CR-400 chroma meter was used to measure the colour of four *Liluva* powder samples (spray-/freeze-dried powder of pea soaking water and spray-/freeze-dried powder of pea cooking water) after calibration with a CR-A43 calibration plate. First, the powder samples were homogenized and weighed (3 g per sample) before they were poured into transparent containers. Then, the colour parameters of *Liluva* samples, including L* (lightness), a* (greenness to redness), and b* (blueness to yellowness), were measured. The surfaces of samples were touched by the light tube directly. Samples were homogenized manually after each detection, and the colour of *Liluva* powder samples were measured in triplicates.

2.7. Cake Preparation

Sponge cakes were made based on the recipe of Mustafa and collaborators [19]. Briefly, 110 mL of split yellow pea cooking water or spray-dried split yellow pea cooking water was mixed with 3 g of apple cider vinegar for 7 min by using a Brabantia BBEK1092 stand mixer. The mixer was started at low speed, and the maximum speed was set when the solution became foamy. After 7 min, 130 g of icing sugar was added to the mixer and mixed for 3 min at the maximum speed. Afterwards, the blend of 130 g of plant flour and 7 g of baking powder was manually mixed with the creamy foam. During mixing, flour was added into the foam three times. After adequately mixing, 110 g of each batter was weighted and poured into the baking pan and was baked at 180 °C for 15 min in a preheated Turbofan oven (Moffat Ltd., model E32 M, Rolleston, IN, USA). The baked cakes were cooled to room temperature and cut into small square-like pieces before the sensory test.

2.8. Sensory Analysis

Modified from the method inferred by Sveinsdottir and collaborators [20], 20 untrained participants from Lincoln University were involved in the sensory test of sponge cake. Participants were asked to taste two sponge samples (one contained raw split yellow pea cooking water; the other contained reconstituted spray-dried split yellow pea cooking water) in the sensory room (with individual booths) at Lincoln University, New Zealand. Freeze-dried samples were not considered since they were not food grade. Soaking water samples were not tested since they contained less protein than cooking water, thus making them less suitable for egg replacement. The small, square-like sponge cake samples were put into small plastic containers (without lids). Samples were labelled with digital codes that were in random order. Water and crackers were also provided for participants to clean their mouths to avoid product carry-over. All the participants were asked to evaluate the appearance, aroma, texture, taste, and overall preference of the sponge cake samples by using a 9-point hedonic scale (1—dislike extremely; 5—neither dislike nor like; 9—like extremely).

2.9. Data Analysis

All data was calculated and presented as average $\pm$ standard deviation by using Excel, Microsoft 365. Statistical analysis was performed by Minitab version 19. One-way analysis of variance (ANOVA) was applied to the analysis of colour, foaming ability, and emulsifying activity. Analysis of the preliminary sensory test was conducted by ANOVA using the general linear model (GLM) procedure and a post-hoc Tukey's honest significant difference (HSD) test ($p < 0.05$).

3. Results and Discussion

3.1. Proximate Composition

According to Table 1, spray-dried split yellow pea cooking water had higher dry matter, protein, and fibre contents, but lower amounts of soluble carbohydrates and minerals than spray-dried split yellow pea soaking water. Previous studies about the compositions of freeze-dried split yellow pea cooking and soaking water presented slightly different profiles, where freeze-dried pea soaking water showed higher levels of protein (31.7%), soluble carbohydrates (36.5%), and minerals (13.8%) but lower fibre content (18%) than freeze-dried pea cooking water (28.2% of protein, 25.2% of soluble carbohydrates, 9.1% of minerals, and 34.7% of fibre) [5,21]. When comparing different drying methods under the same type of *Liluva*, spray-dried water demonstrated higher fibre but lower soluble carbohydrate and mineral levels than freeze-dried water [5,21]. In terms of the protein content, spray-dried soaking water presented a lower amount of protein than freeze-dried soaking water, while spray-dried cooking water conversely showed higher protein content than freeze-dried cooking water [5,21].

Table 1. Proximate composition of spray-dried pea soaking and cooking water powders. Different letters in the same row represent statistical difference ($p < 0.05$).

Nutrients (g/100 g)	Spray-Dried Pea Soaking Water Powder	Spray-Dried Pea Cooking Water Powder
Moisture content	7.15 $\pm$ 0.07 [a]	4.86 $\pm$ 0.14 [b]
Protein	25.16 $\pm$ 0.08 [a]	34.63 $\pm$ 0.43 [b]
Soluble carbohydrates	25.17 $\pm$ 1.70 [a]	17.44 $\pm$ 0.92 [b]
Insoluble carbohydrates	32.43 $\pm$ 1.82 [a]	35.99 $\pm$ 0.98 [b]
Minerals	10.08 $\pm$ 0.11 [a]	7.08 $\pm$ 0.06 [b]

In pea soaking and cooking raw water, the content of proteins composes about 30% of the dry matter [2]. Similar protein concentrations were observed in pea powders, with 25.2% in spray-dried soaking powder, 34.6% in spray-dried cooking powder, 31.7% in freeze-dried soaking powder, and 28.2% in freeze-dried cooking powder. This is possibly because freeze-drying does not involve heating and therefore does not denature the protein. While spray-drying applies heat to the peas' protein, the treatment time might be too short to denature proteins [22]. In terms of the higher protein content in spray-dried pea cooking water (34.6%), it is possibly because the low soluble fibre content in spray-dried pea cooking powder increased its percentage of protein content.

Additionally, as reported by Serventi [2], the content of soluble carbohydrates in split yellow pea raw soaking water was 0.69 g/100 g. In other words, the content of soluble carbohydrates made up to 37% of the dry matter. Additionally, the dry matter in split yellow pea raw cooking water also contained a high percentage of soluble carbohydrates at around 30% [2]. Compared with the results shown in Table 1, spray-drying decreased the content of soluble carbohydrates in pea soaking and cooking water from about 37 to 25.2% and 30 to 17.4% as inferred by Shishir and Chen [23]. Sugars have low molecular weight and glass transition temperature, so the mobility of sugar molecules increases with heat treatment at above 20 °C. Therefore, the reason for the lower content of soluble carbohydrates may be that the sugar in split yellow pea water sticks to the dryer when it comes across heat; thus, the recovery yield of the material decreases [23].

3.2. Free Amino Acids

Twenty-one free amino acids found in pea raw water and spray-/freeze-dried powder were measured and analysed (Table 2). The content of some amino acids in pea samples exceeded the detecting limits, such as glutamic acid in all samples and aspartic acid in the spray-dried and freeze-dried soaking water powder. Additionally, some of the samples contained amino acids that exceeded the quantification limits, such as cysteine in all samples except those of pea soaking water and freeze-dried pea cooking water powder.

Table 2. Free amino acid profiles of different pea soaking and cooking water. Different letters represent statistical difference ($p < 0.05$).

Amino Acid	Soaking Water			Cooking Water		
	Raw (μM)	Spray-Dried (μM)	Freeze-Dried (μM)	Raw (μM)	Spray-Dried (μM)	Freeze-Dried (μM)
Asp	106.33 ± 5.21 [b]	*	*	291.49 ± 4.33 [a]	320.68 ± 14.42 [a]	125.36 ± 2.96 [b]
Glu	*	*	*	*	*	*
Cys	**	**	94.30 ± 0.74 [b]	**	**	163.97 ± 5.09 [a]
Asn	*	749.26 ± 16.53 [e]	1422.57 ± 72.20 [d]	4428.08 ± 23.66 [a]	4144.78 ± 18.95 [b]	2058.07 ± 14.58 [c]
Ser	342.56 ± 8.24 [a]	221.54 ± 0.95 [b]	183.88 ± 7.58 [cd]	197.30 ± 1.91 [c]	168.69 ± 2.61 [d]	136.06 ± 4.04 [e]
Gln	246.78 ± 9.60 [a]	58.57 ± 2.30 [c]	101.47 ± 4.26 [b]	21.71 ± 7.57 [d]	13.97 ± 0.36 [d]	9.48 ± 1.02 [d]
His	128.58 ± 1.63 [b]	192.50 ± 1.36 [a]	86.43 ± 10.99 [c]	127.60 ± 0.29 [b]	121.75 ± 1.75 [b]	100.43 ± 6.53 [c]
Gly	*	*	558.31 ± 5.44 [a]	288.65 ± 5.18 [d]	379.22 ± 4.58 [c]	420.11 ± 16.28 [b]
Thr	376.86 ± 10.36 [b]	542.50 ± 20.09 [a]	267.82 ± 1.80 [d]	325.46 ± 7.05 [c]	261.43 ± 6.07 [d]	238.35 ± 12.24 [d]
Arg	*	422.27 ± 0.07	*	*	*	*
Ala	389.20 ± 5.18 [c]	730.06 ± 0.19 [a]	306.10 ± 1.53 [e]	392.26 ± 0.74 [bc]	409.71 ± 3.57 [b]	343.02 ± 8.98 [d]
Tau	*	*	*	609.93 ± 1.04 [b]	643.41 ± 2.50 [a]	*
Tyr	136.34 ± 1.79 [a]	132.00 ± 0.08 [a]	101.95 ± 0.86 [c]	108.27 ± 0.51 [b]	112.18 ± 0.03 [b]	99.97 ± 2.66 [c]
Val	218.58 ± 2.98 [b]	300.00 ± 1.28 [a]	187.71 ± 0.30 [c]	136.62 ± 1.09 [e]	139.43 ± 1.61 [e]	159.75 ± 4.41 [d]
Met	48.41 ± 0.84 [bc]	75.07 ± 0.18 [a]	28.75 ± 0.15 [d]	51.31 ± 0.20 [b]	48.89 ± 0.39 [bc]	47.60 ± 1.62 [c]
Try	49.39 ± 0.58 [c]	68.18 ± 0.23 [a]	34.65 ± 0.15 [d]	63.31 ± 0.94 [b]	64.81 ± 0.54 [b]	62.10 ± 1.11 [b]
Phe	93.89 ± 1.03 [d]	178.84 ± 0.32 [a]	100.00 ± 0.58 [bc]	97.19 ± 0.04 [cd]	103.92 ± 1.03 [b]	98.97 ± 2.26 [c]
Ile	97.44 ± 1.42 [c]	142.49 ± 0.69 [a]	101.76 ± 0.10 [b]	64.00 ± 0.26 [d]	66.89 ± 1.24 [d]	63.82 ± 1.35 [d]
Lys	197.00 ± 0.72 [bc]	185.34 ± 1.56 [cd]	206.75 ± 0.52 [ab]	171.85 ± 1.65 [e]	176.33 ± 5.96 [de]	215.73 ± 4.32 [a]
Leu	123.54 ± 0.94 [b]	189.15 ± 2.18 [a]	133.58 ± 0.98 [b]	125.19 ± 0.32 [b]	125.88 ± 7.53 [b]	137.06 ± 3.41 [b]
Pro	356.46 ± 7.78 [a]	350.79 ± 0.84 [a]	251.89 ± 13.09 [b]	208.06 ± 3.01 [bc]	232.64 ± 17.66 [b]	173.52 ± 14.97 [c]

* means that the amount of amino acid exceeded the detecting limits. ** means that the amount of amino acid exceeded the quantification limits.

Nosworthy and co-authors [1] reported that the contents of methionine (around 0.22% of dry matter) and cysteine (about 0.25% of dry matter) in yellow peas were limited, while lysine was abundant at around 1.69% of dry matter. This can be linked to the high content of albumin proteins in yellow peas (14 g/100 g), which are known to have many sulphur-containing amino acids and lysine in the protein sequence, which results in high lysine content [2]. Therefore, the ratio of lysine and methionine/cysteine and in pea raw water and dried powder is important.

According to Table 2, the ratio of lysine and methionine in freeze-dried pea powder (around 7.19 µM for soaking powder and about 4.53 µM for cooking powder) was higher than that in spray-dried pea powder (around 2.47 µM for soaking powder and approximately 3.61 µM for cooking powder) and pea raw water (with about 4.07 µM for soaking water and around 3.35 f µM or cooking water). This means that freeze-drying increases the content of lysine and/or decreases methionine's content in split yellow peas. Additionally, the ratio of lysine and methionine was almost twice as high in pea raw soaking water than in spray-dried soaking water powder at about 4.07 µM and 2.47 µM, respectively. Spray-dried pea cooking water powder had a slightly higher ratio of lysine and methionine compared to pea raw cooking water (approximately 3.61 µM and 3.35 µM, respectively). As mentioned by Brishti and collaborators [24], lysine is an amino acid that is active in the occurrence of the Maillard reaction. This might explain the lower content of lysine in spray-dried powder, as spray-drying is a thermal treatment that induces the chemical reaction of amino acid and sugars in a material.

3.3. Protein Molecular Weight Distribution

The protein composition of pea raw water and dried powder diluted samples were visualized by SDS-PAGE (Figure 1). In general, the protein composition (estimated by protein bands in the 1D gel) of split yellow pea raw soaking water was similar to its spray-dried or freeze-dried powder diluted water samples. Similarly to pea soaking water samples, split yellow pea raw/spray-dried/freeze-dried cooking water samples showed no significant differences in the protein bands on the 1D gel, though differences in intensity between the raw water and dried water are noticeable. This illustrates that the overall protein profile in peas was not significantly affected by drying.

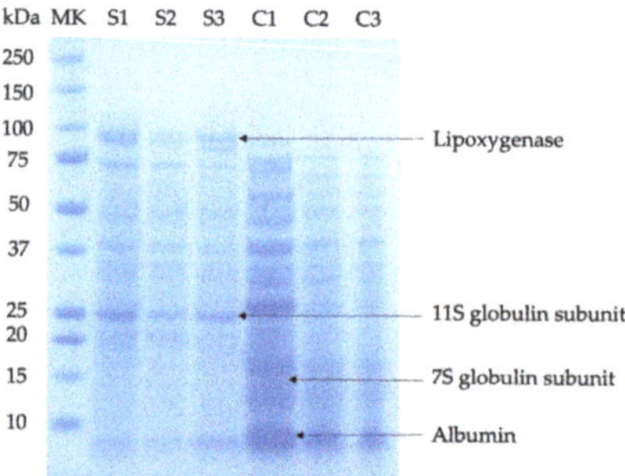

Figure 1. Representative SDS-PAGE gel of the samples studied. The left lane indicates molecular weight (KDa). MK—molecular markers; S1—split yellow pea raw soaking water; S2—split yellow pea spray-dried soaking water; S3—split yellow pea freeze-dried soaking water; C1—split yellow pea raw cooking water; C2—split yellow pea spray-dried cooking water; C3—split yellow pea freeze-dried cooking water.

A comparison of the expected molecular weight of some proteins of interest with literature was made. According to Buhl and collaborators. [16], lipoxygenase corresponded with the molecular weight of 99 kDa, and albumin was located around 10 to 12 kDa. The 7S and 11S globulin units were represented in protein bands around 15 kDa and 25 kDa, respectively [24]. In the gel, low molecular weight proteins such as albumin were more intense in split yellow pea cooking water samples than in split yellow pea soaking water samples. The reason for the higher intensity of albumin protein in pea cooking water might be the high nitrogen loss of split yellow peas, which resulted from the exposure of peas' starchy, proteinaceous endosperms into boiling water in the cooking process [2].

Additionally, split yellow pea soaking water samples contained some large molecular weight proteins that split yellow pea cooking water samples did not have, such as lipoxygenase. The disappearance of lipoxygenase in pea cooking water samples might be because of cooking, which is a heat treatment that might denature the enzymes [2]. In terms of globulins, it is a high content protein in legume seeds and act as storage proteins [2]. Similar to the findings shown by Brishti and coworkers [24] (i.e., that mung bean proteins contained lower levels of globulins), the content of the globulins in split yellow peas was low, as the gel bands were relatively very light in colour.

3.4. Particle Size

Regardless of the application of drying treatments or not, the particle sizes in split yellow pea cooking water were larger than in split yellow pea soaking water (Table 3). The reason is that more insoluble carbohydrates were lost during the pea boiling process compared to the soaking process. According to Serventi [2], cellulose, hemicellulose, and pectin are the insoluble polysaccharides in peas. This explains why the particle sizes in pea cooking water were larger than that in pea soaking water.

Table 3. The particle sizes of raw/spray-dried/freeze-dried split yellow pea soaking/cooking water.

Ingredient	Physical State	Dx (10)	Dx (50)	Dx (90)
Pea Soaking Water	Raw	3.59	18.1	83.0
	Spray-dried	4.91	17.8	48.6
	Freeze-dried	3.38	22.2	96.1
Pea Cooking Water	Raw	41.2	150.0	380.0
	Spray-dried	17.4	58.3	128.0
	Freeze-dried	103	345	737.0

This is also demonstrated in Figure 2, where spray-drying is shown to have significantly decreased the size of large particles to below 100 μm, with the peak particle size at about 20 μm. In addition, the majority of the particles in spray-dried pea powder diluted water were distributed from about 1 to 50 μm. Additionally, freeze-drying did not much change the size of small particles in pea soaking water. However, more large-size particles were formed in split yellow pea soaking water after freeze-drying. Similar results were found in pea cooking water (Figure 3). With the application of spray-drying, more particles in spray-dried split yellow pea cooking water were distributed in the smaller size classes compared with those in split yellow pea raw cooking water. On the other hand, more large-size particles were formed in split yellow pea cooking water after freeze-drying.

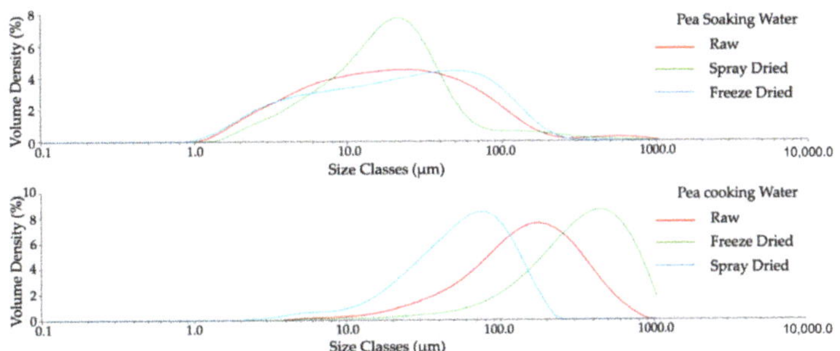

Figure 2. Particle size distribution of pea soaking water (**above**) and pea cooking water (**below**) in different physical states: raw, spray-dried and freeze-dried.

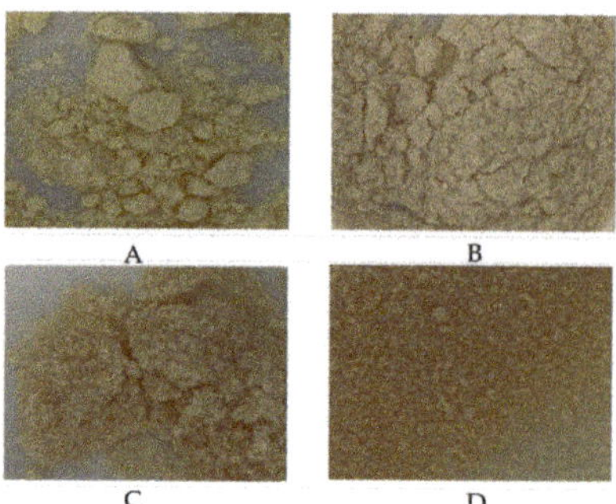

Figure 3. Pictures of the pea water powders. (**A**) Spray-dried pea soaking powder, (**B**) Spray-dried pea cooking powder, (**C**) Freeze-dried pea soaking powder, (**D**) Freeze-dried pea cooking powder.

The results of spray-dried yellow pea samples are in line with the findings reported by del Rio and collaborators [25], who applied spray-drying to the protein isolates of yellow pea and decreased the proteins' particle size. The heat treatment and atomization of spray-drying could be the reason for the breaking down of the particles. Brishti and co-authors [24] also described that freeze-drying resulted in the highest particle size of mung bean protein isolates compared to other drying methods, such as spray-drying and oven drying. This is possibly because of the aggregation of particles during the production of ice crystals in freeze-drying [24]. Joshi and collaborators (2011) [21] obtained similar results in lentil protein isolates as well.

3.5. Colour of Powders

In general, there were significant differences among most of the powder samples in terms of lightness, redness, and yellowness. Spray-drying lightened the colour of split yellow pea soaking and cooking water more significantly than freeze-drying. The freeze-dried powder of split yellow pea soaking or cooking water, on the other hand, exhibited redder and yellower colour than the spray-dried powder of the same water material. No colour differences were observed between freeze-dried split yellow pea soaking water powder and freeze-dried split yellow pea cooking water powder.

Among four powder samples (Table 4), spray-dried powder of split yellow pea soaking water showed the lightest colour, followed by spray-dried powder of split yellow pea cooking water, at 92.1 and 90.3, respectively. Freeze-dried powder of split yellow pea soaking and cooking water had no significant difference in lightness at 81.9 and 81.0, respectively. Freeze-dried powder of split yellow pea cooking water had the reddest and yellowest colour compared to other powder samples, with 4.31 for the red colour and 22.6 for the yellow colour.

Table 4. The colour parameters (L*—lightness, a*—redness, and b*—yellowness) of spray-dried/freeze-dried split yellow pea soaking/cooking powder. Different letters represent statistical difference ($p < 0.05$).

Samples		Lightness (L*)	Redness (a*)	Yellowness (b*)
Soaking water	Spray-dried	92.1 ± 0.7 [a]	−1.39 ± 0.05 [d]	19.9 ± 0.3 [c]
	Freeze-dried	81.9 ± 1.0 [c]	2.36 ± 0.10 [b]	20.5 ± 0.1 [b]
Cooking water	Spray-dried	90.3 ± 0.2 [b]	1.37 ± 0.03 [c]	15.7 ± 0.3 [d]
	Freeze-dried	81.0 ± 0.3 [c]	4.31 ± 0.09 [a]	22.6 ± 0.1 [a]

Similar results were found by Brishti and co-authors [24], who indicated that spray-dried mung bean protein isolate powder had lighter, less red and yellow colour than its freeze-dried powder. Lentils' protein isolates were also shown darker, redder and yellower colour after freeze-drying than after spray-drying [21]. The reason for the lighter colour of the spray-dried powder was because spray-drying broke down the particles of mung bean protein isolates; thus, more light was refracted because of their larger surface area [24]. This finding indicates that light colour is correlated to the particle size of the materials. Furthermore, the higher browning index of freeze-drying compared to spray-drying also explains the lighter colour of spray-dried samples [24].

The yellow colour of yellow peas is mainly contributed by carotenoids [26]. It is possible that freeze-drying, which is a not a heat treatment, preserves the carotenoids in split yellow peas. Degradation and isomerization may occur during heat treatment, which would reduce the yellow colour of the samples [27]. Moreover, the particles may aggregate in freeze-drying [24]. The larger particle size may be the reason for the darker, yellower, and redder colour of freeze-dried powder.

3.6. Sensory Quality

In the sensory analysis of sponge cakes, sponge cakes made with split yellow pea raw cooking water showed no significant difference compared to sponge cakes made with the diluted water of split yellow pea spray-dried cooking powder in all sensory attributes, including appearance, aroma, taste, texture, and overall preference (Table 5). This result indicated that spray-drying did not greatly impact the sensory profile of split yellow pea cooking water.

Table 5. Sensory profile (appearance, aroma, taste, texture, and overall preference) of sponge cakes containing split yellow pea raw cooking water (raw) or reconstituted pea cooking water powder (spray-dried). Different letters represent statistical difference ($p < 0.05$).

Recipe	Appearance	Aroma	Taste	Texture	Overall Preference
Raw	6.65 ± 0.99 [a]	6.05 ± 1.03 [a]	6.50 ± 1.47 [a]	6.85 ± 1.09 [a]	6.50 ± 1.36 [a]
Spray-dried	6.55 ± 0.89 [a]	6.25 ± 1.21 [a]	6.40 ± 1.47 [a]	6.30 ± 1.30 [a]	6.60 ± 1.73 [a]

Our research group previously reported that pea raw cooking water contained about 30 g/100 g of protein and around 30 g/100 g of soluble fibre, while spray-dried pea cooking powder diluted water had higher protein (34.6 g/100 g) but lower soluble fibre content (17.4 g/100 g) [2]. With higher protein and lower soluble fibre content, food products might be drier because of the unbalance of water distribution in food. The aroma score reported in the current study is in disagreement with a study by Avellone and coworkers [28], who investigated the effects of the spray-drying technique on wine's quality. They indicated that the spray-drying process caused significant reductions of aroma compounds in wines. However, in the current study, the liking scores of product aroma showed no significant difference. This suggests that spray-drying could be an ideal technique to preserve the aromatic property of *Liluva*.

4. Conclusions

In summary, spray-drying decreased the content of soluble fibre and lysine in split yellow pea water due to sugar loss and Maillard reaction. However, spray-drying and freeze-drying did not greatly affect the protein content or protein profile of pea water, as shown by the SDS-PAGE gel (showed similar protein bands). Compared to pea raw water, the particles in pea spray-dried powder diluted water samples were mostly related to smaller size classes, while freeze-dried powder diluted water samples were the opposite, highlighting the influence of the different drying mechanisms. Aside from particle size, colour was influenced by the drying method as spray-dried powder, due to its smaller particle sizes, refracted more light. The reduction in the red and yellow colour of spray-dried powder compared to freeze-dried powder also illustrated that heat treatment might degrade some of the pigments in split yellow peas.

In the preliminary sensory test, no significant difference was found in the sensory profiles of pea raw water and powder samples. Thus, spray-drying can be used to dry *Liluva* from peas without greatly influencing peas' properties. Future experiments are required to determine the mineral profiles of pea dried powders and to investigate their emulsifying activity to further confirm the effects of drying methods on pea composition and sensory profiles.

Author Contributions: Conceptualization, H.T.C. and L.S.; methodology, Z.F. and L.S.; investigation, W.C. and H.T.C.; resources, W.C., H.T.C. and Z.F.; data curation, W.C. and L.S.; writing—original draft preparation, W.C.; writing—review and editing, Z.F., E.M. and L.S.; visualization, L.S.; supervision, L.S.; project administration, L.S. All authors have read and agreed to the published version of the manuscript.

Funding: This research received no external funding. The research fund was offered by Lincoln University in support of Master's dissertations under the course called FOOD660.

Institutional Review Board Statement: The study was conducted according to the guidelines of the Declaration of Helsinki, and approved by the Ethics Committee of Lincoln University (protocol code 2020-60, approved on 11 December 2020).

Informed Consent Statement: Informed consent was obtained from all subjects involved in the study.

Acknowledgments: The authors would like to thank the company Dry Food NZ Ltd. for performing spray-drying of samples. Furthermore, the authors would like to thank Jiao Zhang and Roger Cresswell for the analysis of total nitrogen, Rosy Tung for the ash and carbohydrate analyses, and Jenny Zhao for the free amino acid quantification.

Conflicts of Interest: The authors declare no conflict of interest.

Appendix A

Table A1. Spray-drying parameters of split yellow pea water.

	Cooking Water	Soaking Water
Input (kg)	11.80	27.44
Output (kg)	0.25	0.13
Drying period	1 h 15 min	1 h
Target recovery (%)	5	2
Actual recovery (%)	2.12	0.46
Inlet air temperature (°C)	179	179
Outlet air temperature (°C)	80	94.6
Spray frequency (Hz)	170	180
Screw pump frequency (Hz)	23	10

References

1. Nosworthy, M.G.; Franczyk, A.J.; Medina, G.; Neufeld, J.; Appah, P.; Utioh, A.; Frohlich, P.; House, J.D. Effect of processing on the in vitro and in vivo protein quality of yellow and green split peas (*Pisum sativum*). *J. Agric. Food Chem* **2017**, *65*, 7790–7796. [CrossRef] [PubMed]
2. Serventi, L. *Upcycling Legume Water: From Wastewater Food Ingredients*, 1st ed.; Springer Nature: Cham, Switzerland, 2020.
3. Serventi, L. Soaking water composition. In *Upcycling Legume Water: From Wastewater Food Ingredients*, 1st ed.; Springer Nature: Cham, Switzerland, 2020; pp. 27–39.
4. Serventi, L. Cooking water composition. In *Upcycling Legume Water: From Wastewater Food Ingredients*, 1st ed.; Springer Nature: Cham, Switzerland, 2020; pp. 73–85.
5. Huang, S.; Liu, Y.; Zhang, W.; Dale, K.J.; Liu, S.; Zhu, J.; Serventi, L. Composition of legume soaking water and emulsifying properties in gluten-free bread. *Food Sci. Technol. Int.* **2018**, *24*, 232–241. [CrossRef]
6. Barbosa, J.; Teixeira, P. evelopment of probiotic fruit juice powders by spray-drying: A review. *Food Rev. Int* **2016**, *33*, 335–358. [CrossRef]

7. Liu, Y.; Zhang, Z.; Hu, L. High efficient freeze-drying technology in food industry. *Crit Rev. Food Sci Nutr* **2021**, 1–19. [CrossRef]

8. O'Sullivan, J.J.; Norwood, E.-A.; O'Mahony, J.A.; Kelly, A.L. Atomisation technologies used in spray-drying in the dairy industry: A review. *J. Food Eng.* **2019**, *243*, 57–69. [CrossRef]

9. AACC International. Moisture-Air Oven Methods, drying at 103 °C. Methods 44-15.02. In *Approved Methods of Analysis*, 11th ed.; Paul, S.T., Ed.; American Association of Cereal Chemists: Washington, MN, USA, 2000.

10. Jermyn, M.A. A new method for determining ketohexoses in the presence of aldohexoses. *Nature* **1956**, *177*, 38–39. [CrossRef]

11. Pollock, C.J.; Jones, T. Seasonal patterns of fructan metabolism in forage grasses. *New. Phytol.* **1979**, *83*, 9–15. [CrossRef]

12. AOAC. *Official Methods of Analysis of AOAC International*, 16th ed.; AOAC International Publ.: Arlington, TX, USA, 1995.

13. AOAC. *Official Methods of Analysis of AOAC International*, 18th ed.; AOAC International Publ.: Arlington, TX, USA, 1995.

14. Heems, D.; Luck, G.; Fraudeau, C.; Verette, E. Fully automated precolumn derivatization, on-line dialysis and high-performance liquid chromatography analysis of amino acids in food, beverages and feedstuff. *J. Chromatpogr. A* **1998**, *198*, 9–17. [CrossRef]

15. Carducci, C.; Birarelli, M.; Leuzzi, V.; Santagata, G.; Serafini, P.; Antonozzi, I. Automated method for the measurement if amino acids in ureine by high performance liquid chromatography. *J. Chromatogr. A* **1996**, *729*, 173–180. [CrossRef]

16. Buhl, T.F.; Christensen, C.H.; Hammershøj, M. Aquafaba as an egg white substitute in food foams and emulsions: Protein composition and functional behavior. *Food Hydrocolloid* **2019**, *96*, 354–364. [CrossRef]

17. Govoreanu, R.; Saveyn, H.; Van der Meeren, P.; Nopens, I.; Vanrolleghem, P.A. A methodological approach for direct quantification of the activated sludge floc size distribution by using different techniques. *Water Sci. Technol.* **2009**, *60*, 1857–1867. [CrossRef] [PubMed]

18. Yagiz, Y.; Balaban, M.O.; Kristinsson, H.G.; Welt, B.A.; Marshall, M.R. Comparison of Minolta colorimeter and machine vision system in measuring colour of irradiated Atlantic salmon. *J. Sci. Food Agric.* **2009**, *89*, 728–730. [CrossRef]

19. Mustafa, R.; He, Y.; Shim, Y.Y.; Reaney, M.J.T. Aquafaba, wastewater from chickpea canning, functions as an egg replacer in sponge cake. *Int. J. Food Sci. Technol.* **2018**, *53*, 2247–2255. [CrossRef]

20. Sveinsdottir, K.; Martinsdottir, E.; Thorsdottir, F.; Schelvis, R.; Kole, A.; Thorsdottir, I. Evaluation of Farmed Cod Products by a Trained Sensory Panel and Consumers in Different Test Settings. *J. Sens. Stud.* **2010**, *25*, 280–293. [CrossRef]

21. Stantiall, S.E.; Dale, K.J.; Calizo, F.S.; Serventi, L. Application of pulses cooking water as functional ingredients: The foaming and gelling abilities. *Eur. Food Res. Technol.* **2018**, *244*, 97–104. [CrossRef]

22. Joshi, M.; Adhikari, B.; Aldred, P.; Panozzo, J.F.; Kasapis, S. Physicochemical and functional properties of lentil protein isolates prepared by different drying methods. *Food Chem.* **2011**, *129*, 1513–1522. [CrossRef]

23. Shishir, M.R.I.; Chen, W. Trends of spray-drying: A critical review on drying of fruit and vegetable juices. *Trends Food Sci. Tech.* **2017**, *65*, 49–67. [CrossRef]

24. Brishti, F.H.; Chay, S.Y.; Muhammad, K.; Ismail-Fitry, M.R.; Zarei, M.; Karthikeyan, S.; Saari, N. Effects of drying techniques on the physicochemical, functional, thermal, structural and rheological properties of mung bean (*Vigna radiata*) protein isolate powder. *Food Res. Int.* **2020**, *138*, 109783. [CrossRef]

25. del Rio, A.R.; Opazo-Navarrete, M.; Cepero-Betancourt, Y.; Tabilo-Munizaga, G.; Boom, R.M.; Janssen, A.E. Heat-induced changes in microstructure of spray-dried plant protein isolates and its implications on in vitro gastric digestion. *LWT* **2020**, *118*, 108795. [CrossRef]

26. Shreenithee, C.R.; Prabhasankar, P. Effect of different shapes on the quality, microstructure, sensory and nutritional characteristics of yellow pea flour incorporated pasta. *J. Food Meas. Charact.* **2013**, *7*, 166–176. [CrossRef]

27. Lin, C.H.; Chen, B.H. Stability of carotenoids in tomato juice during processing. *Eur. Food Res. Technol.* **2005**, *22*, 274–280. [CrossRef]

28. Avellone, G.; Salvo, A.; Costa, R.; Saija, E.; Bongiorno, D.; Di Stefano, V.; Calabrese, G.; Dugo, G. Investigation on the influence of spray-drying technology on the quality of Sicilian Nero d'Avola wines. *Food Chem.* **2018**, *240*, 222–230. [CrossRef]

 *foods*

Article

Reinvigorating Modern Breadmaking Based on Ancient Practices and Plant Ingredients, with Implementation of a Physicochemical Approach

Vasileia Sereti [1], Athina Lazaridou [1,*], Costas G. Biliaderis [1] and Soultana Maria Valamoti [2,3]

[1] Laboratory of Food Chemistry and Biochemistry, Department of Food Science and Technology, School of Agriculture, Aristotle University of Thessaloniki, P.O. Box 235, 54124 Thessaloniki, Greece; vasisere@agro.auth.gr (V.S.); biliader@agro.auth.gr (C.G.B.)

[2] Laboratory for Interdisciplinary Research in Archaeology (LIRA), Department of Archaeology, School of History and Archaeology, Aristotle University of Thessaloniki, 54124 Thessaloniki, Greece; sval@hist.auth.gr

[3] Center for Interdisciplinary Research and Innovation, Aristotle University of Thessaloniki (CIRI-AUTH), Balkan Center, Buildings A & B, 10th km Thessaloniki-Thermi Rd, P.O. Box 8318, 57001 Thessaloniki, Greece

* Correspondence: athlazar@agro.auth.gr

Citation: Sereti, V.; Lazaridou, A.; Biliaderis, C.G.; Valamoti, S.M. Reinvigorating Modern Breadmaking Based on Ancient Practices and Plant Ingredients, with Implementation of a Physicochemical Approach. *Foods* **2021**, *10*, 789. https://doi.org/10.3390/foods10040789

Academic Editors: Luca Serventi, Charles Brennan and Rana Mustafa

Received: 28 February 2021
Accepted: 5 April 2021
Published: 7 April 2021

Publisher's Note: MDPI stays neutral with regard to jurisdictional claims in published maps and institutional affiliations.

Abstract: In this study, the potential use of ancient plant ingredients in emerging bakery products based on possible prehistoric and/or ancient practices of grinding and breadmaking was explored. Various ancient grains, nuts and seeds (einkorn wheat, barley, acorn, lentil, poppy seeds, linseed) were ground using prehistoric grinding tool replicas. Barley-based sourdough prepared by multiple back-slopping steps was added to dough made from einkorn alone or mixed with the above ingredients (20% level) or commercial flours alone (common wheat, spelt, barley). Sieving analysis showed that 40% of the einkorn flour particles were >400 μm, whereas commercial barley and common wheat flours were finer. Differential scanning calorimetry revealed that lentil flour exhibited higher melting peak temperature and lower apparent enthalpy of starch gelatinization. Among all bread formulations tested, barley dough exhibited the highest elastic modulus and complex viscosity, as determined by dynamic rheometry; einkorn breads fortified with linseed and barley had the softest and hardest crust, respectively, as indicated by texture analysis; and common wheat gave the highest loaf-specific volume. Barley sourdough inclusion into einkorn dough did not affect the extent of starch retrogradation in the baked product. Generally, incorporation of ancient plant ingredients into contemporary bread formulations seems to be feasible.

Keywords: prehistoric grinding practices; ancient grain flours; breadmaking; starch gelatinization; dough rheology; bread quality parameters

1. Introduction

Recent archaeological research has revealed a wide range of plant materials, preserved mainly as charred plant remains, which have been most likely used for preparation of human food [1–6]. In certain cases, actual food remains from prehistoric Europe have been conserved through charring among the burnt debris in cooking areas or houses destroyed by fire [4]. The study of these findings, together with a good knowledge of food preparation techniques, as reflected in the remains of cooking installations, pots and grinding equipment found at prehistoric sites, allows glimpses into past culinary practices and the nutritional benefits of specific ways of food preparation; moreover, such remains allow for the exploration of an evolutionary continuum in the use of plant foods from prehistoric times to the present.

The archaeobotanical record reveals that a wide range of cereals and pulses constituted the staple foods of prehistoric European communities [7,8]. In the Neolithic and Bronze Age of Greece, the glume wheats hold a dominant position among the cereal species presumably used for food and included three species: einkorn, emmer and 'new glume' wheat (*Triticum. monococcum, T. dicoccum* and *T. timopheevi*, respectively). During the Bronze Age, spelt wheat (*T. spelta*) also appears for the first time at the end of the 3rd millennium B.C., while the free-threshing wheats (common wheat, *Triticum aestivum* and durum wheat, *Triticum durum*) had a more limited occurrence in prehistoric times [9]. Nowadays, the glume wheats, together with certain pseudocereal species, are widely referred to as 'ancient grains', whereas those described as 'modern' (naked) wheats (*T. aestivum* and *T. durum*) have already been grown since the Neolithic period. These free-threshing cereals, over time, became the dominant cereals grown for food in Western Asia and Europe at the expense of the glume wheats, perhaps because the latter have a lower productivity per cultivated area, a higher bulk volume when stored in their glumes and a greater labour input required for dehusking before milling [10,11]. An additional factor may have been the low quality of their storage proteins (i.e., inferior gluten aggregation properties and inadequacy to form a strong hydrated network structure), attributed to the low amount of high molecular weight polymeric gluten fractions, resulting in difficulties in dough handling and poor breadmaking performance [12]. Apart from the above wheat species, several other plant-derived materials could have been used for food preparation in prehistoric times, i.e., barley (*Hordeum vulgare*) grain and lentil (*Lens culinaris*) seeds, as well as seed of acorn nut (*Quercus* sp.), linseed (*Linum usitatissimum*) and poppy seed (*Papaver somniferum*) transformed into flour by grinding [4,6].

The transformation of cereals and other plant-derived food materials would have taken place with the facilities available at the time, which included stone grinding and pounding implements [13], cooking installations and pots [14]. The limited availability of remains of actual plant foods preserved in the archaeological record include different types of processed cereals, such as porridges and breads, although a clear distinction between these preparations is not straightforward [9] unless complete loaves or large bread fragments were sometimes preserved, presumably in the context of specific ritual or funerary events [3,15,16]. Precooked ground cereal foods reported in the archaeobotanical record have been identified from the end of the 3rd millennium B.C. in Mesimeriani Toumba, and they could correspond to a prehistoric bulgur or trachanas [4,17]. These actual food remains from a distant past provide a glimpse into prehistoric culinary practices, yet the recipes that led to such food remains are poorly understood; therefore, an attempt is being made to systematically investigate these aspects under the project of PlantCult [18].

Consumer interest in ancient wheat species (e.g., spelt wheat, einkorn wheat) with regard to their use in bakery products has recently emerged, particularly because these grains are rich sources of bioactive components and hence suitable for producing high value food products with enhanced nutrient content and health benefits, especially when used in the form of whole flours [19–21]. For instance, einkorn wheat seems to have nutritional properties that could distinguish it from common wheat varieties, although its rheological dough behavior may be inferior for breadmaking. Actually, einkorn wheat has a high content of ash, protein and essential amino acids, as well as various phenolic antioxidants [22–24]. Furthermore, another ancient cereal grain, barley, is an important source of cereal β-glucans, which are soluble dietary fibers, well-known for their hypocholesterolemic and hypoglycemic action [25]; in fact, barley flour is the main ingredient of several traditional bakery products made on the Greek island of Crete, such as barley rusks ('Dakos'), that exist as staple food of the Mediterranean diet [26,27]. Other non-cereal raw materials, such as seeds from legumes, have been recently used to fortify breads due to their high content in carbohydrates, dietary fibers, vitamins, minerals, phytochemicals, and particularly proteins with a better amino-acids profile, ensuring a balanced diet when consumed in combination with cereals [28]. Additionally, seeds from nuts, such as acorns, have been incorporated into breads, mostly because of their antioxidant potential [29]. Oilseeds, such as linseed and their fractions, have been also engaged in modern breadmaking practices for nutritional enhancement of bakery products with soluble dietary fibers, essential amino acids and essential fatty acids, with the aim of reducing starch digestibility rates and improving atherogenic risk factors [30,31].

In this context, fortification of bakery items with ancient grains and seeds could have an impact to the wellbeing of individuals and contribute to the reduction of risk factors and/or management of chronic diseases, such as diabetes type 2 and cardiovascular diseases. Furthermore, adopting ancient breadmaking practices could give rise to new bakery products with improved nutritional attributes, health benefits and minimally processed food items, without additives and improvers, thereby satisfying the new consumer's preferences and emerging life-trends related to 'clean labelling' of food products, healthy diets, and wellbeing.

In the present study, experimental breads were prepared under controlled conditions using plant ingredients identified in the prehistoric record and resembling some of the prehistoric and/or ancient breadmaking practices (ingredients, grinding tools, leavening agents) in an effort to establish in ongoing studies [32] a relevant foundation for the recognition of past culinary practices, as reflected in the archaeobotanical charred remains of breads produced from cereals, legumes, nuts and oilseeds. The current work aimed to evaluate the physicochemical characteristics of flours, doughs and breads made from these plant materials (grains, seeds, nuts), following their reduction into smaller particles (meal/flour) by prehistoric replica grinding tools and employing food preparation practices that were likely to be employed in prehistoric and/or ancient times. Besides simulating some aspects of prehistoric and/or ancient cuisine, the ultimate goal of our research is to use this information in the design and development of future bakery products with improved nutritional attributes and increased consumer acceptability.

2. Materials and Methods

2.1. Flours

Five different plant raw materials—einkorn wheat (*T. monococcum*), barley (*Hordeum vulgare*), acorn (*Quercus* sp.), lentil (*Lens culinaris*), poppy seed (*Papaver somniferum*) and linseed (*Linum usitatissimum*)—were ground using prehistoric grinding tool replicas constructed for the PLANTCULT project (Figure 1), as described by Bofill et al. [33]. Specifically, einkorn grains were ground by three different types of grinding tools made from three different grinding stone materials (andesite, sandstone and granite) with three different grinding ways: (a) a grinding slab with a handstone of the "overhanging" type used in a back-and-forth reciprocal motion; (b) a grinding slab with a small handstone used in a back-and-forth reciprocal motion; (c) a grinding slab with a small handstone used in a circular and free-motion action. The rest of the grains and seeds were ground by the tool made from sandstone with the second way of grinding. The einkorn flour used for breadmaking and the study of the flour, dough and bread physicochemical properties was a mixture of the 9 flour samples ground by the three different stones and grinding ways mixed in equal amounts; this mixture was used as a representative flour sample prepared by the prehistoric grinding tool replicas. Two commercial flours from common wheat (*T. aestivum*), type T70 (white flour), and spelt wheat (*T. spelta*), whole flour, were also employed for breadmaking, whereas a commercial barley whole flour was also used for breadmaking and sourdough preparation. All the above three commercial flours were organic, provided by a local supplier (Doumos, Irinis Garden, Aridaia, Pella, Greece).

2.1.1. Particle Size Distribution

Flour particle size distribution was determined by sieve analysis using 100 g of a weighed sample, which was passed through a series of sieves with pore sizes from top to bottom: 4-mesh (0.96 mm), 5-mesh (0.80 mm), 6-mesh (0.65 mm), 8-mesh (0.50 mm), 10-mesh (0.40 mm), 16-mesh (0.24 mm), 20-mesh (0.20 mm), 30-mesh (0.13 mm), 40-mesh (0.10 mm) and 70-mesh (0.063 mm). For particle size analysis of einkorn flour, all of the above sieves were used, whereas for the commercial flours of barley and common wheat, only the last six sieves were employed. After shaking for 15 min by a mechanical vibratory shifter, the amount of flour retained on each sieve (WS) was weighed, and the contents of retained fractions were calculated as the percentage of the initial (total, WT) flour weight as follows:

$$\text{Retained fraction (\%)} = (WS/WT) \times 100 \tag{1}$$

Figure 1. Grinding cereals for experimental foods in the context of the European Research Council (ERC) PlantCult project.

The cumulative percent passing through the sieves was calculated by subtracting the cumulative percent of retained fractions from 100%.

$$\text{Cumulative percent Passing \%} = 100\% - \text{Cumulative Retained \%} \tag{2}$$

The particle size parameters of d_{50} (median diameter) and d_{90} that are commonly used in the classification of ground materials were estimated, representing 50% and 90%, respectively, of the particles with diameters smaller than the specified values.

2.1.2. Calorimetric Study of Starch Gelatinization Properties

The starch gelatinization properties of flours and flour mixtures (Table 1) used for breadmaking were studied by Differential Scanning Calorimetry (DSC) using a PL DSC-Gold calorimeter (Polymer Labs. Ltd., Epsom, UK). All flours obtained from the prehistoric grinding tools were passed through a coarse sieve (~1mm) to remove the coarse particles of husk and bran and thus increase the content of starch in the samples. Aqueous slurries of the sieved samples (about 25–30mg) containing 35% w/w solids were sealed hermetically into DSC aluminum pans. The pans were heated from 8 to 120 °C at a heating rate of 5 °C/min; samples were heated under a continuous flow of dry N_2 gas (20 mL/min) to avoid moisture condensation during measurement. Three specimens from each flour preparation were analyzed by DSC. Parameters estimated from the DSC thermographs were the onset (T_o) and peak (T_m) starch gelatinization temperature, as well as the apparent starch gelatinization enthalpy (ΔH) calculated from the area of the endothermic (melting) peak, following calibration of the calorimeter with indium.

2.2. Sourdough

2.2.1. Sourdough Preparation

The sourdough used for breadmaking was prepared for project PLANTCULT by the traditional method of spontaneous back-slopped sourdough (Type I); all the preparation steps are presented in detail in Figure 2. For the first fermentation step, a mixture of barley and einkorn wheat flour (barley:einkorn 1:1) with grape must (grape must:flour mixture 1:1 w/w) was used, following a recipe provided in a modern sourdough preparation book for the wider public [34]. The use of leavened bread is mentioned in the Hippocratic corpus, while the use of a diluted form of grape juice or of the by-product of alcoholic fermentation of grapes for breadmaking are mentioned in 'Geoponika', dated to the 10th c. A.D. [35]. Following fermentation at room temperature for 24 h, the mother sourdough was propagated five times by back-slopping steps using the barley and einkorn flour blend mixed with grape must, resulting in a mature mother sourdough. These fermentation steps were followed by further multiple (9) back-slopping steps at room temperature using barley flour, water and an amount of the mature sourdough; this mature sourdough was stored at 5 °C until it was used for breadmaking. Seven back-slopping steps using barley flour for the sourdough 'refreshment' were performed every 24 h to propagate the mature sourdough before breadmaking (Figure 2); fermentation during refreshments was carried out in closed vessels under controlled laboratory conditions in an incubator (Sanyo Incubator, MIR-154, SanyoElectric Co. Ltd., Ora-Gun, Gunma, Japan) at 30 °C for 24 h. The dough yield (DY) was defined as:

$$DY = (\text{Flour weight} + \text{Water weight}) \times 100/\text{Flour Weight} \qquad (3)$$

and it was 200 for these back-slopping steps during the making of barley-based flour sourdough.

Table 1. Bread formulations.

Sample Symbol	Flour Formulation	Sourdough (% *w/w* Flour Basis)
Control	Einkorn mixture [1] (100%)	-
Einkorn	Einkorn mixture (100%)	20
Einkorn + Spelt	Mixture of einkorn mixture (80%) with spelt (20%)	20
Einkorn + Acorn	Mixture of einkorn mixture (80%) with acorn (20%)	20
Einkorn + Lentil	Mixture of einkorn mixture (80%) with lentil (20%)	20
Einkorn + Barley	Mixture of einkorn mixture (80%) with barley (20%)	20
Einkorn + Poppy seed	Mixture of einkorn mixture (80%) with poppy seed (20%)	20
Einkorn + Linseed	Mixture of einkorn mixture (80%) with linseed (20%)	20
Spelt	Commercial spelt (100%)	20
Barley	Commercial barley (100%)	20
Common wheat	Commercial common wheat (*Triticum aestivum*) (100%)	20

[1] Mixture of the 9 flour samples ground by the three different stones (prehistoric grinding tool replicas) and three grinding ways mixed in equal amounts.

2.2.2. Microbiological and Physicochemical Characteristics of Sourdough

For the determination of colony forming units (CFU/g) in mature mother sourdough and the final back-slopped barley-based sourdough used for breadmaking, 10 g portions of sourdough were homogenized with 90 mL NaCl (0.99%, *w/v*), followed by decimal dilutions in the same saline solution. Enumeration of microorganisms was carried out using the pour plate technique. Lactic acid bacteria were enumerated on De Man Rogosa and Sharpe agar (MRS agar) (Merck KGaA, Darmstadt, Germany) containing natamycin, following incubation at 30 °C for 3 days, while for enumeration of yeasts, Yeast Malt Agar (YM agar) (Sigma-Aldrich, St. Louis, MO, USA), containing tartaric acid was used after incubation at 25 °C for 5 days.

For pH and total titratable acidity (TTA) determinations, 10 g of sourdough were mixed with 90 mL of sterile distilled water, and the resultant suspension was kept under continuous stirring during both measurements. The pH was measured with a Bante 210 pH meter (Bante Instruments Co., Shanghai, China), and the TTA was determined by titration of the sourdough suspension with 0.1 N NaOH until a final pH of 8.4; the TTA values were expressed as mL of 0.1 N NaOH/g of sourdough.

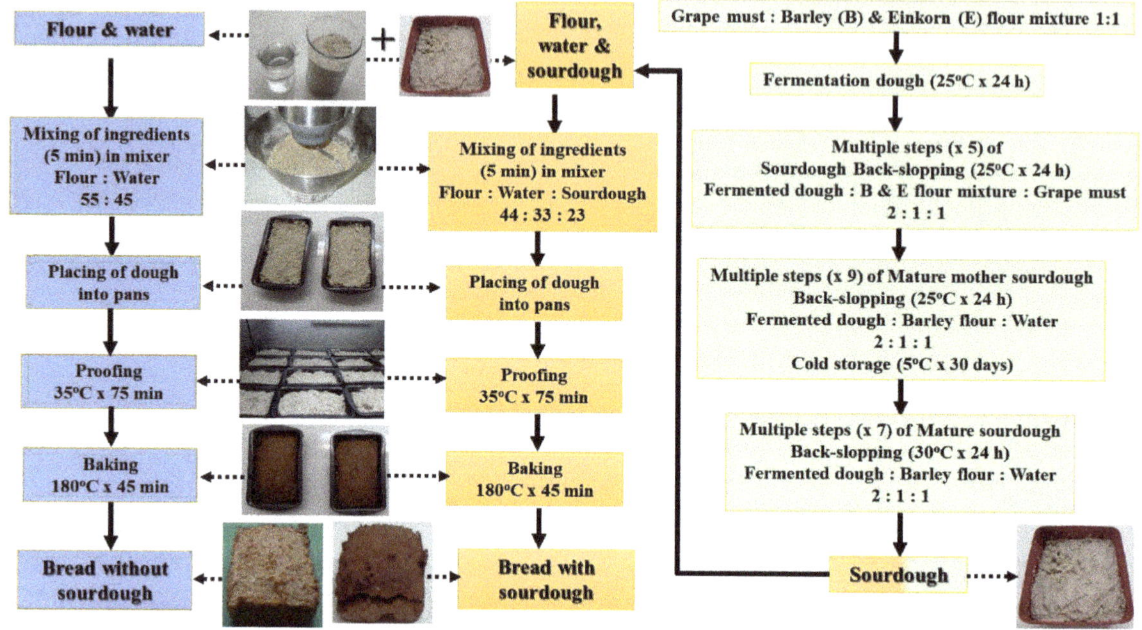

Figure 2. Flow chart of the protocol followed for mature sourdough production and breadmaking without and with sourdough usage; the flour (or flour mixture) formulation used for breadmaking without and with sourdough is given in Table 1.

2.3. Doughs

2.3.1. Dough Preparation

For the study of dough rheological properties, flours obtained with the prehistoric grinding tools were passed through a coarse sieve (~1 mm) to remove all coarse particles of husk and bran and thus increase their homogeneity and improve their water retention capacity. Dough samples from all tested flours used for breadmaking (Table 1) without sourdough were prepared by mixing the flours with tap water (flour:water 50:50 w/w), and followed by hand-kneading for 5 min. Before rheological testing, all dough samples were wrapped with a plastic membrane to avoid water loss and rested at room temperature for 20 min, allowing uniform moisture distribution and dough matrix relaxation before testing. The dough preparation procedure for each formulation was repeated in triplicate.

2.3.2. Dough Rheology

Oscillatory measurements of doughs were performed by a rotational Physica MCR 300 rheometer (Physica Mess-technic GmbH, Stuttgart, Germany) using a parallel plate geometry (50 mm diameter and 2 mm gap) with a solvent trap to avoid moisture loss during measurements; the plate had a sanded surface to prevent slippage of the measuring fixture. The temperature was regulated at 20 °C by a controlled temperature peltier system (TEZ 150P/MCR) with an accuracy of ±0.1 °C. After loading, the dough sample was left to further rest in the geometry for 15 min prior to measurement. Oscillatory measurements were performed, at which the storage (or elastic) modulus, G′; loss modulus, G″; and complex viscosity, η*, at a strain level of 0.01%, were monitored over an angular frequencies range of 0.1–50 Hz. The data of the rheological measurements were analyzed with the supporting rheometer software US200 V2.21.

2.4. Breads

2.4.1. Bread Formulations

Breads were made from einkorn, spelt, barley and common wheat flours alone as well as from flour mixtures of einkorn with another grain or seed flour, i.e., spelt, acorn, lentil, barley, poppy seed and linseed (einkorn: other flour in an 80:20 ratio), with the addition of sourdough; all bread formulations are shown in Table 1. Einkorn was employed as the primary flour (larger proportion in the mixture), since it was a very important species used in prehistoric Greece [36–38]; moreover, it has previously been found that it can have a relatively acceptable breadmaking performance, especially when sourdough is included to an einkorn bread formulation [39–41]. Additionally, a control bread formulation made from einkorn flour alone without sourdough was prepared for comparison. The back-slopped barley-based sourdough (originating from the mature mother sourdough) was used for breadmaking (Figure 2) at the level of 20% w/w on a flour basis (DY 200); i.e., the amount of flour from sourdough was 20g in 100g of total flour in the bread formulation. The level of added water in all dough formulations, including the sourdough water amount, was 80% (flour basis); the same water level was also added to the control sample.

2.4.2. Breadmaking Process

For the breadmaking process, flours obtained with the prehistoric grinding tools were used without removal of the coarse particles of husk and bran in order to closely resemble the prehistoric practices. Actually, pieces of cereal bran and husk embedded in a cereal-based amorphous matrix (gelatinized starch) have been identified in several archaeological sites in Southeastern Europe, indicating that breadmaking and/or porridge making from cereal grains were possibly widespread food preparation methods in prehistoric times, by employing relatively simple processing tools and practices [9].

The flow chart presented in Figure 2 describes the major steps of breadmaking followed in the current study. Breads were prepared by mixing sourdough, flour and water for 15 min with a mixer (KMM023, Kenwood Major Titanium, Kenwood Ltd., Havant, UK) at medium speed and room temperature. The composite dough (150 g) was placed in pans (three pans for each dough formulation) and proofed under controlled temperature and relative humidity (RH) conditions in an incubator (Sanyo Incubator, MIR-154, SanyoElectric Co. Ltd., Ora-Gun, Gunma, Japan) at 35 °C and 100% relative humidity (RH) for 75 min. Baking was performed in an oven (air-o-stream combi oven, Electrolux Professional SpA, Pordenone, Italy) under controlled temperature (180 °C) and RH (100%) conditions for 45 min; the exhaust valve of the oven was opened in the last 10 min of baking to remove excess humidity from the chamber. The breadmaking procedure for each bread formulation was repeated in triplicate. Analyses of quality parameters of the breads were performed after cooling them down to room temperature for 1.5 h, and the reported values of tested quality attributes are mean values from the three breadmaking processes.

2.4.3. Bread Quality Characteristics

Large deformation mechanical properties of the experimental breads were examined by a puncture test using a Texture Analyser (TA-XT2i, Stable Micro systems, Godalming, Surrey, UK) calibrated with a 5 kg load cell. The bread loaves were compressed with a spherical probe (0.635 cm diameter) up to rupture of the crust at a crosshead speed of 0.4 mm/s. Hardness of the crust was taken as the peak force of the force-displacement curve; two values obtained from the same loaf were averaged, and two loaf breads from each breadmaking process were averaged into one replicate.

The volume of the bread loaves was determined with a homemade volume meter made from plexiglass and based on the rapeseed displacement method [42]. Bread loaves were weighted, and the specific volume was calculated as the ratio of volume/bread weight. For each bread formulation, the measurement was carried on three loaves (one from each breadmaking process).

2.5. Microscopy and Changes in Starch Physical State during Breadmaking and Bread Storage

Suspensions in ethanol of einkorn and commercial barley flours, barley sourdough, einkorn dough without and with sourdough, as well as crumb of fresh einkorn bread without and with sourdough were prepared under stirring. Aliquots of suspensions were taken by aspiration and examined using an Olympus BX51 microscope (Japan) equipped with dry lenses, a microscope digital camera Olympus DP70 and the Olympus micro DP70 software. The microscopic observation was carried out after staining of starch with iodine solution in bright field and cross polarized light in order to observe the starch granules and the changes in starch microstructure during dough preparation and breadmaking in the presence or absence of sourdough. At least ten captures for each sample were taken.

Changes in starch physical state upon breadmaking and bread storage were also examined by DSC analysis of the crumb of fresh and staled einkorn bread without and with sourdough; for the staling events in the starch matrix, bread loaves were stored in sealed polypropylene bags at 3 °C for 6 days. Before analysis, crumb samples from fresh and stored bread were lyophilized and then ground into fine powder using liquid nitrogen. Aqueous slurries of the lyophilized samples (about 25–30 mg) containing 35% w/w solids were hermetically sealed into DSC aluminum pans and heated from 8 to 120 °C at a heating rate of 5 °C·min^{-1}. Three crumb specimens of each bread preparation from all the breadmaking processes were tested by DSC. The onset melting temperature (T_o^{RET}), the peak (T_m^{RET}) melting temperature and the apparent melting enthalpy (ΔH^{RET}) of the retrograded starch were determined.

2.6. Statistical Analysis

All physicochemical parameters of flours were tested in triplicates. Mean values of dough parameters, loaf specific volume and crumb retrogradation parameters were the average from three dough specimens, bread loaves and crumb samples, respectively (one from each dough or bread making procedures). For crust hardness, firstly, values from two different points of the same loaf were averaged, and the mean values presented in this study were obtained from two different breads that averaged into one replicate of the three breadmaking repetitions.

Statistical analyses were performed by the IBM SPSS statistical software (version 23.0, IBM Corp., Armonk, NY, USA). All parameters of flour, dough and bread properties were analyzed by a one-way ANOVA, according to a generalized linear model, examining the effect of flours and their mixtures at all breadmaking stages. Differences between mean values were compared using the Tukey's test at a = 0.05 significance level.

3. Results and Discussion

3.1. Flour Properties

Analysis of sieving classifies the flour particles by size as well by shape. The particles of flour are usually spherical-like, such that their diameters correspond to the sides of the square sieve opening. Particle size cumulative distribution curves of einkorn, barley and common wheat flour are presented in Figure 3. Sieve analysis of flour particle size distribution showed that 40% of the particles of einkorn flour had a size > 400 μm, while the sizes of all particles of commercial barley and common wheat flour were smaller than this size. Specifically, the d_{50} and d_{90} values of einkorn, barley and common wheat flour were 287, 228, 123 μm and 879, 313, 252 μm, respectively. Thus, einkorn flour exhibited the largest particle size at all particle size distribution ranges among the three tested flours since einkorn grains were ground by prehistoric grinding tool replicas (stone grinding), in contrast with the other two commercial flours, which were finely ground by industrial mills. Additionally, between the two commercial flours, barley flour had larger particles as it was a whole flour compared with those from common wheat, which was a white flour (70% extraction rate); it is well-known that whole flours include bran and thus have higher mean particle size compared with flours originating mostly from the endosperm of cereal grains (white flours).

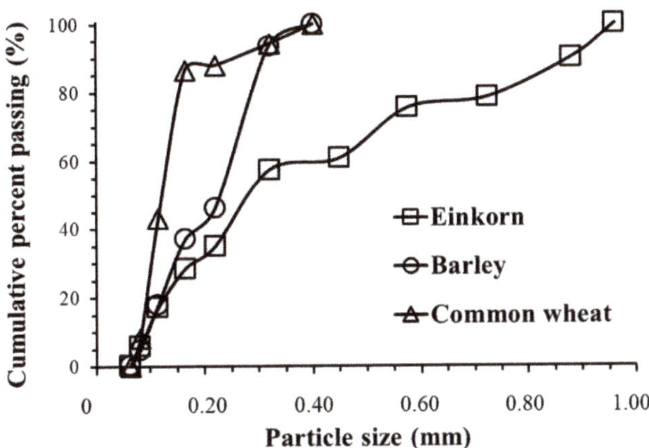

Figure 3. Cumulative distribution of particle size of einkorn, barley and common wheat flours.

According to previous studies, the particle size of flours significantly affects the rate of water absorption during dough making, as fine particles absorb water faster due to their greater surface area [43]. Smaller flour particles from some starchy grains such as quinoa were also found to have an impact on starch gelatinization properties [44], i.e., the finer flours exhibited lower starch gelatinization temperatures. Moreover, it has been shown that flour particle size largely affects cereal flour dough rheological behavior, with doughs from coarser barley flour (d_{50} = 350 μm) exhibiting increased stiffness and resistance to deformation and flow compared with that of a fine (d_{50} = 200 μm) barley flour preparation [27]. However, in our study, the effect of particle size on dough rheological and thermal properties cannot be clearly unraveled, since the different tested flours originated from different grain species, and their compositional differences can certainly have a stronger impact on flour and dough functional properties. It is worth noting that flour particle size can have a significant effect on the starch digestibility of bakery products and thus can have an impact on postprandial glycemic responses. Our previous studies using an in vitro assay simulating the human digestion process have shown a lower starch degradation by digestive enzymes for rusks made from a coarse barley flour compared with the products made from a fine flour (37% vs. 53% after 5 h of digestion) [27]; therefore, coarse flours such as those ground by stone mills, as employed in the present study, can lead to a better attenuation of glucose blood levels compared with products made by commonly used wheat fine flours. Moreover, whole ancient grain flours are rich sources of dietary fiber and other bioactive compounds, such as antioxidants [21–23].

The DSC thermograms of the slurries (35% w/w solids) of tested flour samples showed the typical endothermic peak of starch gelatinization at around 56.0–80.0 °C (Figure 4), which is the usual temperature range at which this phase transition occurs, at similar water levels to those used in the present study [45]. The endothermic peak is attributed to absorbed thermal energy, resulting in the breaking of the hydrogen bonds between adjoining starch polymeric chains existing in double helical conformations. The swelling of starch granules in heated aqueous dispersions usually starts at a temperature corresponding to the onset temperature (T_o) of this endothermic transition and the disruption (melting) of starch molecular orders (mostly double helical structures of amylopectin) upon gelatinization occurs at the peak temperature (T_m) (Figure 4); the area under the endothermic peak expresses the apparent melting enthalpy (ΔH), reflecting the net amount of heat, required for the disruption of short- and long-range molecular orders in the starch granules of the heated sample.

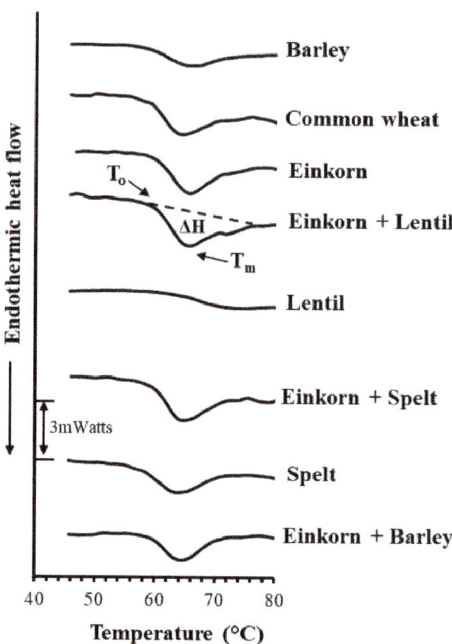

Figure 4. Representative Differential Scanning Calorimetry (DSC) thermographs of slurries (35% *w/w*) of the flour samples used for breadmaking (heating rate 5 °C/min): T_o, onset starch gelatinization temperature; T_m, peak starch gelatinization temperature; ΔH, apparent starch gelatinization enthalpy; notation of samples as in Table 1.

Lentil flour dispersions exhibited the highest T_o (64.7 °C) and T_m (74.0 °C) and the lowest ΔH (0.9 mJ/mg) values among all tested flour samples, implying higher resistance of this legume starch towards gelatinization and a relatively lower amount of double helical structures (primarily involving the amylopectin component) compared with those of cereal starches (Table 2 and Figure 4). It is well known that the thermal transition temperatures of cereal and legume flours differ among different species and are influenced by water, protein and amylose content, level and type of helical structures in the starch granules, distribution of amylopectin branch chains, and the presence of monoacyl lipids, which can complex with amylose chains into single helical structures during starch gelatinization [45,46]. Amylopectin plays a major role in starch granule crystallinity; however, in the case of high amylose content starch, the melting temperature of crystalline regions increases, the endotherm broadens and there is also a change in the apparent gelatinization enthalpy of the starch component [47]. Therefore, the differences in T_o, T_m and ΔH values among lentil seed and cereal grain flours observed in the current study could be attributed to the higher amylose content of starch, ~30–40%, and the lower total starch concentration of flour, ~50%, for legumes than typical cereal grains (einkorn, spelt and barley), which have ~20–25% and 60–70%, respectively [23,46,48–52]. Among cereal grains, spelt showed the lowest T_o temperature (56.2 °C) followed by the common wheat (57.6 °C), while T_o values for flours of barley, einkorn and its mixture with lentil, spelt and barley displayed higher values, ranging from 58.9 to 59.4 °C (Table 2). For cereal flours and flour mixtures with einkorn as their major component, the T_m values were similar, varying in the narrow range of 61.6–65.7 °C, while their gelatinization enthalpy ranged between 3.2 to 5.6 mJ/mg of flour (Table 2). It seems that the gelatinization properties of flour mixtures were not largely influenced by enrichment with the secondary flour, being similar to those of the base flour (einkorn), since fortification with the secondary flour in the flour mixtures was in a small proportion (20% *w/w*). Moreover, barley flour showed significantly lower apparent enthalpy and higher onset temperature of gelatinization compared with common wheat flour (Table 2), probably due to the presence of a higher amount of non-starch polysaccharides (cell wall materials from bran and endosperm) in the whole barley flour. In accordance with our findings, Tester and Sommerville [53] demonstrated that the presence of non-starch polysaccharides limited water availability and reduced the leaching of amylose from starch and hence the swelling factor of starch granules during gelatinization, resulting in an increase in the apparent T_o and a reduction in ΔH.

Table 2. Starch gelatinization parameters of slurries (35% *w/w*) of the flour samples used for breadmaking; heating rate 5 °C/min.

Samples [1]	T_o (°C) [2]	T_m (°C) [2]	ΔH [2] (mJ/mg of Flour)
Barley	58.91 (±0.06) c [3]	65.70 (±0.21) a	3.22 (±0.09) b
Common wheat	57.56 (±0.79) b	61.61 (±1.66) a	5.55 (±0.35) c
Einkorn	59.27 (±0.08) c	64.29 (±0.14) a	4.27 (±1.67) bc
Einkorn + Lentil	58.97 (±0.72) c	65.56 (±0.11) a	5.60 (±0.57) c
Lentil	64.74 (±0.24) d	73.98 (±0.39) b	0.93 (±0.46) a
Einkorn + Spelt	59.14 (±0.17) c	64.86 (±0.52) a	5.54 (±0.93) c
Spelt	56.15 (±0.26) a	63.42 (±1.51) a	4.55 (±0.42) bc
Einkorn + Barley	59.44 (±0.16) c	65.10 (±1.00) a	4.57 (±0.59) bc

[1] Notation of samples is given in Table 1. [2] T_o: onset starch gelatinization temperature, T_m: peak starch gelatinization temperature and ΔH: apparent starch gelatinization enthalpy. [3] Values followed by the same letter for the same column are not significantly different ($p > 0.05$, Tukey's test).

3.2. Microbiological and Physicochemical Characteristics of Sourdough

The predominant microflora in sourdoughs are lactic acid bacteria (LAB), while the number of yeasts is limited [54]. In our study, the number of cells of lactic acid bacteria were much higher (9.8 log CFU/g) than that of yeasts (6 log CFU/g) in the sourdough added to bread dough. The numerous refreshment steps of the original sourdough aimed to establish a final sourdough preparation for breadmaking with increased LAB cell density and a rather suppressed yeast population. Indeed, it appeared that the conditions of the multiple refreshment steps did not allow propagation of yeasts and mostly resulted in the domination of LAB. The pH of the final sourdough dropped to 3.8, whereas the TTA reached a value of 11.7 mL NaOH 0.1 N/10 g of sourdough; these values can be considered as indices of a well-developed sourdough system [55]. Other researchers reported similar LAB and yeast cell densities, pH values, and TTA levels for spontaneous back-slopped barley sourdoughs, as well as for spontaneously fermented durum-wheat-based sourdough when must grape was employed as an added ingredient to properly 'drive' the fermentation process by providing fermentable sugars and competitive microflora, as well as to prevent undesirable microbial deviations [56,57].

3.3. Dough Rheological Properties

The rheological properties of flour doughs are influenced by many factors, such as dough ingredients (composition), temperature, water uptake and type of mixing, but the most important is the type of flour used [58]. Figure 5a illustrates three representative mechanical spectra of einkorn, barley and common wheat doughs. All dough formulations showed the typical solid, elastic-like behavior of wheat and non-wheat-based doughs, with the elastic modulus being greater than the loss modulus over the whole frequency range and both moduli being slightly dependent on frequency [59,60]. The G′ and η* values that were obtained from the frequency sweep test ranged between 2080–33400 Pa and 76.8–1214.0 Pa·s, respectively (Figure 5b). The barley dough exhibited significantly ($p < 0.05$) higher elastic modulus and complex viscosity values compared with all doughs made from any wheat species flour (einkorn, spelt and common wheat), suggesting that the barley dough was the most resistant to deformation and flow. Barley non-starch polysaccharides, such as β-glucans and arabinoxylans, can provide a composite dough with increased structural strength, stiffness and viscosity because of their ability to bind large amounts of water [61–63]. On the other hand, the different wheat species, both 'modern' (common wheat) and 'ancient grains' (einkorn and spelt wheat), produced doughs with similar rheological parameters ($p > 0.05$).

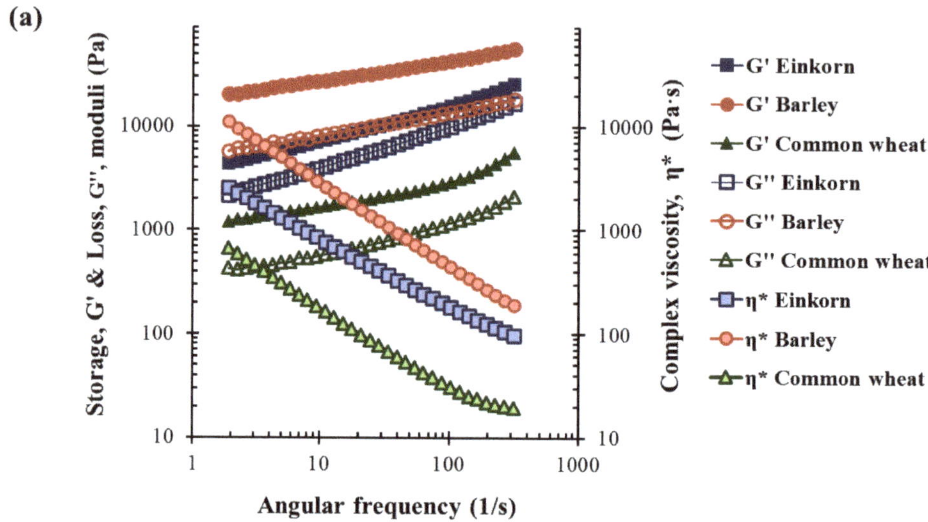

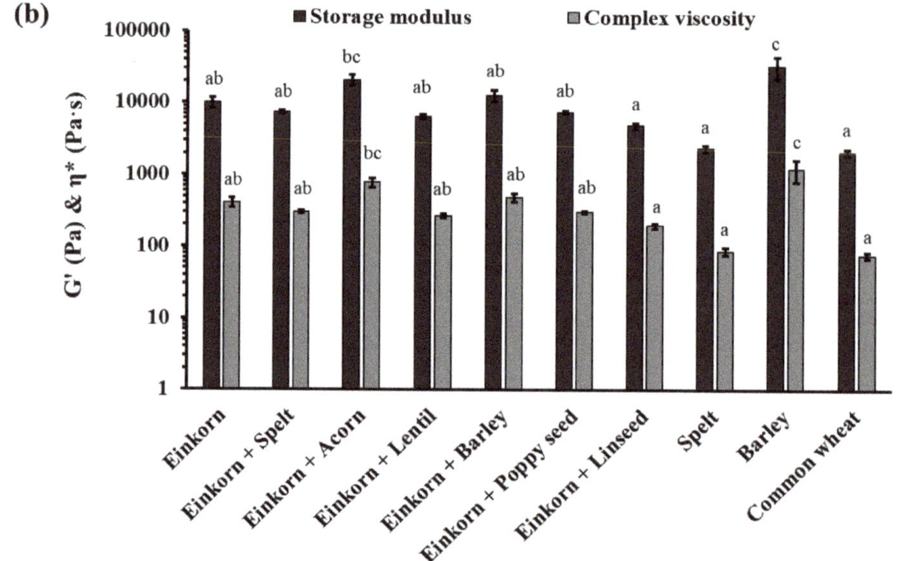

Figure 5. Representative mechanical spectra of doughs (flour:water 50:50 w/w) of the flour samples used for breadmaking (**a**) and the derived storage modulus, G′, and complex viscosity, η*, at 30 1/s angular frequency (**b**); strain 0.01%, 20 °C. Values followed by the same letter for the each specified rheological parameter are not significantly different ($p > 0.05$, Tukey's test); notation of samples as in Table 1.

The supplementation of einkorn wheat dough with spelt, barley, acorn, lentil, poppy-seed and linseed flour did not result in any significant ($p > 0.05$) change in its rheological behavior (Figure 5b). Nevertheless, the incorporation of linseed into einkorn wheat dough led to a small decrease of dough elasticity, probably due to its lubricating action, while inclusion of acorn flour resulted in a slight strengthening of the dough; thus, the most elastic and viscous einkorn doughs were those supplemented by acorn flour, as indicated by the respective values of their rheological parameters. Our findings are in accordance with those of Beltrão Martins et al. [64] and Korus et al. [65] who studied the influence of acorn flour on rheological properties of gluten-free dough; these researchers found that the incorporation of acorn flour into gluten-free dough formulations resulted in an increase of G′ values.

It is worth noting that high dough elasticity and viscosity implies high resistance of dough to deformation (high G′ value) and flow (high η* value) and cannot necessarily be related to improved dough and bread textural properties [66]. High viscosity and elasticity of a wheat flour dough could also be linked with limited dough expansion and insufficient retention of the incorporated gas cells during mixing and/or from sourdough fermentation, thereby resulting in a more compact crumb macrostructure and lower bread volume.

3.4. Evaluation of Bread Quality Characteristics

Quality characteristics of breads, such as appearance, loaf volume and bread texture, are major determinants of the product acceptability by consumers. Despite the expected positive health implications of 'ancient' cereal grains consumption, their involvement in bread production usually leads to doughs characterized by high tenacity and low extensibility, while the resultant breads exhibit reduced loaf volume [67,68]. However, a small number of spelt, emmer and einkorn cultivars were found to have favorable gluten characteristics for good baking performance [12]. In the current study, the breads made were based on 'ancient' grains enriched with other 'ancient' plant derived ingredients—which were often used in food products, including breads, in the prehistoric and ancient past by humans [3,4,6–9,15,16]—by employing stone grinding tools [13,33] similar to those used in prehistoric times and sourdough as the only leavening agent; leavened bread was likely used in ancient cuisine as well [35].

The appearance of the loaf cross-sections of all bread formulations is given in Figure 6. Macroscopically, it was shown that the crumb macrostructure of ancient grain-based breads was inferior compared with that of the commercial common wheat flour. Thus, breads from einkorn and its mixtures with other plant materials had uneven gas cell distributions exhibiting large pores. The latter implies gas cell coalescence during breadmaking, probably due to the weak protein gel network formed upon dough mixing and kneading, as well as the presence of much larger amounts of bran particles in the formulation compared with common wheat bread; bran particles can also weaken the gluten network by dilution, competition for water absorption and interruption of its continuity. Additionally, bread from commercial barley flour showed a relatively more cohesive and compact crumb macrostructure than the other formulations, which is possibly attributed to the presence of non-starch polysaccharides (arabinoxylans and β-glucans) that can largely increase the viscosity and elasticity of the dough, resulting in prevention of extensive dough raising; this observation is in agreement with the findings from the mechanical spectra of dough formulations, in which barley dough showed the highest G′ and η* values (Figure 5) among all the tested preparations. Moreover, bread enriched with acorn flour exhibited the darkest crumb and crust color among the samples due to the presence of this dark-coloured ingredient (Figure 6), indicative of the presence of a high concentration of phenolic compounds in the raw material.

Sourdough inclusion in einkorn bread formulation did not seem to affect the crust hardness (Figure 7a); thus, the crust hardness of einkorn bread with sourdough did not significantly differ ($p > 0.05$) from that of the control bread (einkorn without sourdough). On the other hand, incorporation of linseed into the einkorn bread decreased ($p < 0.05$) the crust hardness significantly; thus, einkorn bread fortified with linseed had the lowest crust hardness among all tested samples. In agreement with our findings, Marpalle et al. [69] have also observed that bread's softness increased with increasing level of flaxseed flour added in fortified breads. Moreover, poppyseed inclusion in einkorn bread formulation resulted in a softer crust than the other einkorn-based breads, although this effect was not significant ($p > 0.05$) (Figure 7a). Soft crust texture of bread formulations fortified with these oilseeds could be attributed to the high fat content of linseed [70–73] and poppyseeds [74,75] (~40–45% fat), with the lipids acting as lubricants, decreasing the crust hardness of the final product. Instead, the inclusion of stone-ground barley in einkorn bread formulation resulted in the highest crust hardness value among all samples (Figure 7a).

Figure 6. Appearance of a cross-section of breads; all breads made with sourdough except control; notation of samples as in Table 1.

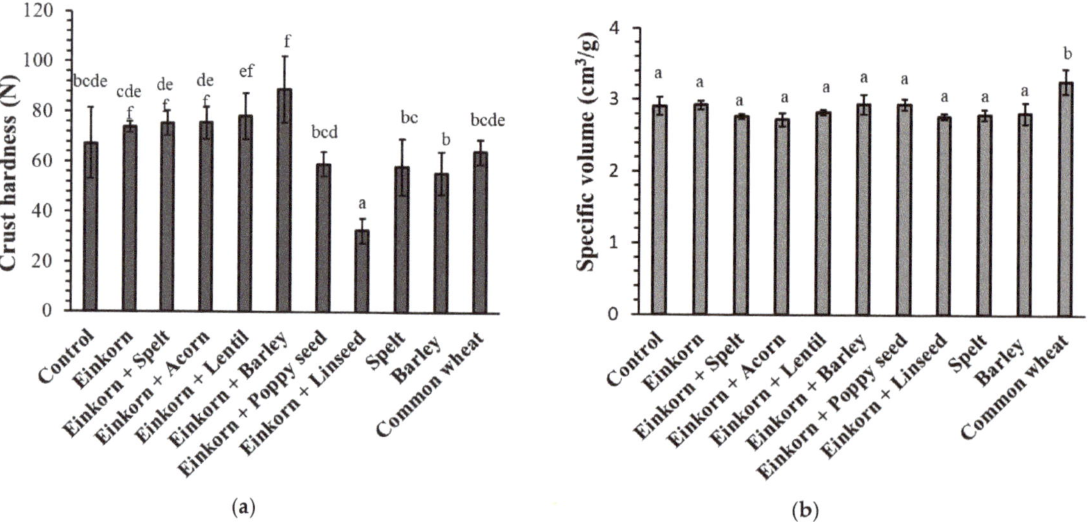

Figure 7. Crust hardness (**a**) and specific volume (**b**) of breads; control formulation is einkorn bread without sourdough. Values followed by the same letter are not significantly different ($p > 0.05$, Tukey's test); notation of samples as in Table 1.

Loaf volume is commonly considered as the most important indicator of bread quality. The bread from common wheat flour had significantly higher loaf specific volume compared with all other bread samples (Figure 7b). The higher volume is ascribed to the unique viscoelastic properties of gluten in common wheat flour resulting in the development of a strong protein cross-linked network in dough, which leads to retention of gas cells during proofing and baking. The presence of high amounts of bran particles or non-starch polysaccharides in einkorn, spelt and barley-based breads, on the other hand, can lead to a loaf volume reduction, since these carbohydrate polymeric materials can compete with proteins and starch for water absorption-retention and thus interrupt the continuity of a well-developed protein gel network during dough mixing, as well as the gelatinized starch in the dough continuous phase throughout baking; the presence of all these non-starch polysaccharides (soluble as well as insoluble particles) also weakens the continuity of the composite gluten-starch network formed upon baking of the dough. Similarly, Geisslitz et al. [12], comparing breads made by 'ancient' wheat species (einkorn, spelt and emmer) to those of common wheat and durum wheat, reported that among the five wheat species, the common wheat flour gave the highest loaf volume. According to these researchers, a high molecular weight glutenin subfraction, namely glutenin macropolymer (GMP), is positively correlated with dough water absorption and bread volume, pointing to a strong impact of protein quality (glutenin fraction) on breadmaking performance; among the different species of wheat, the GMP contents of common wheat, spelt and einkorn were ~0.8, 0.6 and 0.3 g/100g of whole meal flour, respectively.

3.5. Starch Physical State of Fresh and Stored Bread

Iodine staining of the starch granules of einkorn and barley flour revealed the typical bimodal size distribution and the characteristic oval and round shape of wheat and barley starch granules (Figure 8a-left). Additionally, optical birefringence of both flours was evidenced in the cross-polarized micrographs, reflecting the ordered structures in the starch granules at a molecular level (Figure 8a, right).

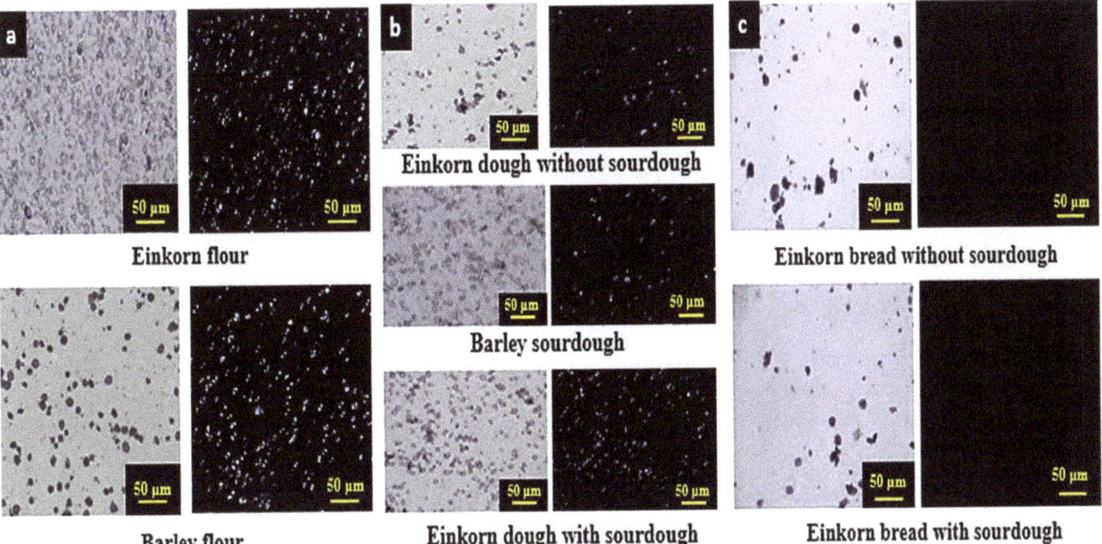

Figure 8. Optical micrographs of starch granules from einkorn and barley flours (**a**), barley sourdough and einkorn dough with and without sourdough (**b**), and crumb of einkorn fresh bread with and without sourdough (**c**), stained by iodine solution and observed under bright-field light (left pictures) and cross-polarised light (right pictures).

As expected, shape, integrity, size and birefringence of einkorn and barley starch granules were preserved in the dough and sourdough preparations (Figure 8b). It is worth noting that some starch granules in the einkorn control dough (without sourdough) seemed to be clustered (Figure 8b, top). Other researchers studying the dough microstructure have found, by staining both protein and starch, that these components are not evenly distributed in the dough, and there are some regions of the dough where several starch granules are gathered [76]; it seems that the gluten network fills the space between the water-fused starch granules. In contrast, starch granules are more evenly distributed in barley sourdough (Figure 8b, middle); most likely, the network of the barley storage protein is weaker and does not hinder a homogeneous distribution of starch granules into the water phase. A more homogeneous distribution of starch granules was also favored in the einkorn dough specimens when barley sourdough was added (Figure 8b, bottom); possibly, sourdough dilutes and interrupts the continuous native gluten gel network in the einkorn dough system.

During baking, the rise of temperature up to about 95 °C in the crumb and the level of added water to the dough allows the gelatinization of starch to occur. Therefore, starch granules appeared larger than those in flour and dough due to extensive swelling and loss of their oval or round shape; nevertheless, the swollen granules still retained their identity (Figure 8c, left). The loss of granular integrity is probably attributed to the melting of starch crystallites and leaching of starch molecules (mostly amylose) from the swollen granules, which largely occur upon starch gelatinization. Moreover, the presence of sourdough in einkorn bread did not seem to affect the starch microstructure in the crumb. As expected, bread crumb exhibited almost no birefringence, since starch granule swelling upon gelatinization was accompanied by loss of the ordered structure of starch molecules, i.e., the melting of the crystalline domains initially present in native starch granules (Figure 8c, right).

Heat-moisture mediated disruption of the ordered structures in granular starch (gelatinization) is generally a prerequisite for its utilization because it changes the rheological properties of the system and has a major influence on the functionality and digestibility of starch-containing products [45], e.g., gelatinized starch largely contributes to the formation of a fine porous crumb structure in bread. However, formation of new structures (intra- and intermolecular associations) upon cooling and storage of starch systems, named as retrogradation, may be detrimental to end-product quality (texture changes), i.e., starch retrogradation significantly contributes to the hardening of bread crumb upon the staling of bread and other bakery items [77]. Starch retrogradation involves reassociation of the polymeric chains, creation of a new molecular order (mostly double helices of the amylopectin outer chains) among starch chains, and crystallization of double helical aggregates. The most common method to monitor these phenomena and to probe the development of the various structural domains in a starchy matrix is calorimetry.

As previously mentioned, the DSC thermograms of einkorn flour slurries (35% *w/w* solids) showed the typical endotherm peak of starch gelatinization at around 55–80 °C, with the peak starch gelatinization temperature (T_m) occurring at about 65 °C, while the enthalpy (ΔH) required for the disruption (melting) of the ordered structures in native starch is estimated by the area under this peak (Figures 4 and 9). This endothermic peak is not found in the DSC thermograms of fresh bread crumb (Figure 9) because of disordering (gelatinization) of starch molecules upon baking; this observation is consistent with the total loss of granule birefringence viewed by cross-polarized light microscopy of freshly prepared bread (Figure 8c, right). However, an endothermic peak (staling endotherm) at a lower temperature range (38–55 °C), with a peak temperature of around 50 °C, eventually appears on the DSC thermograms of bread crumb (35% *w/w* solids) after storage for long time (6 days) (Figure 9). The endothermic peak of staled bread is attributed to the melting of the retrograded amylopectin fraction (re-organized double helices of the outer short chains in the amylopectin molecules); this starch component retrogrades slowly upon storage of starchy aqueous systems, and the area under the endothermic peak continuously increases with storage time [45]. Specifically, this endothermic transition represents the melting of retrograded amylopectin, with ΔH^{RET}, T_o^{RET} and T_m^{RET} corresponding to the apparent melting enthalpy and the onset and peak temperature, respectively, for melting of the re-ordered starch chains (Figure 9). The inclusion of sourdough in bread formulation did not significantly ($p > 0.05$) affect these three parameters; the respective values of ΔH^{RET}, T_o^{RET} and T_m^{RET} were found to be 2.0 ± 0.6 mJ/mg (of dry bread), 38.0 ± 0.3 °C and 49.1 ± 0.6 °C for stored einkorn bread without sourdough, and 2.7 ± 0.7 mJ/mg (of dry bread) 38.1 ± 0.3 °C and 48.5 ± 0.9 °C for bread samples with sourdough.

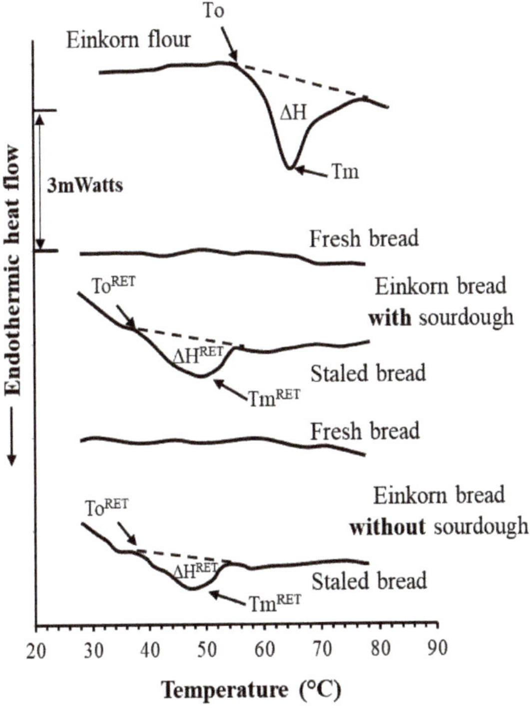

Figure 9. Representative DSC thermographs of slurries (35% w/w) of einkorn flour and crumb of fresh and staled (stored at 3 °C for 6 days) bread without and with sourdough (heating rate 5 °C/min). T_o^{RET}, onset of melting temperature of retrograded starch; T_m^{RET}, peak melting temperature of retrograded starch; and ΔH^{RET}, melting enthalpy of retrograded starch.

In a previous study, the evolution of ΔH^{RET} in common wheat bread (crumb) upon storage (0–5 days) was examined, and it was found that the typical endothermic peak of melting of retrograded starch appeared at approximately 40–60 °C with T_m^{RET} at 51 °C, and the ΔH^{RET} value for the 5th day of storage was 2.7 mJ/mg of dry bread [78]; these results are in agreement with the findings of the current study. Therefore, bread made from einkorn wheat seemed to follow a similar starch retrogradation behavior to that made from common wheat. This may imply that replacement of common wheat with einkorn in bread formulations may not have an adverse impact on the shelf-life of such a composite bakery product.

4. Conclusions

In the present study, the physicochemical properties of flours, doughs and breads made by using ancient plant ingredients (grain, nuts and seeds) as raw materials and adopting prehistoric grinding tool replicas for flour milling and sourdough making by multiple back-slopping steps as a bread leavening process were investigated. The research findings indicated that the particle size of einkorn flour ground with prehistoric-like stone mills was largely higher than it was for commercial fine flours of barley and common wheat. The starch gelatinization properties among cereal-based (spelt, barley, common wheat) flours were similar, while lentil flour had higher gelatinization temperature and lower enthalpy values. Overall, variations in the thermal, rheological and textural properties of the tested flours, doughs and bread formulations were observed among the different plant genera. More specifically, the barley flour made more elastic and viscous doughs compared with those containing any of the various wheat species used (einkorn, spelt and common wheat); additionally, inclusion of acorn into einkorn-based dough formulation resulted in similar rheological behavior with that of barley dough. Among the different wheat species tested, no significant variations in dough rheological properties were noticed. Similarly, the various tested cereal genera (barley and wheat species) did not differ in bread crust texture characteristics. However, fortification of einkorn bread with stone-grounded barley and linseed flour resulted in an increase and decrease in crust hardness, respectively. As expected, common wheat exhibited the highest loaf-specific volume among all bread formulations. On the other hand, the addition of barley-based sourdough into einkorn dough formulation did not affect either the textural properties and loaf volume, nor the extent of starch retrogradation of the final baked product. Overall, the use of 'ancient' plant materials in making of sourdough bread seems to be a promising method for delivering a 'clean labelling' feature in bakery items, satisfying the consumer's demand for healthy and naturally produced breads, as well as exhibiting quality attributes comparable to breads made from conventional wheat flours.

Author Contributions: Conceptualization, S.M.V.; methodology, A.L., C.G.B. and S.M.V.; investigation, V.S.; resources, S.M.V., A.L. and C.G.B.; data curation, V.S. and A.L.; writing—original draft preparation, V.S.; writing—review and editing, A.L., C.G.B. and S.M.V.; supervision, S.M.V. and A.L.; funding acquisition, S.M.V. All authors have read and agreed to the published version of the manuscript.

Funding: This research was funded by the European Research Council project "PlantCult: identifying the food cultures of ancient Europe", under the European Union's Horizon 2020 research and innovation program (grant agreement 682529, consolidator grant 2016–2021) and the APC was funded by the same grant (grant agreement 682529, consolidator grant 2016–2021).

Institutional Review Board Statement: Not applicable.

Informed Consent Statement: Not applicable.

Acknowledgments: We are grateful to Danai Chondrou, Tassos Bekiaris, Ismini Ninou, Maria Bofill and Toni Palomo who helped with the design and construction of the grinding stones that were used to grind the modern reference material. We are also grateful to Pavlos Lathiras and Eleftheria Almasidou for grinding the specimens used in this study. We also thank Vasiliki Kapetanaki for Figure 1.

Conflicts of Interest: The authors declare no conflict of interest.

References

1. Arranz-Otaegui, A.; Carretero, L.G.; Ramsey, M.N.; Fuller, D.Q.; Richter, T. Archaeobotanical evidence reveals the origins of bread 14,400 years ago in northeastern Jordan. *Proc. Natl. Acad. Sci. USA* **2018**, *115*, 7925–7930. [CrossRef]
2. Carretero, L.G.; Wollstonecroft, M.; Fuller, D.Q. A methodological approach to the study of archaeological cereal meals: A case study at Çatalhöyük East (Turkey). *Veg. Hist. Archaeobot.* **2017**, *26*, 415–432. [CrossRef] [PubMed]
3. Primavera, M.; Heiss, A.G.; Valamoti, M.S.; Quarta, G.; Masieri, M.; Fiorentino, G. Inside sacrificial cakes: Plant components and production processes of food offerings at the Demeter and Persephone sanctuary of Monte Papalucio (Oria, southern Italy). *Archaeol. Anthrop. Sci.* **2019**, *11*, 1273–1287. [CrossRef]
4. Valamoti, S.M.; Samuel, D.; Bayram, M.; Marinova, E. Prehistoric cereal foods from Greece and Bulgaria: Investigation of starch microstructure in experimental and archaeological charred remains. *Veg. Hist. Archaeobot.* **2008**, *17*, 265–276. [CrossRef]

5. Heiss, A.G.; Antolín, F.; Berihuete Azorín, M.; Biederer, B.; Erlach, R.; Gail, N.; Griebl, M.; Linke, R.; Lochner, M.; Marinova, E.; et al. The Hoard of the Rings. "Odd" annular bread-like objects as a case study for cereal-product diversity at the Late Bronze Age hillfort site of Stillfried (Lower Austria). *PLoS ONE* **2019**, *14*. [CrossRef] [PubMed]

6. Valamoti, S.M. Plant food ingredients and 'recipes' from Prehistoric Greece: The archaeobotanical evidence. In *Plants and Culture: Seeds of the Cultural Heritage of Europe*; Morel, J.P., Mercuri, A.M., Eds.; Centro Europeo per i Beni Culturali Ravello Edipuglia: Bari, Italy, 2009; pp. 25–38.

7. Zohary, D.; Hopf, M.; Weiss, E. Cereals. In *Domestication of Plants in the Old World: The Origin and Spread of Domesticated Plants in Southwest Asia, Europe, and the Mediterranean Basin*, 4th ed.; Oxford University Press: Oxford, UK, 2012; pp. 20–74.

8. Zohary, D.; Hopf, M.; Weiss, E. Pulses. In *Domestication of Plants in the Old World: The Origin and Spread of Domesticated Plants in Southwest Asia, Europe, and the Mediterranean Basin*, 4th ed.; Oxford University Press: Oxford, UK, 2012; pp. 75–99.

9. Valamoti, S.M.; Marinova, E.; Heiss, A.G.; Hristova, I.; Petridou, C.; Popova, T.; Michou, S.; Papadopoulou, L.; Chrysostomou, P.; Darcque, P.; et al. Prehistoric cereal foods of southeastern Europe: An archaeobotanical exploration. *J. Archaeol. Sci.* **2019**, *104*, 97–113. [CrossRef]

10. Longin, C.F.H.; Ziegler, J.; Schweiggert, R.; Koehler, P.; Carle, R.; Wuerschum, T. Comparative study of hulled (einkorn, emmer, and spelt) and naked wheats (durum and bread wheat): Agronomic performance and quality traits. *Crop Sci.* **2015**, *56*, 302–311. [CrossRef]

11. Peña-Chocarro, L.; Zapata Peña, L.; Emilio González Urquijo, J.; Ibáñez Estévez, J.J. Einkorn (*Triticum monococcum* L.) cultivation in mountain communities of the western Rif (Morocco): An ethnoarchaeological project. In *From Foragers to Farmers*; Fairbairn, A., Weiss, E., Eds.; Oxbow Books: Oxford, UK, 2009; pp. 104–111.

12. Geisslitz, S.; Wieser, H.; Scherf, K.A.; Koehler, P. Gluten protein composition and aggregation properties as predictors for bread volume of common wheat, spelt, durum wheat, emmer and einkorn. *J. Cereal Sci.* **2018**, *83*, 204–212. [CrossRef]

13. Valamoti, S.M.; Chondrou, D.; Bekiaris, T.; Ninou, I.; Alonso, N.; Bofill, M.; Ivanova, M.; Laparidou, S.; McNamee, C.; Palomo, A.; et al. Plant foods, stone tools and food preparation in prehistoric Europe: An integrative approach in the context of ERC funded project PLANTCULT. *J. Lithic Stud.* **2020**, *7*, 1–21. [CrossRef]

14. Dimoula, A.; Tsirtsoni, Z.; Yiouni, P.; Stagkidis, I.; Ntinou, M.; Prevost-Dermarkar, S.; Papadopoulou, E.; Valamoti, S.M. Experimental investigation of ceramic technology and plant food cooking in Neolithic northern Greece. *STAR Sci. Technol. Archaeol. Res.* **2019**, *5*, 269–286. [CrossRef]

15. Heiss, A.G. Bread. In *Archaeology of Food: An Encyclopedia*; Metheny, K.B., Beaudry, M.C., Eds.; Rowman & Littlefield: Lanham, MD, USA, 2015; Volume 1, pp. 70–75.

16. Heiss, A.G.; Pouget, N.; Wiethold, J.; Delor-Ahü, A.; Le Goff, I. Tissue-based analysis of a charred flat bread (galette) from a Roman cemetery at Saint-Memmie (Dép. Marne, Champagne-Ardenne, north-eastern France). *J. Archaeol. Sci.* **2015**, *55*, 71–82. [CrossRef]

17. Valamoti, S.M.; Petridou, C.; Berihuete-Azorín, M.; Stika, H.-P.; Papadopoulou, L.; Mimi, I. Deciphering ancient 'recipes' from charred cereal fragments: An integrated methodological approach using experimental, ethnographic and archaeological evidence. *J. Archaeol. Sci.* **2021**, *128*, 105347. [CrossRef]

18. Valamoti, S.M.; Jacomet, S.; Stika, H.P.; Heiss, A.G. The PLANTCULT Project: Identifying the plant food cultures of ancient Europe. *Antiquity* **2017**, *91*, e9. [CrossRef]

19. Abdel-Aal, E.M.; Hucl, P.; Sosulski, F.W.; Bhirud, P.R. Kernel, milling and baking properties of spring-type spelt and einkorn wheats. *J. Cereal Sci.* **1997**, *26*, 363–370. [CrossRef]

20. Brandolini, A.; Hidalgo, A.; Moscaritolo, S. Chemical composition and pasting properties of einkorn (*Triticum monococcum* L. subsp. *monococcum*) whole meal flour. *J. Cereal Sci.* **2008**, *47*, 599–609. [CrossRef]

21. Cooper, R. Re-discovering ancient wheat varieties as functional foods. *J. Tradit. Complement. Altern. Med.* **2015**, *5*, 138–143. [CrossRef]

22. Ruibal-Mendieta, N.L.; Delacroix, D.L.; Mignolet, E.; Pycke, J.M.; Marques, C.; Rozenberg, R.; Delzenne, N.M. Spelt (*Triticum aestivum* ssp. spelta) as a source of breadmaking flours and bran naturally enriched in oleic acid and minerals but not phytic acid. *J. Agric. Food Chem.* **2005**, *53*, 2751–2759. [CrossRef]

23. Arzani, A.; Ashraf, M. Cultivated ancient wheats (*Triticum* spp.): A potential source of health-beneficial food products. *Compr. Rev. Food Sci. Food Saf.* **2017**, *16*, 477–488. [CrossRef] [PubMed]

24. Hidalgo, A.; Brandolini, A. Nutritional properties of einkorn wheat (*Triticum monococcum* L.). *J. Sci. Food Agric.* **2014**, *94*, 601–612. [CrossRef] [PubMed]

25. Baik, B.K.; Ullrich, S.E. Barley for food: Characteristics, improvement, and renewed interest. *J. Cereal Sci.* **2008**, *48*, 233–242. [CrossRef]

26. Lazaridou, A.; Marinopoulou, A.; Matsoukas, N.P.; Biliaderis, C.G. Impact of flour particle size and autoclaving on β-glucan physicochemical properties and starch digestibility of barley rusks as assessed by in vitro assays. *Bioact. Carbohydr. Diet. Fibre* **2014**, *4*, 58–73. [CrossRef]

27. Lazaridou, A.; Marinopoulou, A.; Biliaderis, C.G. Impact of flour particle size and hydrothermal treatment on dough rheology and quality of barley rusks. *Food Hydrocoll.* **2019**, *87*, 561–569. [CrossRef]

28. Boukid, F.; Zannini, E.; Carini, E.; Vittadini, E. Pulses for bread fortification: A necessity or a choice? *Trends Food Sci. Technol.* **2019**, *88*, 416–428. [CrossRef]

29. Beltrão Martins, R.; Gouvinhas, I.; Nunes, M.C.; Alcides Peres, J.; Raymundo, A.; Barros, A.I. Acorn flour as a source of bioactive compounds in gluten-free bread. *Molecules* **2020**, *25*, 3568. [CrossRef] [PubMed]

30. Sęczyk, Ł.; Świeca, M.; Dziki, D.; Anders, A.; Gawlik-Dziki, U. Antioxidant, nutritional and functional characteristics of wheat bread enriched with ground flaxseed hulls. *Food Chem.* **2017**, *214*, 32–38. [CrossRef] [PubMed]

31. Watkins, T.R.; Tomeo, A.C.; Struck, M.L.; Palumbo, L.; Bierenbaum, M.L. Improving atherogenic risk factors with flax seed bread. In *Developments in Food Science*; Charalambous, G., Ed.; Elsevier: Amsterdam, The Netherlands, 1995; Volume 37, pp. 649–658. [CrossRef]

32. Petridou, C. Plant Foods from Prehistoric Greece: An Interdisciplinary Approach. Ph.D. Thesis, School of History and Archaeology, Aristotle University of Thessaloniki, Thessaloniki, Greece. unpublished, thesis in preparation.

33. Bofill, M.; Chondrou, D.; Palomo, A.; Procopiou, H.; Valamoti, S.M. Processing plants for food: Experimental grinding within the ERC-project PLANTCULT. *J. Lithic Stud.* **2020**, *7*, 1–26. [CrossRef]

34. Karagiannis, G. *Fysiko Prozymi*; Psihalos Publications: Athens, Greece, 2015. (In Greek)

35. Valamoti, S.M.; Fyntikoglou, V.; Symponis, K. *Plant Foods of Ancient Greece: An Inquiry into Ancient Texts and Plant Remains until the Roman Conquest*; University Studio Press: Thessaloniki, Greece, unpublished, book in preparation.

36. Jones, G. Charred grain from late bronze age Gla, Boiotia. *Annu. Br. Sch. Athens* **1995**, *90*, 235–238. [CrossRef]

37. Valamoti, S.M. The plant remains from Makriyalos: Preliminary observations. In *Neolithic Society in Greece*; Hastead, P., Ed.; Sheffield Academic Press: Sheffield, UK, 1999; pp. 136–138.

38. Valamoti, S.M. *Plants and People in Late Neolithic and Early Bronze Age Northern Greece: An Archaeobotanical Investigation*; British Archaeological Reports: Oxford, UK, 2004; Volume 1258.

39. Borghi, B.; Castagna, R.; Corbellini, M.; Heun, M.; Salamini, F. Breadmaking quality of einkorn wheat (*Triticum monococcum* ssp. monococcum). *Cereal Chem.* **1996**, *73*, 208–214.

40. Corbellini, M.; Empilli, S.; Vaccino, P.; Brandolini, A.; Borghi, B.; Heun, M.; Salamini, F. Einkorn characterization for bread and cookie production in relation to protein subunit composition. *Cereal Chem.* **1999**, *76*, 727–733. [CrossRef]

41. Lomolino, G.; Morari, F.; Dal Ferro, N.; Vincenzi, S.; Pasini, G. Investigating the einkorn (*Triticum monococcum*) and common wheat (*Triticum aestivum*) bread crumb structure with X-ray microtomography: Effects on rheological and sensory properties. *Int. J. Food Sci. Technol.* **2017**, *52*, 1498–1507. [CrossRef]

42. AACC International. 10–05.01 Rapeseed displacement method. In *Approved Methods of the American Association of Cereal Chemists*, 11th ed.; American Association of Cereal Chemists International: St. Paul, MN, USA, 2010.

43. Oladunmoye, O.O.; Aworh, O.C.; Maziya-Dixon, B.; Erukainure, O.L.; Elemo, G.N. Chemical and functional properties of cassava starch, durum wheat semolina flour, and their blends. *Food Sci. Nutr.* **2014**, *2*, 132–138. [CrossRef] [PubMed]

44. Ahmed, J.; Thomas, L.; Arfat, Y.A. Functional, rheological, microstructural and antioxidant properties of quinoa flour in dispersions as influenced by particle size. *Food Res. Int.* **2019**, *116*, 302–311. [CrossRef] [PubMed]

45. Biliaderis, C.G. Structural transitions and related physical properties of starch. In *Starch*, 3rd ed.; BeMiller, J., Whistler, R., Eds.; Academic Press: Cambridge, MA, USA, 2009; pp. 293–372. [CrossRef]

46. Koehler, P.; Wieser, H. Chemistry of Cereal Grains. In *Handbook on Sourdough Biotechnology*; Gobbetti, M., Gänzle, M., Eds.; Springer: Boston, MA, USA, 2013; pp. 11–45.

47. Flipse, E.; Keetels, C.J.A.M.; Jacobsen, E.; Visser, R.G.F. The dosage effect of the wildtype GBSS allele is linear for GBSS activity but not for amylose content: Absence of amylose has a distinct influence on the physico-chemical properties of starch. *Theor. Appl. Genet.* **1996**, *92*, 121–127. [CrossRef]

48. Biliaderis, C.G.; Maurice, T.J.; Vose, J.R. Starch gelatinization phenomena studied by differential scanning calorimetry. *J. Food Sci.* **1980**, *45*, 1669–1674. [CrossRef]

49. Biliaderis, C.G.; Grant, D.R.; Vose, J.R. Structural characterization of legume starches. I. Studies on amylose, amylopectin and β-limit dextrins. *Cereal Chem.* **1981**, *58*, 496–502.

50. Ranhotraa, G.S.; Gelrotha, J.A.; Glasera, B.K.; Lorenzb, K.J. Nutrient Composition of Spelt Wheat. *J. Food Compos. Anal.* **1996**, *9*, 81–84. [CrossRef]

51. Sotomayor, C.; Frias, J.; Fornal, J.; Sadowska, J.; Urbano, G.; Vidal-Valverde, C. Lentil Starch Content and its Microscopical Structure as Influenced by Natural Fermentation. *Starch* **1999**, *51*, 152–156. [CrossRef]

52. Xu, M.; Jin, Z.; Simsek, S.; Hall, C.; Rao, J.; Chen, B. Effect of germination on the chemical composition, thermal, pasting, and moisture sorption properties of flours from chickpea, lentil, and yellow pea. *Food Chem.* **2019**, *295*, 579–587. [CrossRef] [PubMed]

53. Tester, R.F.; Sommerville, M.D. The effects of non-starch polysaccharides on the extent of gelatinisation, swelling and α-amylase hydrolysis of maize and wheat starches. *Food Hydrocoll.* **2003**, *17*, 41–54. [CrossRef]

54. Gobbetti, M.; Corsetti, A.; Rossi, J. The sourdough microflora. Interactions between lactic acid bacteria and yeasts: Metabolism of carbohydrates. *Appl. Microbiol. Biotechnol.* **1994**, *41*, 456–460. [CrossRef]

55. Corsetti, A. Technology of Sourdough Fermentation and Sourdough Applications. In *Handbook on Sourdough Biotechnology*; Gobbetti, M., Gänzle, M., Eds.; Springer: Boston, MA, USA, 2013; pp. 85–103. [CrossRef]

56. Harth, H.; Van Kerrebroeck, S.; De Vuyst, L. Community dynamics and metabolite target analysis of spontaneous, backslopped barley sourdough fermentations under laboratory and bakery conditions. *Int. J. Food Microbiol.* **2016**, *228*, 22–32. [CrossRef] [PubMed]

57. Minervini, F.; Celano, G.; Lattanzi, A.; De Angelis, M.; Gobbetti, M. Added ingredients affect the microbiota and biochemical characteristics of durum wheat type-I sourdough. *Food Microbiol.* **2016**, *60*, 112–113. [CrossRef]

58. Mirsaeedghazi, H.; Emam-Djomeh, Z.; Mousavi, S.M.A. Rheometric measurement of dough rheological characteristics and factors affecting it. *Int. J. Agric. Biol.* **2008**, *10*, 112–119.

59. Lazaridou, A.; Duta, D.; Papageorgiou, M.; Belc, N.; Biliaderis, C.G. Effects of hydrocolloids on dough rheology and bread quality parameters in gluten-free formulations. *J. Food Eng.* **2007**, *79*, 1033–1047. [CrossRef]

60. Weipert, D. The benefits of basic rheometry in studying dough rheology. *Cereal Chem.* **1990**, *67*, 311–317.

61. Brennan, C.S.; Cleary, L.J. Utilisation Glucagel® in the β-glucan enrichment of breads: A physicochemical and nutritional evaluation. *Food Res. Int.* **2007**, *40*, 291–296. [CrossRef]

62. Izydorczyk, M.S.; Hussain, A.; MacGregor, A.W. Effect of barley and barley components on rheological properties of wheat dough. *J. Cereal Sci.* **2001**, *34*, 251–260. [CrossRef]

63. Sharma, P.; Kotari, S.L. Barley: Impact of processing on physicochemical and thermal properties—A review. *Food Rev. Int.* **2017**, *33*, 359–381. [CrossRef]

64. Beltrão Martins, R.; Nunes, M.C.; Ferreira, L.M.M.; Peres, J.A.; Barros, A.I.R.N.A.; Raymundo, A. Impact of acorn flour on gluten-free dough rheology properties. *Foods* **2020**, *9*, 560. [CrossRef]

65. Korus, J.; Witczak, M.; Ziobro, R.; Juszczak, L. The influence of acorn flour on rheological properties of gluten-free dough and physical characteristics of the bread. *Eur. Food Res. Technol.* **2015**, *240*, 1135–1143. [CrossRef]

66. Van Bockstaele, F.; De Leyn, I.; Eeckhout, M.; Dewettinck, K. Non-linear creep-recovery measurements as a tool for evaluating the viscoelastic properties of wheat flour dough. *J. Food Eng.* **2011**, *107*, 50–59. [CrossRef]

67. Cappelli, A.; Cini, E.; Guerrini, L.; Masella, P.; Angeloni, G.; Parenti, A. Predictive models of the rheological properties and optimal water content in doughs: An application to ancient grain flours with different degrees of refining. *J. Cereal Sci.* **2018**, *83*, 229–235. [CrossRef]

68. Guerrini, L.; Parenti, O.; Angeloni, G.; Zanoni, B. The bread making process of ancient wheat: A semi-structured interview to bakers. *J. Cereal Sci.* **2019**, *87*, 9–17. [CrossRef]

69. Marpalle, P.; Sonawane, S.K.; Arya, S.S. Effect of flaxseed flour addition on physicochemical and sensory properties of functional bread. *LWT Food Sci. Technol.* **2014**, *58*, 614–619. [CrossRef]

70. Piłat, B.; Zadernowski, R. Physicochemical characteristics of linseed oil and flour. *Pol. J. Nat. Sci.* **2010**, *25*, 106–113. [CrossRef]

71. Sharma, H.P.; Sharma, S.; Nema, V.P. Physico-chemical and functional properties of flour prepared from native and roasted whole linseeds. *J. Pharmacogn. Phytochem.* **2020**, *9*, 1428–1433.

72. Sudha, M.L.; Begum, K.; Ramasarma, P.R. Nutritional characteristics of linseed/flaxseed (*Linum usitatissimum*) and its application in muffin making. *J. Texture Stud.* **2010**, *41*, 563–578. [CrossRef]

73. Yadav, A.; Singh, S.; Yadav, A.; Aman, Z.; Yadav, R.K. Physico-chemical composition of two linseed flour. *J. Pharmacogn. Phytochem.* **2018**, *7*, 57–59.

74. Nergiz, C.; Ötles, S. The proximate composition and some minor constituents of poppy seeds. *J. Sci. Food Agric.* **1994**, *66*, 117–120. [CrossRef]

75. Singh, K.K.; Mridula, D.; Rehal, J.; Barnwal, P. Flaxseed: A potential source of food, feed and fiber. *Crit. Rev. Food Sci.* **2011**, *51*, 210–222. [CrossRef]

76. Hug-Iten, S.; Handschin, S.; Conde-Petit, B.; Escher, F. Changes in Starch Microstructure on Baking and Staling of Wheat Bread. *LWT Food Sci. Technol.* **1999**, *32*, 255–260. [CrossRef]

77. Gray, J.A.; Bemiller, J.N. Bread staling: Molecular basis and control. *Compr. Rev. Food Sci. Food Saf.* **2003**, *2*, 1–21. [CrossRef] [PubMed]

78. Vouris, D.G.; Lazaridou, A.; Mandala, I.G.; Biliaderis, C.G. Wheat bread quality attributes using jet milling flour fractions. *LWT Food Sci. Technol.* **2018**, *92*, 540–547. [CrossRef]

 *foods*

Article

Diversifying the Utilization of Maize at Household Level in Zambia: Quality and Consumer Preferences of Maize-Based Snacks

Emmanuel Oladeji Alamu [1,2,*], **Bukola Olaniyan** [2] **and Busie Maziya-Dixon** [2]

[1] International Institute of Tropical Agriculture (IITA), Southern Africa Research and Administration Hub (SARAH) Campus, P.O. Box 310142, Chelstone, Lusaka 10101, Zambia

[2] International Institute of Tropical Agriculture (IITA), PMB 5230, Ibadan 20001, Nigeria; oluwatoyinolaniyan086@gmail.com (B.O.); B.Maziya-Dixon@cgiar.org (B.M.-D.)

[*] Correspondence: o.alamu@cgiar.org; Tel.: +260-97633-8710

Abstract: This study evaluated the nutritional, antinutritional properties, and consumer preferences of five maize-based snacks at the household level. The physical, nutritional, and antinutritional properties were analyzed with standard laboratory methods, while a structured questionnaire was used for the data collection on consumer preferences of the maize products. There were significant ($p < 0.05$) differences in the proximate parameters of the maize snack samples. Antinutritional properties among maize snacks all fell within the permissible range. Respondents from all districts showed no significant ($p > 0.05$) differences in maize chin-chin variants' and maize finger variants' except for Serenje and Mkushi districts where maize chin-chin and maize finger showed significant ($p < 0.05$) differences in their sensory ratings. However, across districts, the most rated maize finger variant was the spiced 100% maize finger. In conclusion, maize-based snacks enriched with soybean flour have proven nutritious with a reasonable acceptability level.

Keywords: maize snacks; nutritional characterization; consumer preferences

Citation: Alamu, E.O.; Olaniyan, B.; Maziya-Dixon, B. Diversifying the Utilization of Maize at Household Level in Zambia: Quality and Consumer Preferences of Maize-Based Snacks. *Foods* **2021**, *10*, 750. https://doi.org/10.3390/foods10040750

Academic Editors: Luca Serventi, Charles Brennan and Rana Mustafa

Received: 16 March 2021
Accepted: 26 March 2021
Published: 1 April 2021

1. Introduction

Maize serves as the nutritional backbone in central, southern, and eastern Zambia [1]. It is the main staple, providing 52% of the local population's daily calorie intake [2]. Maize is regarded as an economic and political crop in Zambia due to enabling maize production across the country [3]. Approximately 80% of Zambian smallholder households and 20% of commercial farmers grow maize. Many produce other crops such as cassava, grown mainly as a food security crop and soybean as a cash crop. However, there has been substantial financial input from the Government of Zambia in maize production. In 2006, Zambia ranked 13th among 51 countries prominent for maize production in Africa, with a total of 0.865 million tons, and this increased to 3.607 million tons by 2016 [4]. According to FAOSTAT [5], Zambia ranked eighth among the top ten maize producing countries in sub-Saharan Africa [6]. Despite government interventions in the maize sector, there were continuous fluctuations in productivity due to prolonged dry seasons and short rainy seasons [4].

In the bid to increase maize yield in Zambia, improved varieties prominent for grain yield, dry matter, and water usage efficiency in areas of low and erratic rainfall were introduced to farmers [1,7]. This was made possible through the collaborative research of HarvestPlus, International Maize and Wheat Improvement Center (CIMMYT), IITA, the Zambia Agriculture Research Institute (ZARI), and other partners who developed maize varieties with qualities such as high yield, disease resistance, consumer acceptability, and high provitamin A carotenoid content [1,8]. The outcome made a positive impact on the food security status of farm households [1]. However, maize's household utilization has been limited to traditional products such as "Nshima" and various local beverages that are not nutritious enough for household consumption [9]. The National food price data reveals that, as the overall cost of food is reducing, foods high in nutrients are more expensive than staple foods [10]. Hidden hunger and undernutrition have been public health issues in Zambia among children and adults alike [11,12]. A recent study suggested that close to 60% of inpatients at a teaching hospital were at nutritional risk, as pre-admission nutritional status was an associated factor [13].

Furthermore, to combat undernutrition/hidden hunger, most especially Fe and Zn deficiencies, there is a need for swift action such as food-to-food fortification. Regular staples consumed without adequate protein have contributed to low dietary diversity, so utilizing the available and preferred staple to create diversity in nutrient-dense foods at an affordable cost is imperative. However, maize is notorious for phytic acid, which binds with proteins and essential minerals such as Fe and Zn [14]. Nevertheless, different processing methods usually reduce it to safe levels for humans; furthermore, new maize germplasms are specifically bred to be low in phytic acid [15]. Thus, product development using maize and legumes (e.g., soybean) can diversify maize's household utilization.

Legumes are one of the world's most valuable food supply sources, particularly in developing countries where food, energy, and nutrients are of utmost concern [16]. They have been recognized as an essential protein source with maximum advantageous bioactive compounds such as minerals and fat-soluble vitamins, particularly soybean [17]. They are well-laden with sulfur-containing amino acids; the most prominent are Lysine and Tryptophan, mostly not found in cereals. They also possess digestible protein [17,18]. Soybean is unique due to its ability to adjust to many soils and climates and its nitrogen fixative ability [19]. This ability qualifies it as an ideal rotational crop for nitrogen fixation for maize crops, especially in a country such as Zambia.

Snacks are convenience foods that have been around for a long time, but demand increased with urbanization and population [19]. Adebowale and Komolafe [20] reported varieties of snacks and dishes produced from maize familiar to the Nigerian populace. Most snacks are made from cereals, but several findings have exposed their low nutrient concentration [21], which can be made up for by fortification or blending with legumes such as soybean [22], groundnut [16], African yam bean [17], cowpea [18], and pigeon pea [23]; also, defatted coconut has been used in the fortification of maize deep-fried snacks [20].

Zambia produces maize on a relatively large scale, so most of the time, the country exports the surplus to neighboring countries. Diversifying maize for snack production at the household level will enhance the maize value chain and improve the nation's nutritional status. Thus, the study aimed at evaluating the physical, nutritional, and antinutritional properties and consumer preferences of five maize-based snacks produced from maize and soybean.

2. Materials and Methods

2.1. Materials

Maize grains, soybean grains, salt, sugar, margarine, baking powder, and vegetable oil were purchased at local Zambian markets.

2.2. Processing Maize Grain to Flour

Maize flour was produced as described by Adeola et al. [24]. The maize grains were dried, sorted, and cleaned to remove stones, dirt, and infested grains. The cleaned maize was then milled using a laboratory hammer mill and passed through a 100-micrometer mesh sieve.

2.3. Processing Soybean to Flour

The soybean flour was prepared using the methods described by Alamu et al. [25]. The grains were cleaned and sorted to eliminate stones and other undesirable materials. The cleaned soybean seeds were roasted slightly at a temperature of about 120 °C for 5 min until the seed coat was loose and easily removed by hand. The roasted seeds were then coarse-milled and winnowed to remove the seed coat. The decorticated soybean was finely milled to 0.5 mm particle size using a laboratory mill (Perten, Hägersten, Sweden) to obtain fine flour. The flour was packaged and appropriately stored before use.

2.4. Maize-Based Products

Five maize-based products: Plain maize finger, spiced maize finger, spiced fortified maize finger, plain maize chin-chin, and fortified maize chin-chin were prepared using 100% high-quality maize flour (HQMF), blended HQMF, and soybean flour (80:20). Plain maize finger and plain maize chin-chin prepared from HQMF were the control in this experimental setup. Table 1 gives a summary of the ingredients weighed in grams for all products.

Table 1. Recipe for maize-based snacks.

Ingredients/Quantity	Products				
	Plain maize finger	Spiced maize finger	Maize-soy finger	Maize chin-chin	Maize-soy chin-chin
Maize flour	500 g	500 g	500 g	500 g	500 g
Soy flour	–	–	125 g	–	125 g
Salt	5 g	5 g	5 g	–	–
Water	625 mL	620 mL	625 mL	250 mL	500 mL
Sugar	–	–	–	100 g	100 g
Baking powder	—	–	–	10 g	10 g
Margarine	–	–	–	40 g	40 g
Eggs	–	–	–	2 medium-sized	2 medium-sized
Onions	–	2 medium-sized	–	–	–

2.4.1. Processing Maize Finger (Kokoro)

A total of 250 g of HQMF was stirred into 625 mL of hot water (95 °C) to form a thick porridge through the maize starch's gelatinization. The porridge was added to the remaining portion of salted 250 g of maize flour to form a sticky dough. This was mixed thoroughly using an electric mixer and then left to cool to room temperature. A small portion of the dough was scooped, smoothened with the palm, and rolled either on the palm or on a smooth surface to form a firm, long-shaped dough. Then it was deep-fried in 2 L of "Ole oil" (Sunflower oil) at the temperature range of 170 to 175 °C for an average of 4.5 min. The color changed from off-white to brown depicting a Maillard reaction, and a long-shaped, crunchy snack was produced.

2.4.2. Processing Maize Finger Fortified with Soy Flour

The processing method for fortified maize finger is the same as plain maize finger. The only difference is the incorporation of 125 g of soy flour added to the remaining portion of salted 250 g of HQMF. Further, the frying time was extended by 30 s, a little longer than for plain maize finger. It took an average of 5 min for the Maillard reaction to be completed. Table 1 explains all the ingredients and quantities added in grams.

2.4.3. Processing Maize Chin-Chin

Five hundred grams of HQMF was mixed with all the dry ingredients in a bowl; 100 g sugar and 10 g baking powder. Forty grams of margarine was added and mixed thoroughly. Two medium-sized eggs with an average weight of 62.2 g and 250 mL of room temperature water were added to form a dough, kneaded to ensure dough uniformity.

The dough was spread on a smooth, clean surface and sprinkled with HQMF to prevent stickiness. A stainless-steel knife was used to cut the dough into small rectangular shapes, which were deep-fried in 2 L of "Ole oil" (Sunflower oil) within a range of 170 to 175 °C as frying progressed. It took an average of 5 min for frying to be completed, and the product was strained and cooled on a clean flat surface; the result was a sweet, crunchy snack.

2.4.4. Processing Maize-Soy Chin-Chin

The processing is the same as for the plain chin-chin. The fortifier, 125 g of High Quality Soy flour (HQSF), was added to the maize flour when all other dry ingredients were added. It took an average frying time of 5 min for the Maillard reaction to be completed when the frying sample became brown.

2.4.5. Processing Spicy Maize Finger (Kokoro)

The processing follows the same as that for maize finger and maize-soy finger. The significant difference is the addition of two medium-sized onions with an average weight of 63.8 g each as a spice.

2.5. Determination of Nutritional and Physico-Chemical Properties of Maize Snacks

The maize finger, maize-soy finger, maize chin-chin, and maize-soy chin-chin were analyzed for moisture, protein, fat, ash, total reducing sugars, total starch, digestible starch, non-digestible starch, amylose, phytate, tannins, pH, bulk density, and color parameters in duplicate

1. Moisture content determination: The pulverized samples were used to determine the moisture content using the method reported by Alamu et al. [25].
2. Ash content determination: The method of AOAC [26] as reported by Alamu et al. [27].
3. Protein content determination: The Kjeldahl method was used to determine the protein content by multiplying the nitrogen value with a conversion factor of 6.25, as described by Alamu et al. [25].
4. Crude fat content determination: The Soxhlet extraction method was used as described by Alamu et al. [27].
5. Digestible starch and total reducing sugar content determination: Digestible starch and total reducing sugar were determined using the Dubois method [28], as reported by Alamu et al. [25].
6. Amylose content determination: The adapted method Williams et al. [29] was used to determine amylose content described by Alamu et al. [25].
7. Carbohydrate content: This was derived by calculating the difference, %CHO = 100—(sum of the percentages of moisture, ash, fat, protein, and crude fiber)

2.6. Determination of Antinutritional Properties of Maize Snacks

1. Phytic acid content: Phytate was determined by the extraction and precipitation of phytic acid according to Wheeler and Ferrel's method [30] as described by Okukpe & Adeloye [31].
2. Tannin content determination: Tannins were determined by the method described by da Silva Lins et al. [32].

2.7. Determination of Functional Properties of Maize Snacks

1. pH determination: This was done using 10 g of pulverized maize finger and maize chin-chin dispersed in 20 mL of deionized water to detect the suspension's pH using a table-top pH meter [33].
2. Bulk density determination: Bulk density was determined using the method recommended by AOAC [26]. The sample (7 g) was placed into a 50 mL graduated measuring cylinder and then tapped gently against the palm until a constant volume was obtained.
3. Color parameters: Color measurements were performed on pulverized samples using a color meter. The color of products was expressed as the average of three L*, a*, and b* readings, where L* stands for brightness, a* redness, a* greenness, b* yellowness,

and b* blueness. A white calibration plate was used to standardize the equipment before color measurements [34].

2.8. Sensory Evaluation and Consumer Preferences of Maize Snacks

The maize finger' quality attributes made from 100% corn flour, spiced 100% maize finger, spiced 80:20% maize-soy finger, 100% maize chin-chin, and 80:20% maize-soy chin-chin were assessed by a 30-member sensory panel. The panelists were well trained on the desired descriptors for all maize products, and all precautions were adhered to strictly before the sensory session. Forms were administered, and panelists were asked to score the samples using the 9-point hedonic scale for taste, color, crispiness, flavor, and overall acceptability. The scores were ranked and subjected to statistical analysis.

The investigation was conducted in each major maize-growing district (Monze, Katete, Serenje, and Mkushi). The areas were selected based on levels of consumption and accessibility. Four hundred and thirteen respondents (413) were available for the survey, with 104 respondents in Monze, 109 in Katete, 108 in Serenje, and 92 in Mkushi. They were all randomly selected. The data was collected using a well-designed questionnaire administered to each respondent by well-trained enumerators. They were adequately informed about the study, and an agreement was reached to obtain their consent. The maize-based products were well coded to avoid a mix-up and presented to the participants randomly according to the method described. The sensory attributes chosen were aroma, appearance, taste, texture, and overall acceptability. The attributes of each product were rated by participants on a 5-point hedonic scale to measure the degree of likeness using qualitative judgments that correspond to 1 = dislike very much, 2 = dislike a little, 3 = neither like nor dislike, 4 = like a little, 5 = like very much. The sensory testing order was such that product appearance was rated first, then aroma, and finally taste and texture. Clean, potable water was supplied to respondents for necessary rinsing of the mouth between one product and another to be precise with each product's sensorial attributes [35].

2.9. Statistical Analysis

The data generated on the proximate, functional, and antinutritional properties were statistically analyzed using IBM SPSS statistical software (Version 21). The data about preference and willingness to consume were subjected to Analysis of Variance (ANOVA) at a 95% level of significance. The differences between means were considered significant at $p < 0.05$ using the Duncan multiple range test.

3. Results and Discussion

3.1. Nutritional Properties of Deep-Fried Maize-Based Snacks

Table 2 shows the results for nutritional and antinutritional properties of deep-fried maize-based snacks. All maize snacks' moisture content was minimal and did not harm the products' quality attributes.

The addition of soy flour and eggs caused an increase of about 50% ash content in the products. It implies that Fe, Zn, and other minerals contents will be markedly higher in the soy-fortified maize products than 100% maize products. All products seem to be high in fat content, mainly due to deep frying. This could have a negative effect on the storability of the products due to unsaturated fatty acid exposure to warm or hot air known as oxidative rancidity [16]. The same trend was observed as reported by [36] that some snacks' high-fat content resulted from the processing techniques that involved the addition of cooking oils and/or deep-frying.

The protein values obtained for the snacks are significantly ($p < 0.05$) different. The addition of legumes to cereals has been scientifically established to improve its protein quality [37–39]. Soy flour is rich in sulfur-containing amino acids such as lysine, which is deficient in maize or eroded. So the addition of soy flour or other legumes enhances maize snacks' nutrient content.

Table 2. Nutritional and Antinutritional properties of deep-fried maize-based snacks.

Products							Parameters							
	MC (%)	Ash (%)	Fat (%)	Protein (%)	Amylose (%)	Amylopectin (%)	Sugar (%)	Starch (%)	TCHO (%)	TDCHO (%)	TNDCHO (%)	Energy Value (Kcal/100 g)	Tannin (mg/g)	Phytate (%)
100% maize kokoro	2.191 [a]	1.611 [c]	15.555 [b]	9.750 [ab]	20.494 [c]	79.506 [b]	1.881 [c]	62.730 [a]	70.893 [b]	64.611 [a]	6.282 [e]	462.565 [b]	1.130 [c]	1.454 [d]
Spiced 100% maize kokoro	2.443 [a]	1.653 [c]	12.852 [c]	9.650 [b]	24.259 [b]	75.741 [c]	2.695 [b]	51.952 [c]	73.402 [a]	54.648 [c]	18.755 [b]	447.874 [c]	1.410 [b]	1.644 [c]
Spiced 20% soy-maize kokoro	2.342 [a]	2.453 [a]	15.431 [b]	10.656 [a]	25.475 [a]	74.525 [d]	3.411 [a]	48.712 [d]	69.117 [c]	52.124 [d]	16.994 [c]	457.976 [b]	1.552 [b]	2.332 [a]
100% maize chin-chin	1.332 [b]	2.340 [b]	18.619 [a]	8.605 [c]	20.114 [d]	79.886 [a]	3.450 [a]	56.126 [b]	69.104 [c]	59.576 [b]	9.528 [d]	478.411 [a]	1.538 [b]	1.236 [e]
20% Soy-maize chin-chin	1.346 [b]	2.455 [d]	18.322 [a]	9.188 [bc]	20.722 [c]	79.278 [b]	3.393 [a]	42.137 [e]	69.689 [c]	45.530 [e]	24.158 [a]	480.407 [a]	1.975 [a]	1.869 [b]
Minimum	1.298	1.447	12.631	8.560	20.076	74.449	1.843	42.137	69.034	45.511	6.066	446.137	1.111	1.234
Maximum	2.598	2.475	18.668	10.938	25.551	79.924	3.488	62.810	73.654	64.653	24.284	480.816	2.050	2.362
Mean	1.931	1.902	16.156	9.570	22.213	77.787	2.966	52.332	70.441	55.298	15.143	465.447	1.521	1.707
Std. deviation	0.529	0.433	2.244	0.735	2.330	2.330	0.645	7.299	1.715	6.847	6.800	13.074	0.291	0.397
Pr > F (Products)	**	***	***	**	***	***	***	***	***	***	***	***	***	***

[a] Parameters were analyzed in duplicate. Mean values in the same column with different letters are significantly different at $p < 0.05$. **, significant at $p < 0.01$; ***, significant at $p < 0.001$. MC = Moisture content; TCHO = Total carbohydrate; TDCHO = Total digestible carbohydrate; TNDCHO = Total non-digestible carbohydrate.

Furthermore, all products' amylose content is highly significant at $p < 0.001$. Research has expounded on the importance of amylose in diets. It positively correlates with resistant starch by slowing down glucose release into the bloodstream, thereby benefiting those managing obesity or hyperinsulinemia [9]. The sugar content across all products has a high significance at $p < 0.001$, with 100% maize finger (kokoro) having the lowest value of 1.88% while 100% maize chin-chin has the highest value of 3.45%. The trend observed is that products fortified with soy flour have an elevated sugar level than their unfortified counterparts. A similar pattern was reported by [24] where the sugar content of kokoro increased as AYF (African yam bean flour) substitution increased. Another possibility is that 100% maize chin-chin records a high sugar content due to sucrose as part of the recipe.

The product that has the highest starch content is the 100% maize finger (62.73%), while 20% soy-maize finger and spiced 20% soy-maize finger have the lowest values of 42.14% and 48.71%, respectively. Maize is predominantly starch accounting for 60–75% of the kernel, and so the starch content is appreciably reduced when substituted with legumes such as soybean. This same trend was reported by [24] when 100% of maize kokoro had the highest starch content than other products substituted with AYF. The total carbohydrate across the products shows no significant ($p > 0.05$) difference except 100% maize finger and its spiced variant. This same observation was reported by [16] and [25], who substituted maize snack (kokoro) with partially defatted groundnut paste and cowpea flour, respectively. The high total carbohydrate values observed for 100% maize finger and its spiced variant could be from starch hydrolysis. This suggests that the 100% maize products (maize finger and chin-chin) record the highest total digestible carbohydrates. This can be explained by their starch structure with high amylopectin, which results in a high degree of branching, a disrupted granular structure of starch, thus increasing its susceptibility to the attack by enzymes and in-vitro digestibility [40]. Although the consumption of 100% maize products may rapidly boost energy levels, especially when an instant burst of energy is required, such as during an endurance sporting event, rapidly digestible carbohydrate is the best choice to make. Therefore, glucose tablets and sports drinks are so popular [40].

Nevertheless, there may also be a tendency to raise blood glucose levels, resulting in hyperglycemia in some individuals, particularly those with impaired glucose tolerance [41]. Alternatively, maize products fortified with soy flour and the spiced 100% maize finger have high values for total non-digestible carbohydrates, with 20% soy-maize chin-chin having the highest value of 24.15%. This qualifies the product to fit into the carbohydrates group known as resistance starch, generally referred to as dietary fiber. This non-digestible carbohydrate is absorbed in the small intestine. It ferments, to some extent, producing short-chain fatty acids with advantages such as improvements in glycemic control, bowel health, and cardiovascular disease prevention [42]. The naturally occurring ones such as verbascose and low molecular weight fructans are found in legumes and onions [40]. This justifies why 20% soy-maize chin-chin has the highest value of 24.15% and the spiced 100% maize finger 18.75%, the second-highest value.

The 20% soy-maize chin-chin has the highest energy value even though there is no significant ($p > 0.05$) energy difference among the products. This agrees with [39], who reported an increase in the energy value with the increasing proportion of pigeon-pea protein concentration in maize flour kokoro. Furthermore, protein and fat are usually associated with high calorific values in food. The relatively high-fat content across the products resulting from deep-frying contributed immensely to the energy value.

The antinutritional properties across the products show no significant ($p > 0.05$) differences for tannin and phytate. The lowest value for tannin was observed in 100% maize finger with a value of 1.13 mg/g, and the highest value observed in 20% soy-maize chin-chin. The products containing soy were observed to have the most tannin concentration, while the highest values for phytate were seen in the spiced 20% soy-maize with a value of 2.33% and the lowest value of 1.23% observed in 100% maize chin-chin. Tannins are known to be notorious for reducing the bioavailability of proteins in humans and animals and notable for their antioxidative and anti-inflammatory characteristics [43]. Phytate also forms complexes with essential minerals and protein in foods, making them unavailable for absorption and deactivating digestive enzymes [19]. Both antinutrients are significantly found in cereals and legumes but usually reduced when subjected to processing such as dehulling, cooking, frying, malting, fermentation, and oven-drying, among others [15]. Hence, the values obtained across all products from Table 2 are considered safe enough for human consumption.3.2. Physicochemical Properties of Deep-Fried Maize-Based Snacks

The maize finger all have bulk densities in a similar range from 0.846 g/mL in spiced 100% maize finger to 0.863 g/mL in spiced 20% soy-maize finger, while the chin-chin samples had almost the same bulk density with a percentage difference of 0.2 g/mL (Table 3). There was no significant ($p > 0.05$) difference between the bulk densities of the products. The bulk density is a vital parameter that defines the ease of packaging and conveyance of particulate foods [9].

Table 3. Physicochemical properties of deep-fried maize-based snack.

Products	pH	Bulk Density (g/mL)	L*	a*	b*
100% maize finger	6.740 [c]	0.853 [a]	5620 [a]	760.0 [a]	3003.5 [a]
spiced 100% maize finger	6.790 [c]	0.846 [a]	5724 [a]	638.5 [a]	3083.5 [a]
Spiced 20% soy-maize finger	6.515 [d]	0.863 [a]	5106 [a]	854.5 [a]	2803.5 [a]
100% maize chin-chin	7.130 [b]	0.780 [a]	5361 [a]	732.5 [a]	2796.5 [a]
20% soy-maize chin-chin	7.430 [a]	0.782 [a]	5374 [a]	719.0 [a]	2720.5 [a]
Minimum	6.510	0.760	4991	389	2693
Maximum	7.440	0.939	5814	907	3339
Mean	6.921	0.825	5437	741	2882
Std. deviation	0.340	0.054	240	141	197
Pr > F (Products)	***	Ns	**	Ns	Ns

[a] Parameters were analyzed in duplicate. Mean values in the same column with different letters are significantly different at $p < 0.05$. ns, not significant at $p < 0.05$; **, significant at $p < 0.01$; ***, significant at $p < 0.001$.

Color is a critical quality attribute in food, influencing consumer choice [44]. The color of the maize snacks (L*, a*, and b* values), which depicts the degree of lightness, redness, and yellowness, shows no significant difference at $p > 0.05$ across all products. However, 100% maize finger have the highest degree of lightness and yellowness values (5724, 5620; 3003.5, 3083.5), respectively, while spiced fortified maize finger have the highest degree of redness (Table 3). The observed trend agrees with the report of Sha et al. [45], whose degree of redness of corn snacks supplemented with soy and chickpea flour increased with supplementation and degree of yellowness was found in corn snacks developed without supplementation of soy or chickpea flour. Furthermore, Anton et al. [38] reported a slight color impact when corn starch-based extruded snacks were fortified with navy beans. At the same time, a little red bean flour addition resulted in evident color changes.

3.2. Demographic Information of the Respondents for Maize Products

Table 4 presents the demographic information and the awareness of the maize chin-chin snacks and their consumption frequency. Significantly few respondents acknowledged knowing about the maize chin-chin snack. Monze District, alongside Serenje District, had an awareness level of 1.94%, recording the lowest awareness level. In comparison, Katete and Mkushi districts had a 2.67% awareness level, the highest level recorded out of a total 9.22% awareness level among the correspondents that claimed to be aware of the products. The highest level of ignorance of the maize chin-chin snack was recorded in Serenje District, where 99% of respondents claimed ignorance of the product. The frequency of consumption of the product was extremely low, just as the level of awareness was equally low, with the lowest being recorded in Monze and Serenje districts.

Table 4. Demographic information of the respondents for maize-based products.

Maize Chin-Chin		Monze	Katete	Serenje	Mkushi
Variables		N (%)	N (%)	N (%)	N (%)
Gender	Female	42 (10.17)	30 (7.26)	61 (14.77)	35 (8.47)
	Male	62 (15.01)	79 (19.13)	47 (11.38)	57 (13.8)
Age (year)	Mean ± SD	35.1 ± 13.27	38.3 ± 12.33	43.8 ± 12.26	41.0 ± 11.42
	Minimum	14	18	21	20
	Maximum	72	105	76	73
Awareness of maize chin-chin N(%) Yes No	Female	8(1.94)	11.6(2.67)	8(1.94)	11(2.67)
	Male	96(23.3)	98(23.73)	99(24.03)	81(19.66)
Maize Kokoro		**Monze**	**Katete**	**Serenje**	**Mkushi**
Variables		N (%)	N (%)	N (%)	N (%)
Gender	Female	59 (14.54)	77 (19.11)	40 (9.93)	59 (14.64)
	Male	43 (10.67)	29 (7.2)	65 (16.13)	31 (7.69)
Age (year)	Mean ± SD	35.6 ± 14.27	38.3 ± 11.24	42.9 ± 12.91	40.7 ± 12.1
	Minimum	14	21	18	15
	Maximum	72	84	79	73
Gender	Female	59(14.54)	77(19.11)	40(9.93)	59(14.64)
	Male	43(10.67)	29(7.2)	65(16.13)	31(7.69)

In contrast, Katete and Mkushi districts recorded a 13.16% daily consumption rate. It was observed that male respondents consumed more of the product, with 59.32% of the total respondents being male. More male respondents consumed the product than females in all the study locations except Serenje District, where awareness is generally low.

There was a very low awareness level of the maize finger' product, with just 12.9% of total respondents claiming knowledge of the product; 2.3% of which are in Mkushi District, where they had the lowest awareness level. The highest awareness level was found in Serenje District. Maize finger snack is least consumed daily in Katete District and Serenje District, having 1% daily consumption frequency. Some 58.31% of total respondents are female and most of them are from the Katete District, while Monze and Mkushi districts share an almost equal percentage of total female respondents.

3.3. Consumer Preference Ratings for Maize Chin-Chin and Maize Finger

Table 5 shows the analysis of variance (ANOVA) results of the effects of gender, product types and, location on the sensory attributes of maize chin-chin and maize finger (Kokoro), respectively. Gender, Product type, and District (Location) had significant effects ($p < 0.001$) on all the sensory attributes for both maize chin-chin and maize finger except appearance and aroma that showed no significant effect of product type and location, respectively. This implies that the sensory attributes ratings were gender-dependent. This follows a similar pattern reported by Alamu et al. [46], who inferred that snack foods' preference depends mostly on biological factors (genetic), which are gender related. To buttress the observation that sensory attributes are gender-dependent, [47] reported that gender significantly affects tenderness, flavor intensity, and overall acceptability on goose meat. The appearance attribute is not essential in rating maize-based snacks studied, and aroma does not affect the rating of maize snacks across the districts. In a study involving consumer acceptability of two variants of maize baobab snacks, all other sensory parameters such as appearance and aroma were not significant in the overall ratings except taste [48]. Taste and texture are the driving sensory attributes observed for these maize snacks. Jaworska and Hoffmann [49] evaluated the relationship between texture and other sensory attributes on potato chips. It was revealed that texture significantly correlated with overall sensory quality and consumer acceptance.

Table 5. ANOVA of consumer preference ratings of maize-based products.

Maize Chin-Chin		Product Attributes				
Source	DF	Appearance	Aroma	Taste	Texture	Overall acceptability
Gender	1	2.2196 **	5.3176 ***	8.1343 ***	6.7947 ***	7.7645 ***
Product	1	1.1634	17.4334 ***	22.7228 ***	17.1441 ***	15.1864 ***
District	3	2.5717 **	1.27899	3.27146 **	3.3450 **	5.2905 ***
Error	821	0.5644	0.6329	0.7062	0.8538	0.5983
Maize Kokoro		**Product Attributes**				
Source	DF	Appearance	Aroma	Taste	Texture	Overall acceptability
Gender	1	5.3334 **	5.5007 ***	19.2687 ***	24.3147 ***	17.0796 ***
Product	2	8.8309	11.9139 ***	47.2613 ***	59.0050 ***	37.2600 ***
District	3	3.0000 **	1.2250	0.6070 **	1.1879 **	3.6325 ***
Error	1202	0.8500	1.0534	2.4767	1.2176	0.9664

, significant at $p < 0.01$; *, significant at $p < 0.001$.

3.4. Consumer Preference Ratings for Maize Chin-Chin Products by District and across Districts

Table 6 shows the consumer preference rating for maize chin-chin by district and across the districts. The appearance, taste, and texture for maize chin-chin are not significantly different at $p < 0.05$ for all locations except Serenje. The appearance ratings of maize chin-chin snacks ranged from 4.8 ± 0.44 for soy-maize chin-chin at Monze District to 4.8 ± 0.51 at Mkushi district for 100% maize chin-chin. It was not significantly different at a value of $p < 0.05$ except in Serenje. The aroma of the various products across the five districts had no significant difference ($p > 0.05$). The aroma is the only attribute that does not significantly differ among the five districts ranging from 4.2 ± 0.82 to 4.7 ± 0.52. The taste attribute was significantly different at a value of $p < 0.05$, ranging from 4.1 ± 0.99 for 100% maize chin-chin at Serenje District to 4.7 ± 0.6 for soy-maize chin-chin at Mkushi District. The overall acceptability ranged from 4.1 ± 1.01 to 4.8 ± 0.44 for 100% maize chin-chin in Serenje District to soy-maize chin-chin product in Mkushi District, respectively, with no significant difference at a value of $p < 0.05$ except Serenje. Moreover, maize chin-chin's overall acceptability for Serenje District is significantly different at $p < 0.05$ compared with other districts. In all the districts, the soy-fortified maize chin-chin variant has the higher overall acceptability ratings except in Katete. Many researchers had reported decreased acceptability of maize snacks when legume fortification increased [18,22], while some reported otherwise [16]. The nature of the legume and the processes subjected to before utilization in fortification may determine the level of product acceptability. For instance, a product fortified with roasted or malted soy flour tends to enjoy higher sensory acceptability than another product fortified with unroasted or unmalted soy flour. This is because roasting or malting decreases the inherent in them to the least, improves flavor and color, and denatures protein, thereby improving digestibility [50].

In Monze and Mkushi, fortified maize chin-chin taste has the highest rating of 4.60 ± 0.84 and 4.70 ± 0.60, respectively. While in Katete and Serenje, the most rated attributes are aroma (4.60 ± 0.74) and appearance (4.80 ± 0.48) for the fortified maize chin-chin. The lowest sensory ratings go to Serenje District; this same trend was observed by [25] in evaluating sensory properties for wheat and cassava chin-chin variants.

There is a significant ($p < 0.05$) difference between the maize products for all sensory attributes except appearance. Nevertheless, fortified maize chin-chin has the highest ratings for all sensory attributes.

Table 7 shows the preferences for maize finger variants according to districts and across the districts. There seems to be no significant difference for sensory parameters in all locations except in Serenje District. All districts had 100% spiced maize finger, the most rated for all sensory parameters, and the next in the rating was the spiced fortified maize finger, which applies to all districts except Mkushi. The maize finger with the least preferred aroma were observed to be the spiced soy-maize finger at the Mkushi District with a value of 3.9 ± 1.21, while the product with the most preferred aroma was spiced 100% maize finger with a value of 4.4 ± 1.00. The spiced soy-maize had the lowest aroma rating due to its characteristic beany flavor [51]. However, the trend observed is that the maize finger product with the lowest overall acceptance level is the 100% maize finger with a mean value of 3.9 ± 0.93, 3.7 ± 1.18, and 3.6 ± 0.89, at Monze, Katete, and Serenje districts, respectively. This strongly suggests that the inclusion of soy flour positively influenced the maize finger' taste. Uzor-Peters et al. [51] reported a similar trend when they fortified maize finger with defatted soya cake flour in different ratios of 1:1, 7:3, and 9:1. The maize-soy finger at a ratio of 9:1 had the highest rating for all sensory attributes.

The most rated maize finger variant is the spiced 100% maize finger, while the next in the rating is spiced soy-maize finger. The spice used on these products is onions (*Allium cepa* L.), a vital vegetable crop used as a spice and food ingredient due to its scent, taste, and intense flavor [52]. It has been reported to be effective against cardiovascular disease, hypolipidemic, anti-hypertensive, anti-diabetic, antithrombotic, and anti-hyperhomocysteinemia effects, and many other biological activities such as antimicrobial, antioxidant, anticarcinogenic, antimutagenic, antiasthmatic, immunomodulatory, and prebiotic activities.

In contrast, the onion's strong flavor successfully masked the beany flavor associated with soy flour used to fortify the spiced soy-maize finger. Thus, the spice's inclusion serves a dual purpose; improvement of sensory attributes and health-promoting activities.

Table 6. Consumer preference ratings for maize chin-chin products by district and across the districts.

District	Sample (Attributes)	N	Appearance Mean ± SD	CV	Aroma Mean ± SD	CV	Taste Mean ± SD	CV	Texture Mean ± SD	CV	OA Mean ± SD	CV
Monze	Soy-maize chin-chin	104	4.80 ± 0.44 [a]	9.2	4.60 ± 0.64 [a]	13.7	4.60 ± 0.84 [a]	18.3	4.50 ± 0.87 [a]	19.42	4.60 ± 0.72 [a]	15.55
	100% maize chin-chin	104	4.80 ± 0.45 [a]	21.5	4.20 ± 0.92 [a]	21.9	4.40 ± 0.79 [a]	18.1	4.10 ± 1.08 [a]	26.29	4.30 ± 0.83 [a]	19.42
Katete	Soy-maize chin-chin	109	4.80 ± 0.46 [a]	15.5	4.60 ± 0.74 [a]	16.2	4.60 ± 0.72 [a]	15.6	4.50 ± 0.75 [a]	16.72	4.60 ± 0.63 [a]	13.6
	100% maize chin-chin	109	4.80 ± 0.47 [a]	15.8	4.30 ± 0.93 [a]	21.6	4.40 ± 0.89 [a]	20.4	4.30 ± 0.98 [a]	22.7	4.50 ± 0.76 [a]	17.14
Serenje	Soy-maize chin-chin	108	4.80 ± 0.48 [b]	19.3	4.30 ± 0.76 [a]	17.5	4.40 ± 0.75 [b]	17	4.10 ± 0.94 [b]	22.55	4.30 ± 0.84 [b]	19.62
	100% maize chin-chin	108	4.80 ± 0.49 [b]	17.2	4.20 ± 0.82 [a]	19.3	4.10 ± 0.99 [b]	24.4	4.10 ± 1 [b]	24.59	4.10 ± 1.01 [b]	24.4
Mkushi	Soy-maize chin-chin	92	4.80 ± 0.50 [a]	15.7	4.70 ± 0.52 [a]	11.1	4.70 ± 0.6 [a]	12.8	4.60 ± 0.68 [a]	14.73	4.80 ± 0.44 [a]	9.21
	100% maize chin-chin	92	4.80 ± 0.51 [a]	15.8	4.20 ± 0.9 [a]	21.6	4.10 ± 1.04 [a]	25.3	4.10 ± 0.99 [a]	24.36	4.30 ± 0.77 [a]	18.04
Ratings across the districts (Attributes)												
	Sample	N	Appearance Mean ± SD	CV	Aroma Mean ± SD	CV	Taste Mean ± SD	CV	Texture Mean ± SD	CV	OA Mean ± SD	CV
	Soy-maize chin chin	413	4.6 ± 0.72 [a]	15.8	4.5 ± 0.69 [a]	15.2	4.6 ± 0.74 [a]	16.3	4.4 ± 0.83 [a]	18.85	4.6 ± 0.7 [a]	15.4
	100% maize chin chin	413	4.5 ± 0.79 [a]	17.7	4.2 ± 0.89 [b]	21	4.2 ± 0.94 [b]	22.1	4.1 ± 1.01 [b]	24.5	4.3 ± 0.86 [b]	19.99
	Total	826	4.5 ± 0.76	16.8	4.4 ± 0.81	18.4	4.4 ± 0.86	19.6	4.3 ± 0.94	21.93	4.4 ± 0.8	17.97

Mean values in the same column with different letters are significantly different at $p < 0.05$; OA = overall acceptability.

Table 7. Consumer preference ratings for maize finger products by district and across the districts.

District	Sample (Attributes)	N	Appearance Mean ± SD	CV	Aroma Mean ± SD	CV	Taste Mean ± SD	CV	Texture Mean ± SD	CV	OA Mean ± SD	CV
Monze	100% maize finger	102	4.4 ± 0.9 [a]	20.21	4.0 ± 0.97 [a]	23.92	3.7 ± 1.11 [a]	29.75	3.5 ± 1.04 [a]	30.06	3.9 ± 0.93 [a]	23.64
	Spiced 100% maize finger	102	4.5 ± 0.92 [a]	20.3	4.3 ± 1.03 [a]	23.98	4.4 ± 0.9 [a]	20.57	4.2 ± 0.99 [a]	23.7	4.5 ± 0.75 [a]	16.83
	Spiced soy-maize finger	102	4.3 ± 0.9 [a]	20.88	4.0 ± 1.07 [a]	26.92	3.9 ± 1.16 [a]	29.98	3.9 ± 1.17 [a]	29.74	4.0 ± 1.05 [a]	26.33
Katete	100% maize finger	106	4.3 ± 0.93 [a]	21.56	3.9 ± 1.27 [a]	32.21	3.6 ± 1.26 [a]	34.93	3.4 ± 1.35 [a]	39.92	3.7 ± 1.18 [a]	31.63
	Spiced 100% maize finger	106	4.6 ± 0.81 [a]	17.46	4.5 ± 0.94 [a]	20.86	4.4 ± 1 [a]	22.88	4.3 ± 1.08 [a]	25.11	4.5 ± 0.94 [a]	20.72
	Spiced soy-maize finger	106	4.4 ± 0.87 [a]	19.79	4.3 ± 1.01 [a]	23.37	4.2 ± 1.04 [a]	24.72	4.3 ± 1.03 [a]	23.82	4.4 ± 0.96 [a]	21.97
Serenje	100% maize finger	105	4.1 ± 0.99 [b]	23.94	4.0 ± 0.93 [a]	23.26	3.5 ± 1.19 [a]	33.97	3.4 ± 1.19 [a]	35.44	3.6 ± 0.89 [b]	24.34
	Spiced 100% maize finger	105	4.4 ± 0.79 b	17.97	4.3 ± 0.86 [a]	20.12	4.3 ± 0.92 [a]	21.4	4.3 ± 0.79 [a]	18.51	4.4 ± 0.78 [b]	17.72
	Spiced soy-maize finger	105	4.2 ± 1.02 b	24.51	4.1 ± 0.97 [a]	23.53	4.2 ± 3.99 [a]	96.09	4 ± 1.02 [a]	25.7	3.9 ± 0.98 [b]	25.35
Mkushi	100% maize finger	90	4.4 ± 0.93 a	21.22	4.3 ± 0.95 [a]	22.17	3.9 ± 1.16 [a]	30.08	3.8 ± 1.08 [a]	28.87	4.2 ± 0.93 [a]	22.25
	Spiced 100% maize finger	90	4.6 ± 0.88 a	19.07	4.4 ± 1.00 [a]	22.77	4.4 ± 0.99 [a]	22.63	4.1 ± 1.2 [a]	29.46	4.4 ± 1.09 [a]	24.67
	Spiced soy-maize finger	89	4.2 ± 1.12 a	26.4	3.9 ± 1.21 [a]	30.87	4 ± 1.25 [a]	31.16	3.9 ± 1.16 [a]	29.66	4.0 ± 1.17 [a]	29.27

Table 7. *Cont.*

Ratings across the districts Attributes		Appearance		Aroma		Taste		Texture		OA	
Sample	N	Mean ± SD	CV	Mean ± SD	CV	Mean ± SD	CV	Mean ± SD	CV	Mean ± SD	CV
100% maize finger	403	4.3 ± 0.94 [b]	21.84	4.1 ± 1.04 [b]	25.77	3.7 ± 1.19 [c]	32.33	3.5 ± 1.18 c	33.98	3.9 ± 1.01 [c]	26.1
Spiced 100% maize finger	403	4.5 ± 0.85 [a]	18.73	4.4 ± 0.96 [a]	21.95	4.4 ± 0.95 [a]	21.8	4.2 ± 1.02 [a]	24.19	4.5 ± 0.89 [a]	20.01
Spiced soy-maize finger	402	4.3 ± 0.98 [b]	22.84	4.1 ± 1.07 [b]	26.15	4.1 ± 2.26 [b]	55.59	4 ± 1.1 [b]	27.29	4.1 ± 1.05 [b]	25.9
Total	1208	4.4 ± 0.93	21.28	4.2 ± 1.04	24.8	4 ± 1.6	39.59	11.7 ± 3.3	85.46	12.5 ± 2.95	72.01

Mean values in the same column with different letters are significantly different at $p < 0.05$. OA = overall acceptability.

4. Conclusions

Maize snacks have proven to contain high nutritional content whose quality can be further improved by fortification with soybean flour with a reasonable level of acceptability. This will help to create a diversity of nutrient-dense foods, thereby shrinking the pool of an under-nourished population. Maize chin-chin fortified with 20% soy flour has the highest acceptability in Monze, Katete, and Mkushi districts. In comparison, the spiced 100% maize finger enjoyed the highest acceptability across all districts except in Mkushi, where the spiced soy-maize finger had the least rating for aroma. This indicates that the soy flour used for fortification may undergo more processing operations or an entirely new processing method to reduce the beany flavor of soybean. Furthermore, two or more legumes may be used to fortify more maize products as protein digestibility and availability will be researched. However, it is essential to know that the Zambians' nutritional status will be considerably upgraded.

Thus, maize in nutritious, healthy snacks in Zambia will benefit the maize value chain's improvement by placing a higher demand on the produce, thereby increasing its economic value and providing job opportunities. There is an urgent need to train farmers and processors to commercialize these relatively new products.

Author Contributions: E.O.A. and B.M.-D. designed the research; E.O.A. conducted the research; B.M.-D. supervised the study; E.O.A. analyzed the data; E.O.A. and B.O. prepared the manuscript, E.O.A., B.O. and B.M.-D. reviewed the manuscript. E.O.A. and B.M.-D. were responsible for the contents of the paper. All authors have read and agreed to the published version of the manuscript.

Funding: This research received no external funding.

Informed Consent Statement: The participants' informed consent was obtained verbally and by signing the informed consent form.

Acknowledgments: The authors acknowledged the support from the African Development Bank (AfDB), CGIAR Maize CRP, CGIAR CRP A4NH, Ministry of Agriculture and Livestock (MoAL), Zambia. In addition, the supports of Prisca Chilese, Ackson Mooya, Jeremiah Hantolo, Sam Ofodile (IITA, Nigeria), and all the field staff were acknowledged.

Conflicts of Interest: The authors declare no conflict of interest.

References

1. Manda, J.; Gardebroek, C.; Kuntashula, E.; Alene, A.D. Impact of improved maize varieties on food security in Eastern Zambia: A doubly robust analysis. *Rev. Dev. Econ.* **2018**, *22*, 1709–1728. [CrossRef]
2. Melkani, A.; Mason, N.M.; Mather, D.L.; Chisanga, B. Smallholder Maize Market Participation and Choice of Marketing Channel in the Presence of Liquidity Constraints: Evidence from Zambia (No. 1879-2020-467); Michigan State University, Department of Agricultural, Food, and Resource Economics, Feed the Future Innovation Lab for Food Security (FSP). 2019. Available online: http://foodsecuritypolicy.msu.edu/ (accessed on 10 September 2020).
3. Timler, C.; Michalscheck, M.; Alvarez, S.; Descheemaeker, K.; Groot, J.C. Exploring options for sustainable intensification through legume integration in different farm types in Eastern Zambia. In *Sustainable Intensification in Smallholder Agriculture: An Integrated Systems Research Approach*; Earthscan: London, UK, 2017; pp. 196–209.
4. Amondo, E.; Simtowe, F.; Erenstein, O. Productivity and production risk effects of adopting drought-tolerant maize varieties in Zambia. *Int. J. Clim. Chang. Strat. Manag.* **2019**, *11*, 570–591. [CrossRef]
5. Duncan, E.G.; O'Sullivan, C.A.; Roper, M.M.; Biggs, J.S.; Peoples, M.B. Influence of co-application of Nitrogen with phosphorus, potassium, and Sulphur on the apparent efficiency of nitrogen fertilizer use, grain yield and protein content of wheat. *Field Crop. Res.* **2018**, *226*, 56–65. [CrossRef]
6. FAOSTAT (2014) FAOSTAT. Available online: http://faostat.fao.org (accessed on 4 August 2020).
7. Mudenda, E.M.; Phiri, E.; Chabala, L.M.; Sichingabula, H.M. Water use efficiency of maize varieties under rain-Fed conditions in Zambia. *Sustain. Agric. Res.* **2017**, *6*. [CrossRef]
8. Menkir, A.; Palacios-Rojas, N.; Alamu, O.; Dias Paes, M.C.; Dhliwayo, T.; Maziya-Dixon, B.; Rocheford, T. *Vitamin A-Biofortified maize: Exploiting Native Genetic Variation for Nutrient Enrichment*; (No. 2187-2019-667); CIMMYT.; IITA.; EMBRAPA.; HarvestPlus; Crop Trust: Bonn, Germany, 2018; pp. 1–3.
9. Alamu, E.O.; Ntawuruhunga, P.; Chileshe, P.; Olaniyan, B.; Mukuka, I.; Maziya-Dixon, B. Nutritional quality of fritters produced from fresh cassava roots, high-quality cassava and soy flour blends, and consumer preferences. *Cogent Food Agric.* **2019**, *5*, 1677129.

10. Harris, J.; Chisanga, B.; Drimie, S.; Kennedy, G. Nutrition transition in Zambia: Changing food supply, food prices, household consumption, diet and nutrition outcomes. *Food Secur.* **2019**, *11*, 371–387. [CrossRef]
11. Farràs, M.; Chandwe, K.; Mayneris-Perxachs, J.; Amadi, B.; Louis-Auguste, J.; Besa, E.; Swann, J.R. Characterizing the metabolic phenotype of intestinal villus blunting in Zambian children with severe acute malnutrition and persistent diarrhea. *PLoS ONE* **2018**, *13*, e0192092. [CrossRef]
12. Maila, G.; Audain, K.; Marinda, P.A. Association between dietary diversity, health and nutritional status of older persons in rural Zambia. *S. Afr. J. Clin. Nut.* **2019**, 1–6. [CrossRef]
13. Miyoba, N.; Musowoya, J.; Mwanza, E.; Malama, A.; Murambiwa, N.; Ogada, I.; Liswaniso, D. Nutritional risk and associated factors of adult inpatients at a teaching hospital in the Copperbelt province in Zambia; A hospital-based cross-sectional study. *BMC Nutr.* **2018**, *4*, 1–6.
14. Anaemene, D.I.; Fadupin, G.T. Effect of Fermentation, Germination and Combined Germination-Fermentation Processing Methods on the Nutrient and Anti-nutrient Contents of Quality Protein Maize (QPM) Seeds. *J. Appl. Sci. Environ. Manag.* **2020**, *24*, 1625–1630. [CrossRef]
15. Nissar, J.; Ahad, T.; Naik, H.R.; Hussain, S.Z. A review phytic acid: As antinutrient or nutraceutical. *J. Pharm. Phytochem.* **2017**, *6*, 1554–1560.
16. Otunola, E.T.; Sunny-Roberts, E.O.; Adejuyitan, J.A.; Famakinwa, A.O. Effects of addition of partially defatted groundnut paste on some Properties of 'kokoro' (a popular snack made from maize paste). *Agric. Biol. J. N. Am.* **2012**, *3*, 280–286. [CrossRef]
17. Idowu, A.O. Nutrient composition and sensory properties of 'kokoro' (a Nigerian snack) made from Maize and African yam bean flour blends. *Int. Food Res. J.* **2015**, *22*, 739.
18. Abegunde, T.A.; Bolaji, O.T.; Adeyemo, T.B. Quality evaluation of maize chips (Kokoro) fortified with cowpea flour. *Niger. Food J.* **2014**, *32*, 97–104. [CrossRef]
19. Ndife, J.; Abasiekong, K.S.; Nweke, B.; Linus-Chibuezeh, A.; Ezeocha, V.C. Production and comparative quality evaluation of chin-chin snacks from maize, soybean, and orange-fleshed sweet potato flour blends. *FUDMA J. Sci.* **2020**, *4*, 300–307. [CrossRef]
20. Adebowale, O.J.; Komolafe, O.M. Effect of supplementation with defatted coconut paste on proximate composition, physical and sensory qualities of a maize-based snack. *J. Culin. Sci. Technol.* **2018**, *16*, 40–51. [CrossRef]
21. Akoja, S.S.; Adebowale, O.J.; Makanjuola, O.M.; Salaam, H. Functional properties, nutritional and sensory qualities of maize-based snack (kokoro) supplemented with protein hydrolysate prepared from pigeon pea (*Cajanus cajan*) seed. *J. Culin. Sci. Technol.* **2017**, *15*, 306–319. [CrossRef]
22. Adelakun, O.E.; Adejuyitan, J.A.; Olajide, J.O.; Alabi, B.K. Effect of soybean substitution on some physical, compositional, and sensory properties of kokoro (a local maize snack). *Eur. Food Res. Technol.* **2005**, *220*, 79–82. [CrossRef]
23. Adegunwa, M.O.; Adeniyi, O.D.; Adebowale, A.A.; Bakare, H.A. Quality Evaluation of Kokoro Produced from Maize–Pigeon Pea Flour Blends. *J. Culin. Sci. Technol.* **2015**, *13*, 200–213. [CrossRef]
24. Adeola, A.A.; Olunlade, B.A.; Ajagunna, A.J. Effect of pigeon pea or soybean substitution for maize on nutritional and sensory attributes of kokoro. *Ann. Sci. Biotechnol.* **2011**, *2*, 61–66.
25. Alamu, E.O.; Maziya-Dixon, B.; Popoola, I.; Gondwe, T.N.P.; Chikoye, D. Nutritional evaluation and consumer preference of legume fortified maize-meal porridge. *J. Food Nutr. Res.* **2016**, *4*, 664–670.
26. AOAC, A. of Official Analytical Chemists. Coffee and tea. In *Official Methods of Analysis*, 17th ed.; AOAC: Gaithersburg, Maryland, 2000.
27. Alamu, E.O.; Popoola, I.; Maziya-Dixon, B. Effect of Soybean (*Glycine max* (L.) Merr.) flour inclusion on the nutritional properties and consumer preference of fritters for improved household nutrition. *Food Sci. Nutr.* **2018**, *6*, 1811–1816. [CrossRef] [PubMed]
28. Dubois, M.; Gilles, K.A.; Hamilton, J.K.; Rebers, P.A.; Smith, F. A colorimetric method for the determination of sugars. *Nature* **1951**, *168*, 167. [CrossRef]
29. Williams, V.R.; Wu, W.T.; Tsai, H.Y.; Bates, H.G. Varietal differences in amylose content of rice starch. *J. Agric. Food Chem.* **1958**, *6*, 47–48. [CrossRef]
30. Wheeler, E.L.; Ferrel, R.E. A method for phytic acid determination in wheat and wheat fractions. *Cereal Chem.* **1971**, *48*, 312–320.
31. Okukpe, K.M.; Adeloye, A.A. Evaluation of the nutritional and antinutritional constituents of some selected browse plants in Kwara State, Nigeria. *Niger Soc. Exp. Biol. J.* **2019**, *11*, 161–165.
32. da Silva Lins, T.R.; Braz, R.L.; Silva, T.C.; Araujo, E.C.G.; De Medeiros, J.X.; Reis, C.A. Tannin content of the bark and branch of Caatinga species. *J. Exp. Agric. Int.* **2019**, *31*, 1–8. [CrossRef]
33. Rodríguez-Miranda, J.; Ruiz-López, I.I.; Herman-Lara, E.; Martínez-Sánchez, C.E.; Delgado-Licon, E.; Vivar-Vera, M.A. Development of extruded snacks using taro (*Colocasia esculenta*) and nixtamalized maize (*Zea mays*) flour blends. *LWT Food Sci. Technol.* **2011**, *44*, 673–680. [CrossRef]
34. Dauda, A.O.; Kayode, R.M.O.; Salami, K.O. Quality Attributes of Snack made from Maize Substituted with Groundnut. *Ceylon J. Sci.* **2020**, *49*, 21–27. [CrossRef]
35. Altamore, L.; Ingrassia, M.; Chironi, S.; Columba, P.; Sortino, G.; Vukadin, A.; Bacarella, S. Pasta experience: Eating with the five senses-A pilot study. *AIMS Agric. Food.* **2018**, *3*, 493–520.
36. Aletor, O.; Ojelabi, A. Comparative evaluation of the nutritive and functional attributes of some traditional Nigerian snacks and oil seed cakes. *Pak. J. Nutr.* **2007**, *6*, 99–103.

37. Arise, A.K.; Oyeyinka, S.A.; Dauda, A.O.; Malomo, S.A.; Allen, B.O. Quality evaluation of maize snacks fortified with bambara groundnut flour. *Ann. Food Sci. Technol.* **2018**, *19*, 283–291.
38. Anton, A.A.; Fulcher, R.G.; Arntfield, S.D. Physical and nutritional impact of fortification of corn starch-based extruded snacks with common bean (*Phaseolus vulgaris* L.) flour: Effects of bean addition and extrusion cooking. *Food Chem.* **2009**, *113*, 989–996. [CrossRef]
39. Akoja, S.S.; Ogunsina, T.I. Chemical Composition, Functional and Sensory Qualities of Maize-Based Snacks (Kokoro) Fortified with Pigeon Pea Protein Concentrate. *IOSR J. Environ. Sci. Toxicol. Food Technol.* **2018**, *12*, 42–49.
40. Panel on the Definition of Dietary Fiber Staff, Food and Nutrition Board Staff, & Institute of Medicine Staff. *Dietary Reference Intakes: Proposed Definition of Dietary Fiber: A Report of the Panel on the Definition of Dietary Fiber and the Standing Committee on the Scientific Evaluation of Dietary Reference Intakes*; National Academy Press: Washington, DC, USA, 2001.
41. Martin, K.E. Glycaemic Response to Varying the Proportions of Starchy Foods and Non-Starchy Vegetables within a Meal: A Randomized Controlled Trial. Ph.D. Thesis, University of Otago, Dunedin, New Zealand, 2017.
42. Alexander, C.; Swanson, K.S.; Fahey, G.C., Jr.; Garleb, K.A. Perspective: Physiologic importance of short-chain fatty acids from non-digestible carbohydrate fermentation. *Adv. Nutr.* **2019**, *10*, 576–589. [CrossRef]
43. Sharma, K.; Vikas, K.; Jaspreet, K.; Beenu, T.; Ankit, G.R.S.; Yogesh, G.; Ashwani, K. Health effects, sources, utilization and safety of tannins: A critical review. *Toxin Rev.* **2019**, 1–13. [CrossRef]
44. Pathare, P.B.; Opara, U.L.; Al-Said, F.A.J. Colour measurement and analysis in fresh and processed foods: A review. *Food Bioprocess. Technol.* **2013**, *6*, 36–60. [CrossRef]
45. Shah, F.U.H.; Sharif, M.K.; Butt, M.S.; Shahid, M. Development of protein, dietary fiber, and micronutrient enriched extruded corn snacks. *J. Texture Stud.* **2017**, *48*, 221–230. [CrossRef]
46. Alamu, E.O.; Maziya-Dixon, B.; Olaniyan, B.; Pheneas, N.; Chikoye, D. Evaluation of nutritional properties of cassava-legumes snacks for domestic consumption—Consumer acceptance and willingness to pay in Zambia. *AIMS Agric. Food* **2020**, *5*, 500. [CrossRef]
47. Uhlířová, L.; Tůmová, E.; Chodová, D.; Vlčková, J.; Ketta, M.; Volek, Z.; Skřivanová, V. The effect of age, genotype and sex on carcass traits, meat quality and sensory attributes of geese. *Asian Australas. J. Anim. Sci.* **2018**, *31*, 421. [CrossRef] [PubMed]
48. Netshishivhe, M.; Omolola, A.O.; Beswa, D.; Mashau, M.E. Physical properties and consumer acceptance of maize-baobab snacks. *Heliyon* **2019**, *5*, e01381. [CrossRef] [PubMed]
49. Jaworska, D.; Hoffmann, M. Relative importance of texture properties in the sensory quality and acceptance of commercial crispy products. *J. Sci. Food Agric.* **2008**, *88*, 1804–1812. [CrossRef]
50. Kavitha, S.; Parimalavalli, R. Effect of processing methods on proximate composition of cereal and legume flours. *J. Hum. Nut. Food Sci.* **2014**, *2*, 1051.
51. Uzor-Peters, P.I.; Arisa, N.U.; Lawrence, C.O.; Osondu, N.S.; Adelaja, A. Effect of partially defatted soybeans or groundnut cake flours on proximate and sensory characteristics of kokoro. *Afr. J. Food Sci.* **2008**, *2*, 98–101.
52. Seifu, M.; Tola, Y.B.; Mohammed, A.; Astatkie, T. Effect of variety and drying temperature on physicochemical quality, functional property, and sensory acceptability of dried onion powder. *Food Sci. Nutr.* **2018**, *6*, 1641–1649. [CrossRef] [PubMed]